Analytical
Gas Chromatography

Second Edition

Analytical
Gas Chromatography

Second Edition

WALTER JENNINGS
ERIC MITTLEFEHLDT
PHILIP STREMPLE

J & W Scientific
Folsom, California

ACADEMIC PRESS
San Diego London Boston New York
Sydney Tokyo Toronto

This book is printed on acid-free paper. ∞

Academic Press
a division of Harcourt Brace & Company
525 B Street, Suite 1900, San Diego, California 92101-4495, USA
http://www.apnet.com

Academic Press Limited
24-28 Oval Road, London NW1 7DX, UK
http://www.hbuk.co.uk/ap/

Library of Congress Cataloging-in-Publication Data

Jennings, Walter
    Analytical gas chromatography.

    Includes index.
    1. Gas chromatography.   2. Capillarity.   I. Title.
    QD79.C45J458   1987   543'.0896   86-28873
    ISBN 0-12-384357-X

PRINTED IN THE UNITED STATES OF AMERICA
97 98 99 00 01 02 MM 9 8 7 6 5 4 3 2 1

CONTENTS

CHAPTER 3
Sample Injection

CHAPTER 4
The Stationary Phase

CHAPTER 5
Variables in the Gas Chromatographic Process

CHAPTER 6
Column Selection, Installation, and Use

CHAPTER 7
Instrument Conversion and Adaptation

CHAPTER 8
Special Analytical Techniques

PREFACE

Some forty years after the process was first conceived, gas chromatography remains the world's most widely used analytical technique. However, the real expertise of a large proportion of today's chromatographers lies in other fields. Lacking opportunity or time to develop a better understanding of the chromatographic principles that underlie their analyses, they use chromatography merely as a means to an end. Because the technique is so powerful, they are still able, however, to generate useful data. Some are under pressure to produce results using equipment that was purchased specifically to generate those results. Others, including a number of GC/MS analysts in environmental laboratories, often perform their analyses using procedures established by regulatory agencies that often discourage deviations from these methods that would otherwise permit improvements in analytical procedure. The net result—which is certainly understandable—is that a large proportion of users have little real knowledge of the actual variables in the chromatographic process, the interaction between those variables, how they are best controlled, how the quality of analytical results could be improved, and how analysis times can be shortened to facilitate the generation of a greater number of more reliable results on the same equipment.

An analyst with a more comprehensive understanding of chromatographic principles, however, can often improve the quality of the data generated and reduce the analytical time. As the laboratory workload increases, shorter analysis times help accommodate increased demands on the available equipment. In turn, this forestalls the need to purchase an additional chromatograph or another mass spectrometer. Knowledgeable users can also extend the usable lifetime of the equipment. Equally important is the sense of personal satisfaction that comes when a "black-box" approach changes to a real understanding of the process. One of W.J.'s most rewarding experiences occurred when a senior professor in Germany who had been publishing chromatographic papers for some twenty years thanked the instructor at the end of a course and remarked, "I never realized how little chromatography I actually knew."

We have taken pains to prepare this second edition in a form that will permit its use as an instructional college-level text, as a "brush-up" manual that the practitioner can read straight through, and as a reference source for the user who merely wishes to review one subject or area. To fulfill these multiple goals properly, some degree of redundancy has been necessary. The user wishing to quickly review the options for increasing retention factors will be reminded of the roles played by K_c, temperature, and β, with a brief mention of their interrelationships. He or she will then be referred to other sections where each of these topics is discussed in greater detail.

Many of our colleagues have generously contributed suggestions and ideas. We are particularly indebted to Roger Schirmer for assistance in the areas of the physical and chemical nature of glass and fused silica, to Shawn Reese and Roy Lautamo for their contributions in the areas of stationary phases, surface treatments, and deactivations, and to Allen Vickers, Mitch Hastings, and others for invaluable help in applications and other areas.

September, 1996

Walter Jennings
Eric Mittlefehldt
Philip Stremple

ABOUT THE AUTHORS

Complementing a research career that began in 1952, Professor Walter Jennings has taught gas chromatography for over 40 years. His teaching has included graduate instruction and regularly scheduled courses at the University of California, Davis. Beginning about 1970, these instructional activities were supplemented by a series of extra-curricular one- to three-day courses, some of which were open enrollment, while others were restricted, in-house courses specially tailored for a variety of industrial concerns. Eventually these evolved into a comprehensive, fast-paced, continuously updated one-day course that is still presented 30 to 40 times each year at points all over the world. To date, it has been estimated that well over 30,000 chromatographers have attended these courses.

Dr. Eric Mittlefehldt is a Senior Research Scientist and Manager of J&W's Custom Column Shoppe. Eric is an accomplished physical–analytical chemist whose experience extends well beyond the practical aspects of capillary column gas chromatography. Originally trained as a surface scientist, Eric has made significant contributions to the characterization of multicomponent polymeric materials routinely used as stationary phases in gas chromatography. His unusual background brings to this edition a solid foundation in chromatographic theory as well as considerable practical experience with regard to the design and implementation of new stationary phases and column technology.

Dr. Philip Stremple received his Ph.D. at the University of Iowa in 1983 and was a postdoctoral fellow at MIT until 1985. He then spent nine years as a practicing analytical chemist exclusively in the area of gas chromatography for the Clorox Company. His areas of expertise then included the GC analysis of industrial raw materials, fragrance and flavors, finished products, and competitive product analysis. For the past two and a half years Phil has been the manager of applied science at J&W Scientific specializing in applications, technical support, and training.

CHAPTER 1

INTRODUCTION

1.1   General Considerations

In the late 1800s, Mikhail Tswett separated natural pigments into colored zones by percolating plant extracts through adsorbent-packed columns. He later used the word "chromatography" to describe this process [1, 2]. Our use of the word has broadened, and "chromatography" is now used for a number of processes in which the substances to be separated are subjected to equilibrium partitioning between two phases. In most cases, one of those phases is stationary and the other is mobile. The principles of liquid–liquid chromatography (LLC) are employed by the separatory funnel at one end of the spectrum, and by the Craig countercurrent distribution apparatus at the other extreme. Applications of liquid–solid chromatography range from paper—through column—to some forms of thin-layer chromatography.

Their work on liquid–solid chromatography (LSC) earned Nobel Prizes for A. J. P. Martin and R. L. M. Synge. It was in his award address that Martin suggested a gas might be used as the mobile phase in chromatographic processes. Some years later, James and Martin [3] subjected to the passage of ethyl acetate vapor a mixture of fatty acids that had been affixed to an adsorbent. In doing so, they demonstrated the sequential elution of the fatty acids. Coupled with an automated titration system, this generated a graph composed of a series of "steps" depicting the sequential additions of base as each successively eluted acid was neutralized by automated titration.

In 1954, Ray [4] inserted the sensing filament of a thermal conductivity cell, constituting one leg of a Wheatstone bridge, into the outlet of a gas chromatographic column. He generated the first modern-day "chromatogram" where each eluting substance generated a Gaussian-type peak. The schematics and chromatograms that Ray published stimulated a number of workers to enter what promised to become a new and exciting field. Within a decade, some hundreds of individual scientists were engaged in both basic and applied research in gas chromatography. Many of these contributions have been detailed elsewhere (e.g., [5, 6]), but several fundamental steps in the development of modern analytical gas chromatography deserve special mention. These include Golay's invention of the open tubular column [7], Desty's elegantly simple design for a glass-capillary-drawing machine [8], and the concept of a thin-walled fused silica column [9].

When a gas is employed as the mobile phase, either a liquid or a solid can be utilized as the stationary phase. These processes are "gas–liquid chromatography" (GLC) and "gas–solid chromatography" (GSC), respectively. The former has greater general utility and is more widely used, while the latter is especially useful for the separation of highly volatile compounds, including fixed gases (see later chapters). In popular usage, the term "gas chromatography" and the abbreviation "GC" are often applied to both processes.

1.2   A Simplistic Approach

In the process of gas chromatography, a thin film of the stationary phase is confined to the column, and continuously swept by a stream of mobile phase (i.e., carrier gas). The two extremes in column types are packed columns and open tubular columns.

Packed columns are typically 2–5 m long, 1–5 mm in internal diameter (ID) (d_c), and are filled with an "inert" granular support, each particle of which is coated with the stationary phase. As implied by the name, micropacked columns are a smaller version of the packed column, usually having IDs of less than 1 mm, and smaller packing granules. The length of a packed column is practically limited by the pressure drop generated by the resistance it offers to gas flow.

There are three general types of open tubular columns. The most widely used is the wall-coated open tubular (WCOT) column, in which the stationary phase exists in the form of a uniform thin film affixed to the inner periphery of an open tube, the column. In porous layer open tubular (PLOT) columns, a porous layer exists on the inner wall of the column, while the central portion is open. Porosity of that layer is sometimes achieved by chemical means such as etching of the wall per se, and in other cases by deposition of the porous particles from a suspension. The porous layer may serve as a support for a stationary phase, or as the "stationary phase" per se. SCOT (support coated open tubular) columns are a form of PLOT column. Commonly used sorbents include porous polymers, aluminum oxide, and selected zeolites. In some open tubular columns, the d_c may be as large as 0.5–0.75 mm. While these are open tubular columns, they should not be regarded as true "capillaries." We will consider all of the above columns in this book, but the use of the word "capillary" will be restricted to columns whose inner diameters do not exceed 0.35 mm.

Whether it is packed or open tubular, the column, which in the normal GC system is connected to the inlet of the gas chromatograph at one end and to the detector at the other, is adjusted to some suitable temperature and continuously swept with the mobile phase (carrier gas). When a mixture of volatile components is introduced to the inlet end of the column, each solute in that sample engages in a highly dynamic equilibrated partitioning between the stationary phase and the mobile phase in accordance with its distribution constant ($K_c = c_S/c_M$). Let us consider a single band of solute at some one point in time: as the solute molecules in the gas phase are swept forward by the carrier gas, those in the stationary phase are carried downcolumn a finite distance. At that instant, the equilibrium distribution K_c is violated at the rear of the band (where c_S is finite and c_M is zero) and at the front of the band (where c_S is zero and c_M is finite). To reestablish the distribution constant throughout the band, the dominant partitioning is from stationary phase to mobile phase at the rear of the band, and from mobile phase to stationary phase at the front of the band. In other words, the flow of carrier gas disrupts the equilibrium distribution at the front and rear of each chromatographing solute band, causing continuous evaporation at the rear and reestablishment at the front of each solute band as it chromatographs through the column (Fig. 1.1). Because all solutes are injected simultaneously, separation is obviously contingent on differences between the K_c values of the individual solutes. The proportion of a solute that is in the mobile phase at any given time is a function of the "net" vapor pressure of that solute; molecules of those components exhibiting higher vapor pressures partition more toward the mobile phase. They are swept toward the detector more rapidly and are the first solutes eluted from the column. Other solutes exhibit lower vapor pressures, either because they are higher-boiling or because they engage in interactions with the stationary phase that effectively reduce their vapor pressures under the chromatographic conditions employed. Individual molecules of these solutes venture into the mobile phase (carrier gas) less frequently, their concentrations in the mobile phase are lower, and they require longer periods of time to reach the detector; hence separation is achieved.

These superficial considerations lead to two credulous generalities. First, gas chromatography is a volatility phenomenon that is applicable only to materials that can be vaporized; it is not useful in the analysis of proteins or metals per se, although we will, in later chapters, discuss methods of converting some nonvolatile substances to volatile derivatives. Second, because separation depends on partitioning between the two phases, the temperature of the column is critical to the analysis. If the column temperature is too low, the solutes remain largely (or wholly) in the stationary phase, and rarely (or never) enter the mobile phase. They neither separate from each other nor (in the extreme case) elute from the column. If the column temperature is too high, the solutes spend most (or all) of their time

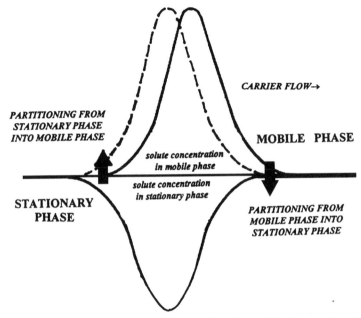

Fig. 1.1. In gas chromatography, each chromatographing solute partitions between the mobile phase (*above the horizontal line*) and the stationary phase (*below the horizontal line*) in accordance with its K_c ($K_c = c_S/c_M$). At time "T," c_S is represented by the curve below, and c_M by the dotted curve above, the horizontal line. Those solute molecules in the mobile phase are continuously swept down stream, and an instant later, at time "$T + t$," c_M is represented by the solid curve above the horizontal line. The K_c has now been violated in the regions marked by the dark arrows. At the left, solutes in stationary phase must migrate to mobile phase to re-establish the K_c; at the right, solutes in mobile phase must partition toward the stationary phase to re-establish the K_c. In traversing the column, each solute band continuously evaporates at the rear and is re-established at the front.

in the mobile phase, rarely (or never) enter the stationary phase, and elute from the column as an unseparated mixture.

It can be useful to (incorrectly) visualize gas chromatography as a stepwise process and to begin by considering the separation of a simple two-component mixture containing (e.g., acetone [boiling point (bp) 57°C]) and ether bp 37°C). If a small amount of that mixture is introduced into a chromatographic column which is continuously swept with carrier gas and held at a temperature where each solute exhibits a suitable vapor pressure, both solutes will immediately partition between the moving gas phase and the immobile stationary phase. All other things being equal, the molecules of the lower-boiling ether that are dissolved in the stationary phase will vaporize before (or more frequently than) the molecules of the higher-boiling acetone [10]. As they enter the mobile gas phase they progress through the column and pass over the virgin stationary phase, where they redissolve. A fraction of a second before the acetone molecules revaporize to be carried downstream again by the carrier gas, the ether molecules move again. Hence the more volatile ether molecules continuously increase their lead over the less volatile acetone molecules, and separation is achieved.

Although this concept may prove helpful in visualizing that a multiplicity of vaporizations and resolutions on the part of the individual solute molecules is one major factor influencing the degree of separation efficiency, it must be stressed that this oversimplification results in an inaccurate picture. For one thing, the chromatographic process is continuous and highly dynamic rather than being a series of discrete steps. At any point in time, some of the molecules of each solute are in the stationary phase and others are in the mobile phase. As the mobile phase moves over virgin stationary phase, some of the mobile-phase-entrained solute molecules dissolve in the stationary phase, while immediately behind the moving front, an equivalent number of dissolved solute molecules vaporize into the mobile phase. Because ether and acetone exhibit different vapor pressures, the ratio

Molecules in stationary phase/molecules in mobile phase

will be larger for the less volatile acetone than for the more volatile ether. Hence, ether is less retained and spends a greater percentage of its transit time in the mobile phase. Hence it will move through the column more rapidly, and at the conclusion of the process a "plug" of ether molecules dispersed in the mobile phase (carrier gas) will emerge to the detector, followed by a second mobile phase "plug" carrying acetone molecules. Under constant chromatographic conditions, the degree of separation (i.e., the distance between the two eluting plugs) will be a function of the solute retention factors and the concentrations (or "sharpness" of the solute plugs).

These concepts are also helpful in emphasizing that the vapor pressure of the solute strongly influences its chromatographic behavior. Solutes undergo no separation in the mobile phase, nor do they undergo separation in the stationary phase. Solute separation is dependent on the differences in solute volatility, which influence the rates (or frequencies) of solute vaporizations and resolutions. This differentiates solute concentrations in the stationary and mobile phases. Hence it is desirable to subject solutes to as many "vaporization steps" as possible (without having other adverse effects; *vide infra*), and this will require that they undergo an equal number of "resolution steps." If the vapor pressures of the solutes are too high, they spend most (or all) of their transit time in the mobile phase and little (or no) separation is achieved. If the vapor pressures are too low, the solutes spend too long in the stationary phase, analysis times become disproportionately long, and sensitivity is also adversely affected (*vide infra*). Column temperature is an obvious method of influencing solute vapor pressures. Another method is through the choice of stationary phase. A "polar" stationary phase reduces the vapor pressures of polar solutes by means of additional solute-stationary phase interactions that may include hydrogen bonding and/or dipole–dipole interactions. These interrelationships are discussed in greater detail in later chapters.

1.3 Simplistic Comparisons of Packed and Open Tubular Columns

Most chromatographers recognize that the open tubular (or "capillary") column is capable of separations that are vastly superior to those obtained on packed columns [11]. Figure 1.2 illustrates separations of an essential oil on a packed column and on two types of open tubular column [12]. One of us (WJ) once experienced instructional difficulties with workers in a developing country who preferred the packed column "because it generated fewer peaks." They viewed with dismay the greater challenge imposed by the large number of peaks generated during the separation of their sample on a more efficient capillary column. The folly of that

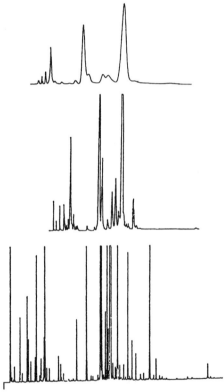

Fig. 1.2. Chromatograms of a peppermint oil on (top) a 6 ft × 1/4 in packed column, (center) a 500 ft × 0.03 in stainless steel open tubular column, and (bottom) a 30 m × 0.25 mm fused silica open tubular (capillary) column. *Adapted from [12], p. 445, and reprinted with permission.*

prejudice is easily emphasized by pointing out that the chromatogram can be made even simpler by omitting the stationary phase. The chromatogram will then consist of a single peak that includes all components.

On the other hand, there are occasions where the required degree of separation can be obtained on a packed column and separation on an optimized open tubular column results in more resolution than is required ("overkill") at the expense of longer analysis times. In situations of this type, some of the superior resolving power of the open tubular column can be traded off to yield equivalent (or improved) separation in a fraction of the analysis time required by the packed column, while generating higher sensitivities in a much more inert system (quantitative reliability is improved). As compared to packed columns and packed column analyses, the open tubular column can also confer distinct cost advantages [13].

These points are illustrated in Fig. 1.3 [14], where the short capillary column delivers separation "equivalent" to that obtained in the much longer packed column analysis. Actually, the capillary resolution is superior. Integrated peak areas from the packed column analysis will include appreciable solvent contributions. The solute peaks are well removed from the solvent in the capillary analysis, and quantitation will be enhanced.

Returning to Fig. 1.2, the striking difference between the two sets of chromatographic results illustrated is best attributed to inequalities in the degree of

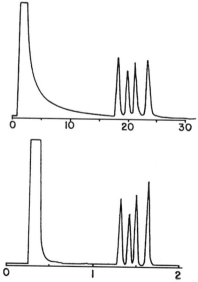

Fig. 1.3. Off-line monitoring of a mixture of methyl benzoates. Top, packed column analysis (25 min), and bottom, 1.7 m × 0.25 mm glass capillary analysis (1.7 min).

randomness exhibited by identical molecules of each individual solute. All identical molecules of each solute exhibit a narrow range of retention times in the bottom chromatogram, but as the ranges of retention times become greater, neighboring peaks exhibit overlap and resolution suffers. These behavioral differences between identical molecules can be attributed to three factors [10]:

1. The packed column offers solute molecules a multiplicity of flow paths: some short, the majority of average length, and some long. Hence identical molecules of each given solute would be expected to spend disparate times in the mobile phase. The open tubular column, on the other hand, has a single flow path, and molecules would be expected to exhibit mobile phase residence times that were much more nearly identical.

2. A similar rationale can be drawn for the randomness of stationary phase residence times. There is much more stationary phase in the packed column, and the film thickness is nonuniform. Thicker regions of stationary phase would be expected to occur in particle crevices and where two or more coated particles come into contact. A solute molecule dissolving in a thinner region of stationary phase would become dispersed and then reemerge to the moving gas phase in a relatively short period. An identical molecule, dissolving in a thicker patch of stationary phase, would take a longer time to reemerge. The times that identical molecules of a given solute spent in the stationary phase would be quite diverse. In the open tubular column, the stationary phase is in a thinner and much more uniform film. Hence, the range of times that identical molecules spent in the stationary phase would be expected to be much narrower.

3. It was previously mentioned that solute volatility (i.e., solute vapor pressures) constitutes an important variable in gas chromatography. The vapor pressure of a solute is an exponential function of the absolute temperature, ergo a minor shift in temperature can have a major effect on vapor pressure. At our present state of instrumental development, solute temperatures are controlled by the temperature of the air in the column oven—the oven air conveys heat to the column wall, the column wall conducts heat to the particles of solid support in contact with the wall, these conduct heat to the stationary phase with which it is coated and to the next particle of solid support, and so on. Packed column support materials, however, are notoriously poor heat conductors. *A temperature range must exist across any transverse section of the packed column.* The range of temperature will be greater for larger-diameter columns and for faster program rates, but even in an isothermal mode, there must be a temperature difference between that packing in contact with the column wall and that packing at the central axis of the column. In an isothermal mode, solute molecules whose flow path is down the center of the packed column will be at a lower temperature, exhibit lower vapor pressures, and spend more of their time in the stationary phase than will identical molecules whose flow paths are closer to the column wall. The problem is exacerbated by

larger diameter-columns and fast heating rates. The fact that these individual flow paths through the column undoubtedly switch back and forth from the central area of the packing to peripheral areas bounded by the column wall does not compensate for this variation. This is one more factor causing identical molecules to exhibit a broadened range of retention times. In the fused silica column, the stationary phase exists as a thin film deposited directly on the inner wall of a tube of very low thermal mass. There should be no temperature variation across any transverse section of the column, provided that the column is not exposed to radiant heat but is heated only by convection. The latter point is an important distinction between the oven requirements for packed and capillary columns and is considered again in a later chapter.

Our goal in chromatography can now be better defined. Gas chromatography should be performed under conditions where (1) solute molecules undergo many interphase transitions (vaporizations and resolutions), (2) identical molecules of each solute exhibit the narrowest possible range of retention times (i.e., the chromatographing band formed by each molecular species is short, hence the standard deviation of the resultant peak is small), and (3) negative contributions (e.g., remixing caused by longitudinal diffusion) must not be permitted to cause excessive degradation of the separation achieved by the processes described above.

1.4   A Simplified Theory of the Chromatographic Process

The primary objective of this book is to hone the practical skills of those using gas chromatography. Practical skills include judicious selection, proper installation, evaluation, and optimized use of state-of-the-art open tubular columns. While a basic comprehension of elementary gas chromatographic theory is essential to attaining these goals, this section is intended as neither a comprehensive nor a rigorous treatment of chromatographic theory. Theoretical considerations have been well covered elsewhere (e.g., [5, 15, 19]). In an attempt to avoid contributing further to the confusion caused by a variety of nonuniform "systems" of nomenclature, the symbols and nomenclature used throughout this discussion are based largely on those suggested by the International Union of Pure and Applied Chemistry (IUPAC) [20] and the American Society for Testing and Materials (ASTM) [21] and are detailed in the Appendix.

A compound subjected to the gas chromatographic process (a "solute") is, on injection into the column, immediately partitioned between the mobile phase and the stationary phase. Its apportionment between the two phases is reflected by the distribution constant K_c, defined as the ratio of the weights of solute per unit volumes of the stationary and mobile phases:

$$K_c = \frac{\text{mass of solute per unit volume stationary phase}}{\text{mass of solute per unit volume mobile phase}} = \frac{W_{i(S)}/V_S}{W_{i(M)}/V_M} \quad (1.1)$$

K_c is a true equilibrium constant, and its magnitude is governed only by the compound, by the stationary phase, and by the temperature. Polar solutes would be expected to dissolve in, disperse through, and engage in intermolecular attractions with polar stationary phases to a much greater degree than would hydrocarbon solutes exposed to the same stationary phase. Logically, the K_c of a polar solute in a polar stationary phase is higher than the K_c of the analogous aliphatic hydrocarbon in the same polar stationary phase. As the temperature of the column is increased, both types of solute exhibit higher vapor pressures and their K_c values (c_S/c_M ratios) decrease, although (in this polar stationary phase) those of the polar solute remain larger than those of the hydrocarbon. Among the members of a homologous series, of course, higher-molecular-weight homologues have lower vapor pressures and higher K_c values.

During its passage through the column, a chromatographing solute spends a fractional part of its total transit time in the stationary liquid phase and the remainder in the mobile gas phase. The mobile phase residence time can be determined by direct measurement. Whenever a solute emerges to the mobile phase, it is transported toward the detector at the same rate as in the mobile phase. Hence, under a given set of conditions, every solute in a given chromatogram *must spend the same length of time in the mobile phase*. This mobile phase residence time can therefore be determined by timing the elution of a substance that never enters the stationary phase, but spends all its time in the mobile phase. Ideally, this could be determined by timing the transit period of an injection of mobile phase (carrier gas), but detection of carrier gas in carrier gas would be impractical. Methane is normally used for this purpose, and although it is recognized that methane does have a discrete stationary phase residence time (particularly in some PLOT columns), it is assumed that this is minuscule and can be ignored in columns of "standard" stationary phase film thickness at reasonable column temperatures.

The column residence time for methane is assigned the symbol t_M, and, as discussed earlier, a solute in the mobile phase is transported toward the detector at the same velocity as the mobile phase. Hence *everything spends t_M time in the mobile phase*. This is the "gas holdup time" (or "gas holdup volume") of the system. The total retention time is equal to the mobile phase residence time t_M, plus the stationary phase residence time. It therefore follows that the stationary phase residence time or the "adjusted retention time" t_R' is

$$t_R' = t_R - t_M \qquad (1.2)$$

Ideally, solute bands will be introduced into the column in such a way that they occupy a very short length of the column (see injection mechanisms in Chapter 3). It is highly desirable that the length of each band increase by a minimum amount as the solute bands traverse the column. Toward this goal, it is desirable that the ranges of retention times exhibited by identical molecules be extremely narrow, i.e., that the standard deviation exhibited by each molecular species is small. As these tight, concentrated bands leave the column, they can be delivered to the detector as narrow, sharp peaks. In actuality, even if the times that identical molecules spend in mobile phase and in stationary phase are precisely the same, other factors such as longitudinal diffusion (occurring primarily in the mobile (gas) phase and to a negligible degree in the stationary phase) cause lengthening of the solute bands during the chromatographic process. The centers of the bands of solutes that have different K_c values will become increasingly separated as they progress through the column, but if the range of retention times exhibited by identical molecules is large or if longitudinal diffusion is excessive, band lengthening may cause the trailing edge of the faster component to interdiffuse with the leading edge of the slower component, resulting in incomplete separation and overlapping peaks. Hence the efficiency with which two components can be resolved is governed not only by their separation factors (see below) but also by the degree of band lengthening that occurs. Insofar as the column is concerned, the separation efficiency is inversely related to the degree of band lengthening. All other things being equal, a minimum degree of band lengthening occurs per unit of column length in a column of high efficiency, and a higher degree of band lengthening occurs per unit of column length in a less efficient column. There also exist extracolumn contributions to band lengthening. These also detract from the separation process and will be considered in later chapters. The term "band broadening" is usually used to describe these phenomena, but as discussed in later chapters, bandwidths are constant and limited by the column diameter. It is really the lengths of those bands that are of concern, because long bands lead to broad peaks.

Inasmuch as both are methods for separating mixtures of volatile compounds, it is not surprising that gas chromatography was promptly compared with the process of fractional distillation, and distillation terminology (i.e., "theoretical plates") was employed (albeit imperfectly; see below) to describe gas chromatographic separation efficiencies.

As detailed above, the separation efficiency of a gas chromatographic column is related to the degree to which a solute band lengthens (which correlates with peak width and affects the standard deviation of the peak, σ) relative to the time the band requires to traverse the column (i.e., its retention time, t_R). The "number of theoretical plates" N is defined as

$$N = (t_R/\sigma)^2 \tag{1.3}$$

where t_R is the time (or distance) from the point of injection to the peak maximum, and σ is the standard deviation of the peak. To avoid the necessity of determining σ, the peak is assumed to be Gaussian (which is usually doubtful; see below), and the problem is simplified. For a Gaussian peak, peak width at base (w_b) is equal to 4.0σ, and peak width at half height (w_h) is equal to 2.354σ (Fig. 1.4). Substitution

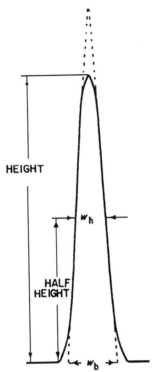

HEIGHT

w_h

HALF HEIGHT

w_b

Fig. 1.4. Characteristics of a Gaussian peak. Peak width at base (w_b, which must be determined by extrapolation from the points of inflection) is equal to 4 standard deviations (σ), and peak width at half height (w_h, which can be directly measured on an on-scale peak) is equal to 2.35 standard deviations (σ).

yields the relationships

$$N = (4t_R/w_b)^2 = 16(t_R/w_b)^2 \tag{1.4}$$

and

$$N = (2.35t_R/w_h)^2 = 5.54(t_R/w_h)^2 \tag{1.5}$$

Because the w_h measurement can be made directly and with greater precision, most workers prefer Eq. (1.5). The same units must, of course, be used for the t_R and w measurements. As mentioned above, these equations assume Gaussian-shaped peaks, and it is very doubtful that the average chromatographic peak is indeed Gaussian. In most cases the deviation is not great, but theoretical plate measurements cannot be applied with any real meaning to peaks that are obviously malformed or asymmetric. Peak asymmetry may be caused by (among other things) reversible adsorption or overloading. These phenomena are discussed in later sections.

Obviously, longitudinal diffusion of components in the column is one factor affecting band lengthening. Lower-molecular-weight solutes (usually characterized by smaller K_c values) would be expected to diffuse to a greater degree per unit time than higher-molecular-weight solutes. Stationary phase diffusivity also affects values of both t_R and w_h (Chapter 5). Consequently, the value of N is a function not only of the column and of the solute but also of the temperature, the type of carrier gas, and the degree to which that gas is compressed (i.e., the pressure drop through the column). These relationships are discussed in later chapters.

The determination of the gas holdup volume t_M by methane injection was discussed briefly above. It is also possible to estimate the value of t_M by calculation [1]. Although t_M is an integral part of t_R and makes a positive contribution to the theoretical plate number, it contributes nothing to the separation process. An empty tube inserted between the injector and the front end of the column would yield larger values of t_M, leading to still larger values for t_R without effecting a change in w_h. This would give a grossly inflated value for the number of theoretical plates, N. This possibility can be eliminated by dealing with an adjusted retention time t'_R:

$$t'_R = t_R - t_M \tag{1.6}$$

This value is used in calculating the number of usable or effective theoretical plates, N_{eff}:

$$N_{eff} = 5.54[(t_R - t_M)/w_h]^2 = 5.54(t'_R/w_h)^2 \tag{1.7}$$

Longer columns (of identical efficiency per unit length) will possess more theoretical plates, although because of complicating factors such as the increased pressure drop, the relationship approximates linearity only at the optimum mobile phase velocity ($\bar{u}_{opt}$; see below).

Efficiencies are sometimes expressed as the number of theoretical plates per meter of column length, e.g., N/m. More often, however, the inverse value—the length of column occupied by one theoretical plate—is used. Once again, distillation terminology is employed, and this is termed the "height equivalent to a theoretical plate." It is usually expressed in millimeters and is given the symbol H:

$$H = L/N \tag{1.8}$$

where L is the column length. Similarly, the height equivalent to one effective theoretical plate is given the symbol H_{eff}:

$$H_{eff} = L/N_{eff} \tag{1.9}$$

Obviously, smaller values of H (or H_{eff}) indicate higher column efficiencies and greater separation potentials. The term H_{min} is used to express the value of H when the column is operating under optimum flow conditions.

Inasmuch as the values of N and N_{eff} are affected by the column temperature, the test compound, the nature of the carrier gas, the carrier gas velocity, the stationary phase film thickness, and the column ID, these parameters must also affect H and H_{eff}.

As previously described, a solute undergoing separation spends a proportion of its transit time in the mobile (gas) phase and the remainder in the stationary (liquid) phase. The sum of these times is, of course, its observed retention time t_R. During the periods when a substance is in the mobile phase, it is moving toward the detector at the same velocity as the carrier gas. Therefore, regardless of their (total) retention times, all substances spend *the same length of time*, equal to t_M, in the mobile phase. The time spent in the stationary phase will therefore be equivalent to the adjusted retention time t'_R. The solute retention factor k is defined as the *amount* (not concentration) of a solute in stationary phase compared to the *amount* of that solute in mobile phase. This is proportional to the *time* the solute spends in stationary phase relative to the *time* it spends in mobile phase:

$$k = t'_R / t_M \qquad (1.10)$$

Theoretical plate numbers (N) are, of course, always larger than effective theoretical plate numbers (N_{eff}), and for a given set of conditions the magnitude of the difference is a function of the retention factor (k):

$$N_{eff} = [k/(k + 1)]N \qquad (1.11)$$

and

$$N = [(k + 1)/k]N_{eff} \qquad (1.12)$$

For a solute whose retention factor is very large (i.e., large k and long retention), t_M is such a minuscule portion of t_R that (for all practical purposes) $t_R = t'_R$. Hence with very large-k solutes, $k \approx (k + 1)$, and $N \approx N_{eff}$.

Figure 1.5 illustrates the effect that the retention factor of the test compound has on the calculated maximum theoretical values of N and N_{eff} for a 50 m × 0.25 mm column. Experimental determinations performed on a series of several solutes rarely exhibit the smooth relationships shown in Fig. 1.5. Part of the reason for this discrepancy lies in the fact that the data for Fig. 1.5 are calculated at the optimum mobile phase velocity for each individual solute, which is a function of the solute retention factor. In many experimental determinations, the mobile phase velocity is constant during the separation of the entire series of solutes. Hence some solutes are chromatographed above (and others below) their optimum velocities. This will have adverse effects on the theoretical plate numbers generated, and the magnitude of those effects will be different for each solute.

Logically, the proportion of the analysis time that a substance spends in the stationary phase relative to the time spent in mobile phase k must be related to its distribution coefficient K_c. This relationship hinges on the relative availability of (i.e., the volumes of the column occupied by) the mobile (gas) and stationary

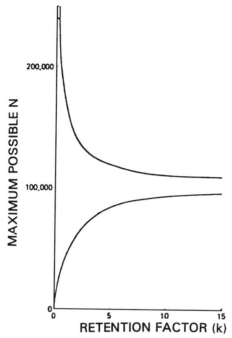

Fig. 1.5. Effect of the solute retention factor (k) on the maximum possible plate numbers for a 50 m × 0.25 mm column. Values calculated at u_{opt} for each value of k. Upper curve, theoretical plates (N); lower curve, effective theoretical plates (N_{eff}).

(liquid) phases (i.e., the column "phase ratio") and is given the symbol β (Table 1.1):

$$\beta = V_M / V_S \tag{1.13}$$

Referring back to Eq. (1.1), it can be seen that the distribution constant K_c can also be defined as

$$K_c = \frac{\text{amount in stationary phase/volume of stationary phase}}{\text{amount in mobile phase/volume of mobile phase}}$$

or

$$K_c = \frac{\text{amount in stationary phase}}{\text{amount in mobile phase}} \times \frac{\text{volume of mobile phase}}{\text{volume of stationary phase}}$$

The latter fraction has just been defined as the phase ratio β, and the former is the same as the retention factor k. From this emerges a relationship that will be utilized later in rationalizing certain injection mechanisms and in understanding the

TABLE 1.1

Phase Ratios of Selected Columns

d (mm)	r (μm)	d_f (μm)	β^a
0.05	25	0.2	63
0.10	50	0.4	63
0.15	75	0.5	75
0.20	100	0.4	125
0.25	125	0.1	625
0.25	125	0.15	416
0.25	125	0.25	250
0.25	125	0.5	125
0.25	125	1.0	63
0.32	160	0.1	800
0.32	160	0.15	533
0.32	160	0.25	320
0.32	160	0.5	160
0.32	160	1.0	80
0.32	160	3.0^b	27
0.32	160	5.0^b	15
0.53	265	1.0	133
0.53	265	1.5	88
0.53	265	3.0^b	44
0.53	265	5.0^b	27

[a] Rounded to the nearest whole number.
[b] These very thick film columns confer other disadvantages and should be restricted to high-diffusivity phases; see Sections 5.12, 5.13, and 6.6.

interrelationships of several parameters in retention and separation characteristics:

$$K_c = \beta k \qquad (1.14)$$

It is apparent that β is a measure of the "openness of the column," and one would expect the phase ratios of open tubular columns to be appreciably larger than those of packed columns, in which the packing not only limits the volume available for mobile phase but also increases the support area over which thicker films of the stationary phase are distributed. The phase ratios of packed columns are usually in the range of about 5–35, whereas in open tubular columns with "normal" (i.e., <0.5 μm) stationary phase film thickness, the values usually fall in the range 50 to 1000. The advent of "bonded" stationary phases has led to open tubular columns with ultrathick (e.g., 3-, 5-, and 8-μm) films, whose phase ratios as may be as low as 10. These special-purpose columns are discussed in later chapters.

The inner surface area of the open tubular column (which at constant film thickness governs V_S) varies directly with column diameter, while the volume of the column (which governs V_M) varies directly with the square of the inner radius, specifically, the distance from the center of the column to the surface of the stationary phase coating. Hence both the diameter of the column and the thickness of the stationary phase film exercise effects on the phase ratio of open tubular columns. For the more commonly used columns, d_f values range from 0.1 to 1.0 μm, and the column diameter is of the order of 200–530 μm. Hence the effect of d_f on the mobile phase volume of the column is usually ignored and the phase ratio is expressed as

$$\pi r^2 / 2r_c \pi d_f \cong r_c / 2d_f \tag{1.15}$$

Inasmuch as the volume of stationary phase is $2r_c \pi d_f$, per unit of column length, the mobile phase volume is more precisely

Total (wall-to-wall) volume − volume occupied by stationary phase

or $\pi r_c^2 - 2r_c \pi d_f$ per unit of column length. A more accurate measure of the phase ratio is therefore

$$\beta = (\pi r^2 - 2r_c \pi d_f)/2r_c \pi d_f = (r - 2d_f)/2d_f \tag{1.16}$$

Because $[(r_c - 2d_f)/2d_f] + 1 = r_c/2d_f$, the values calculated from Eq. (1.15) are exaggerated by one unit. Except for columns of very low phase ratio (e.g., thick film or small diameter), differences between values calculated from Eqs. (1.15) and (1.16) are insignificant. Of even greater import is the fact that because column production tolerances usually approximate ± 0.5 μm for the radius of the tubing and $\pm 5\%$ for d_f, the difference between actual and calculated phase ratios usually exceeds the discrepancy between results derived from the different equations.

The phase ratio of a column can be determined by measuring the k of a solute of known K_c, but the accuracy may be influenced by slight variations in the chemical composition of the stationary phase (as compared to the column used in the initial determination of K_c) or by temperature deviations, both of which would affect K_c. Distribution constants for many of the n-paraffin hydrocarbons on several stationary phases at specific temperatures have been published (e.g., [22–24]). Some of those values are reproduced in Table 1.2. Alternatively, the phase ratio can be determined by comparison with another column of known phase ratio. If it is assumed that the known quantity of stationary phase is uniformly distributed on the inner periphery of open tubular columns coated by a static technique, the phase ratio for that column can be calculated from Eq. (1.15) or (1.16). A test compound can then be chromatographed, and after measuring t_M, k for that compound can be calculated [Eq. (1.9)]. From Eq. (1.14), K_c for the test compound can then be determined, and, as detailed earlier, this value will be the same for that solute on

TABLE 1.2

Selected Distribution Constants

	\multicolumn{9}{c}{Stationary phase}								
	SE-30		OV-210			OV-225		PEG 20M	
Solute	51°C	100°C	48°C	97°C	149°C	97.5°C	148°C	98°C	148.5°C
Hexane	96								
Heptane	229		66						
Octane		87	139						
Nonane			302	45					
Decane			671	80		98			
Undecane				143		183	31		
Dodecane				255	46	348	52		
Tridecane				463	71		82		
Tetradecane					108		134		
Methanol								32	9
Ethanol								35	10
Propanol								64	15
Butanol								113	24
Pentanol								231	38
Hexanol								400	61
Octanol								1192	139
Benzene	161	37	117			42			
Chloroform	102		64			38			

Data selected from Hawkes [23, 24].

any column containing the same stationary phase at that same temperature. When the test compound is then chromatographed on the new column under the same conditions, $K_{c1} = K_{c2}$ and β for the second column can readily be calculated from the relationship

$$\beta_{(2)} = \beta_{(1)}(k_{(1)}/k_{(2)}) \tag{1.17}$$

1.5 Separation of Components

The degree to which two components are separated is a function of (1) the ratio of their retention times and (2) the sharpness of the peaks (N). The ratio of the adjusted retention times of two components, 1 and 2, is termed the separation factor, α. This measures only the separation of the centers of mass of the two peaks:

$$\alpha = t'_{R_{(2)}}/t'_{R_{(1)}} \tag{1.18}$$

From Eq. (1.9) it is readily apparent that the separation factor can also be expressed as

$$\alpha = k_{(2)}/k_{(1)} = K_{c_{(2)}}/K_{c_{(1)}} \tag{1.19}$$

With coeluting solutes, $k_2 = k_1$ and $\alpha = 1.0$. By convention, α is never less than 1.0, so the function of the second (or more retained) solute is always used as the numerator. Solute pairs with large α values can be separated even on low-resolution columns, but as α approaches unity, columns with increasingly larger numbers of theoretical plates are required to achieve separation. Alternatively, of course, one can sometimes select another stationary phase in which the separation factor of those components is larger.

The degree to which two components are resolved is termed resolution R_S:

$$R_S = 2(t_{R(2)} - t_{R(1)})/(w_{b(1)} + w_{b(2)}) \tag{1.20}$$

This definition, however, requires extrapolation to determine the widths of those peaks at their bases. If peaks are assumed to be Gaussian, then, because of the relationship between w_b and w_h discussed earlier in connection with Eq. (1.4) and shown in Fig. 1.4, resolution can also be expressed as

$$R_S = 1.18(t_{R(2)} - t_{R(1)})/(w_{h(1)} + w_{h(2)}) \tag{1.21}$$

Hence resolution (or component separation) is related to the degree of peak broadening *and* retention time differences. As shown in Fig. 1.6, a resolution of 1.0,

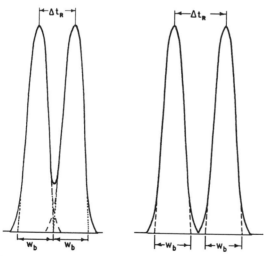

Fig. 1.6. Resolution (R_S) and component separation. With symmetrical peaks, a resolution of 1.5 is usually equivalent to "baseline separation."

while separating idealized peaks, actually results in a considerable degree of over-lap. A resolution of 1.5 will usually achieve baseline separation, but asymmetry, tailing, or gross discrepancies between the sizes of the two peaks can cause complications.

From a knowledge of the separation factor of two compounds, one can closely approximate, under that specific set of conditions, the number of theoretical plates required to achieve a given degree of resolution:

$$N_{req} = 16Rs^2[\alpha/(\alpha - 1)]^2[(k + 1)/k]^2 \qquad (1.22)$$

where k is the retention factor of the second component. This can be rearranged to yield

$$R_S = 1/4\sqrt{N}[(\alpha - 1)/\alpha][k/(k + 1)] \qquad (1.23)$$

Several practical conclusions can be drawn from Eqs. (1.22) and (1.23). First, resolution is proportional to the square root of the number of theoretical plates (or, at $\bar{u}_{opt}$, to the square root of the column length). To double the number of theoretical plates delivered by a 25-m column operating at its optimum carrier gas velocity, a 100-m column of equivalent quality would be required ($25 \times 2^2 = 100$), and the analysis would seem to take four times as long. In actuality, because optimum gas velocities vary inversely with column length (Chapter 5), the longer column has a lower optimum velocity, and analysis on the latter would take more than four times as long, if all other conditions remained constant.

Second, resolution is influenced by both the separation factor α and the retention factor k. This interrelationship is partly responsible for the bewildering array of stationary phases available to the packed column chromatographer. Even at its best, the packed column is capable of delivering only a few thousand theoretical plates. As a result, the packed column chromatographer frequently requires stationary phases of greater selectivity: stationary phases that generate larger α values for those particular solutes under those specific conditions. The open tubular column can achieve much larger theoretical plate numbers. In many cases, a stationary phase unsatisfactory for packed column use (because at that restricted number of theoretical plates, solute separation factors were too small) may deliver complete separation in the open tubular system, accompanied by shorter analysis times and higher sensitivities. These interrelationships may be complicated by the fact that because the phase ratio of the open tubular column is larger, solute retention factors are necessarily smaller [Eq. (1.11)]. Where a very small retention factor is increased, for instance, from 0.01 to 1.0, the factor $[(k + 1)/k]^2$ decreases from 10,000 to 4 and becomes a much less significant multiplier (see Table 1.3). One of the column parameters to be discussed in a later chapter is that of stationary phase film thickness. Thin-film columns are very useful for the analysis of larger and high-boiling solutes but are rarely satisfactory for general-purpose use. Theoretical plate numbers vary indirectly, and retention factors vary directly with the thickness

TABLE 1.3

Values of $[(k + 1)/k]^2$ and N_{req} Corresponding to Different k Values

k	$[(k + 1)/k]^2$	$N_{req}{}^a$	$N_{req}{}^b$
0.01	10,201	162,000,000	44,000,000
0.05	441	7,000,000	1,900,000
0.10	121	1,900,000	527,000
0.15	58.8	930,000	256,000
0.50	9	143,000	39,000
1.0	4	63,500	17,500
3.0	1.8	28,600	7,800
5.0	1.4	22,200	6,100
10.00	1.1	17,500	4,800
50.0	1.05	16,700	4,600

Plate numbers have been rounded and assume $R_S = 1.5$ and $^a\alpha = 1.5$, $^b\alpha = 1.10$.

of the stationary phase film. Under the same conditions, columns with thinner films exhibit higher theoretical plate numbers, but because solutes have smaller retention factors in those columns, separation may be far from satisfactory in spite of the larger plate numbers. As the retention factor of the test compound increases, the factor $[(k + 1)/k]^2$ becomes less and less significant. These relationships are discussed in detail later.

1.6   Effect of Carrier Gas Velocity

A number of valuable concepts are embraced by the van Deemter equation [25], which examines the efficiency of the column (or, after some modification, of the entire system; see later chapters) as a function of the average linear velocity of carrier gas. Gas flows through packed columns are generally described in terms of volumetric flow, (cubic centimeters per minute or milliliters per minute) as measured at the column outlet. With older bubble-type flowmeters, the operator measures the time required for the bubble to traverse a given distance. These were generally replaced by digital-type flowmeters, in which the time between two points is measured by photoelectric sensors. The defect common to all bubble-type meters lies in the fact that readings are increased by the vapor pressure of water. At room temperature, this can yield a 4% error [25]. (The newer family of "bubble-free" flowmeters employs acoustic displacement technology to measure gas flows from 0.1 mL to 1 L per minute. These can be read directly, and require no correction factors).

Gas flow dynamics are much simpler in the open tubular column, and it was soon recognized that solutes are conducted through the column not by the volumetric flow per se, but by the linear velocity of the mobile phase. Volumetric flows are often used to measure and adjust column (i.e., carrier) and extra column (e.g., combustion hydrogen and makeup gas) flows to the detector, but for solute progression through the column, it is preferable to use the "average linear carrier gas velocity" in centimeters per second ($\bar{u}$). This is determined by the injection of a solute for which the detector exhibits good response, and that is assumed to spend zero time in the stationary phase. The elution time of that solute is termed the gas holdup time, t_M, and is used to calculate the average linear gas velocity, $\bar{u}$:

$$\bar{u}_{cm/sec} = L_{cm}/t_{Msec} \tag{1.24}$$

Methane is commonly used with flame ionization detectors (FIDs), methylene chloride functions well for electron capture detection (ECD), in "normal" columns at higher temperatures, and acetonitrile is acceptable for nitrogen/phosphorus detection (NPD), in nitrogen mode. With liquid solutes, the injection is limited to vapors overlying that liquid. Column temperature should be sufficiently high to discourage association of that solute with the stationary phase.

Because volumetric flow is used in calculating detector gas ratios, some users mistakenly attempt to interconvert the two modes of measurement:

$$F = 0.6\pi r_c^2 \bar{u} \tag{1.25}$$

$$\bar{u} = 1.67F/\pi r_c^2 \tag{1.26}$$

where F is the volumetric flow in cubic centimeters per minute, r_c the capillary radius in millimeters, and $\bar{u}$ the average linear carrier gas velocity in centimeters per second. Usually F is measured with a small-volume bubble flowmeter and $\bar{u}$ is determined by methane injection. Such interconversions are obviously inaccurate. F is measured at the column outlet, usually at atmospheric pressure, while $\bar{u}$, when determined as indicated above, is measured above atmospheric pressure. Gases are compressible fluids, and unlike liquid chromatography, where the carrier is a noncompressible fluid, a pressure drop exists through the gas chromatographic column. At the inlet end, pressure and gas density are higher, and linear velocity is lower. At the outlet end, pressure and gas density are lower, and linear velocity is higher. The slope of the velocity curve, and the abruptness with which that curve breaks, are functions of the pressure drop through the column. The average linear gas velocity $\bar{u}$ actually exists at only one point in the column. These topics are discussed in greater detail in Chapter 5.

Column efficiencies are often expressed in terms of the "height equivalent to a theoretical plate," and because $H = L/N$, small values of H denote large values of N and greater separation potentials. The width of the peak is a factor in the determination of N, and as a solute reaches the end of the column and is delivered to the detector, the width of the peak generated by the detector is a function of the time required for the carrier gas to conduct that solute from its position at the end of the column to the detector. This is partly a function of the length of the solute band in the column. It is also a function of the distribution constant of the solute, as discussed earlier. Because the distribution constant is an exponential function of the column temperature, peak widths are affected by the temperature of the column at the time of their elution to the detector (Section 5.7). Obviously, this has an effect on column efficiency measurements based on peak widths, and it complicates efficiency determinations that are conducted under conditions of temperature programming. In most cases, comparative evaluations should employ isothermal conditions.

In its abbreviated form, the van Deemter equation can be written as

$$H = A + B/\bar{u} + C\bar{u} \tag{1.27}$$

where A includes packing and multi-flow-path factors, B is the longitudinal diffusion term, C the resistance to mass transfer from the mobile phase to the stationary phase and from the stationary phase to the mobile phase, and $\bar{u}$ the average linear velocity of the mobile gas phase. Open tubular columns contain no packing and the A term becomes zero, reducing the van Deemter equation to a form known as the Golay equation [7]:

$$H = B/\bar{u} + C\bar{u} \tag{1.28}$$

To achieve the maximum separation potential, the goal is obviously the lowest possible value for H (or H_{eff}), equivalent to the highest possible value of N (or N_{eff}). Elimination of the A term from the van Deemter equation (the Golay equation) is a big step in this direction. As H varies indirectly with the value of $\bar{u}$ in the B term and directly with the value of $\bar{u}$ in the C term, there must exist some optimum value of $\bar{u}$ at which any given system will achieve the highest efficiency for a solute of a given retention factor. This can be calculated or determined graphically, and the interested reader is referred to more general or theoretical references (e.g., [16–18]).

Because plate numbers are a function of the square of the $t_{\text{R}}/w_{\text{h}}$ ratio, H will be decreased by any factors that make this ratio larger and increased by any factors that make it smaller. Although w_{h}, is (in practice) affected by extracolumn factors that include the length of the starting band (i.e., injection efficiency) and the distribution constant (discussed above), if these are held constant w_{h}, reflects lengthening of the band occasioned by its passage through the column (i.e., column efficiency) relative to its transit time (which increases as the solute spends more time in the stationary phase or experiences a greater number of interphase transitions).

One cause of band lengthening is longitudinal diffusion. As discussed in Chapter 5, diffusion constants are of the order of 0.15–0.6 cm^2/sec in mobile phase and approximately 10^{-7} cm^2/sec in the higher diffusivity stationary phases (e.g., polydimethylsiloxane). The contribution of diffusion occurring in the stationary phase is extremely small in comparison with that occurring in the mobile phase, and for most practical purposes stationary phase diffusion can be ignored with respect to band lengthening in open tubular columns. Diffusivity of solute molecules in the gas phase varies inversely with both the density of the carrier gas (i.e., diffusion is greater in hydrogen than in helium than in nitrogen) and the size of the solute molecule (under the same conditions, hexane diffuses more rapidly in helium than does decane). Temperature also has an effect on diffusivity. As the temperature increases, the greater kinetic energy of the individual solute molecules encourages diffusion, but the increased viscosity of the mobile phase discourages diffusion. The gas phase concentration of the solute also plays a role in diffusion; within a given gas under a given set of conditions, a more concentrated vapor plug diffuses more rapidly than does a less concentrated plug of the same vapor.

The two extremes under gas chromatographic conditions would be a stop-flow situation, where band lengthening occasioned by diffusion would be maximal, and a gas velocity sufficiently high to carry the solute band through the column before gas phase diffusion can make a measurable contribution to band lengthening. These considerations are, of course, oversimplifications that assume laminar flow and ignore concentration effects.

The retention time of a solute is governed in part by the time it spends in the stationary phase, specifically, how effectively it utilizes the stationary phase on its route through the column. As an individual solute molecule vaporizes from the stationary phase, it diffuses laterally through the mobile phase (carrier gas) to a new solution site in the stationary phase. During this lateral diffusion it is at the same time being conducted through the column (and past available stationary phase solution sites) by the carrier gas stream. As low carrier gas velocities, a given solute molecule undergoes more resolutions into the stationary phase (hence more vaporizations from the stationary phase) than it does at high velocities. The term "resistance to mass transfer" was coined to measure these dissolution and vaporization processes. "Mass transfer," however, usually implies that special contributions on the part of those molecules at the inter phase (i.e., crossing the interfacial boundary) have been taken into account. Such "surface tension phenomena" almost surely play at least minor roles in gas chromatography, but as currently used, the van Deemter equation treats solute diffusion in the mobile (and in the stationary) phase as the definitive barrier to this step, and "mass transport" seems a more precise and appropriate term.

It should now be apparent that the best results would be achieved under conditions of infinitely slow longitudinal diffusion and infinitely fast lateral diffusion in the mobile phase. Obviously, this will not be possible.

A primary difference between packed and open tubular columns is that the former have thicker and more irregular films of stationary phase, separated by smaller gas volumes. The difference becomes readily apparent if the Golay equation is written so that the two contributions to the "resistance to mass transport" term can be differentiated, that is, expressed as "mass transport from mobile phase to stationary phase" and "mass transport from stationary phase to mobile phase":

$$H = B/\bar{u} + (C_M\bar{u} + C_S\bar{u}) \tag{1.29}$$

Most of the internal volume in the packed column is occupied by the packing. Each particle of packing is coated with stationary phase, and the carrier gas is restricted to the remaining unoccupied volume. Most of the problem in mass transport in the packed column is moving the solute from the stationary phase to the mobile phase. That solute must then transit a comparatively short distance through mobile phase, and is once again in stationary phase. As a result, the C_M term is vanishingly small in comparison to the C_S term. In open tubular columns the reverse is true; with standard film thickness, the solute must transit a relatively short path in stationary phase, and a relatively long path in mobile phase. In the open tubular column, C_S is trivial and can usually be ignored; C_M is limiting. (This generality is true for "normal" open tubular columns. For columns having thick films of stationary phase and/or stationary phases of low diffusivities, C_S becomes increasingly important; see below.) The significance of this distinction and how it affects the choice of stationary phase and carrier gas are considered in Chapter 5.

Again, it can be helpful to (incorrectly) visualize gas chromatography as a stepwise process. As the linear velocity of the carrier gas approaches zero, a vaporizing solute molecule is conducted (longitudinally) through a minimum length of column while it diffuses laterally across the column to its next stationary phase solution site. Under these conditions of very low mobile phase velocity, each solute molecule would experience a maximum number of solution and vaporization steps during its passage through the column, and while retention times would become very large, the value of H—as influenced only by this parameter and ignoring band lengthening occasioned by longitudinal diffusion—would approach zero and be dictated largely by the efficiency of the injection process (Chapter 3). As the velocity of the mobile phase is increased, the length of column through which the vaporizing solute molecule is swept before it can diffuse across the gas volume and dissolve in a new solubility site also increases. Assuming that the rate of transverse diffusion remained constant, the degree to which an average solute molecule is denied access to the stationary phase would be directly proportional to the mobile phase velocity. High carrier gas velocities decrease both the time spent in the stationary phase (fewer solution steps) and the time spent in the mobile phase. Unless the velocity becomes so high that the flow becomes turbulent or solute equilibrations are affected, solute retention factors should remain the same.

In Fig. 1.7, the line representing the resistance to mass transport (lateral diffusion) intersects the ordinate slightly above the abscissa as detailed above. The slope of the line appears to vary directly with column length, but this effect is actually more the result of pressure drop across the column (longer columns operate at higher head pressures) than column length per se. The slope of this

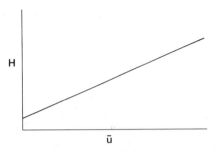

Fig. 1.7. Effect of the average linear carrier gas velocity (u) on column efficiency (H) as influenced only by the "resistance to mass transport" (C) term of the van Deemter (and Golay) equation. Ignoring the longitudinal diffusion term (B), the degree of inter-phase mass transport varies inversely with gas velocity. At lower velocities, N is larger, ergo H is smaller, and at higher velocities, N is smaller and H is larger. Insofar as this factor alone is concerned, N varies indirectly and H varies directly with the gas velocity u.

line also varies directly with column diameter (open tubular columns) and carrier gas density. It varies inversely with column temperature. In other words, anything that inhibits the process of lateral diffusion—lower column temperature (usually, but the resultant decrease in gas viscosity may complicate the net effect of this parameter), higher-k (i.e., higher-molecular-weight, hence larger) solutes, denser carrier gases (longer columns require larger pressure drops, resulting in a denser carrier gas)—or increases the distance over which that diffusion must occur (larger-diameter columns) will produce a steeper slope.

The negative contribution of longitudinal diffusion to separation must also be considered. Component resolution is enhanced by preserving to the largest degree possible the separation achieved by the multiple vaporization steps considered above. A major erosive influence is longitudinal diffusion leading to remixing of partially resolved solute bands. Longitudinal diffusion occurs almost entirely in the mobile gas phase. The band-lengthening contribution of diffusion in the stationary phase is normally minuscule (approximately six orders of magnitude less). Diffusion is a function of time, and the time the solute bands spend in the mobile gas phase is a function of the mobile phase velocity. At a very low mobile phase velocity, the negative contribution of longitudinal diffusion is large, N decreases, and H becomes large. At a high mobile phase velocity, the negative contribution of longitudinal diffusion is small, N remains large, and H is small. Because diffusion is also a function of concentration, the net result of longitudinal diffusion as influenced by the velocity of the mobile phase (ignoring mass transport) is a curve, as shown in Fig. 1.8.

The van Deemter curve [25] represents the sum of the resistance to mass transport term ($H = C_M \bar{u} + C_S \bar{u}$; Fig. 1.8), and the longitudinal diffusion term ($H = B/\bar{u}$; Fig. 1.9), as shown in Fig. 1.10. Experimental curves (see later chapters) vary from the usual theoretical curves in that they continue to curve upward at higher

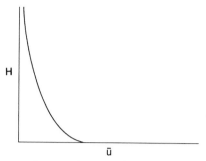

Fig. 1.8. Effect of the average linear carrier gas velocity (u) on column efficiency (H) as influenced by longitudinal diffusion (B term of the van Deemter and Golay equation) alone; resistance to mass transport (C term, shown in Fig. 1.7, above) is not taken into account. See text for discussion.

velocities rather than exhibit a straight line whose slope is dictated by the C term. This discrepancy is related to the fact that at constant outlet pressure, the average linear velocity of the mobile gas phase ($\bar{u}$) can be varied only by varying the inlet pressure of the column, that is, changing the pressure drop. It is the resultant changes in gas velocities and gas densities (which affect gas diffusivities) that are largely responsible for the discrepancies between theoretical and experimental van Deemter curves. The van Deemter treatment is discussed in much greater detail in Chapter 5.

It should be mentioned that optimum velocities vary inversely with the solute retention factor (k), and that this effect is most pronounced at low values of k (Fig. 1.10). The skilled chromatographer usually operates at $\bar{u}_{opt}$ only to prove the system. Most analyses should be performed at higher gas velocities, generating

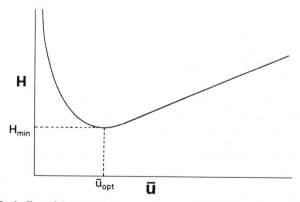

Fig. 1.9. Total effect of the average linear carrier gas velocity ($\bar{u}$) on column efficiency (H), the van Deemter curve represents the summation of the curves shown in Figs. 1.7 and 1.8.

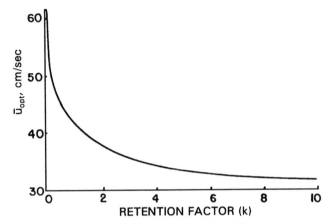

Fig. 1.10. Optimum linear carrier gas velocity as a function of the solute retention factor, k. Note that the effect is extreme only at very low values of k, and becomes much less significant at values of $k = 2$ and higher.

results that are commensurate with the degree of separation required, the efficiency of the column, the range of retention factors embraced by the sample solutes, and the time available for the analysis (see later chapters).

References

1. M. Tswett, *Ber. Dtsch. Bot. Ges.* **24**:316 (1906).
2. M. Tswett, *Ber. Dtsch. Bot. Ges.* **24**:384 (1906).
3. A. T. James and A. J. P. Martin, *Biochem. J.* **50**:697 (1952).
4. N. H. Ray, *J. Appl. Chem.* **4**:21 (1954).
5. Various authors, in *Chromatography* (E. Heftmann, ed). Van Nostrand-Reinhold, Princeton, New Jersey, 1961.
6. Various authors, in *75 Years of Chromatography—A Historical Dialogue* (L. S. Ettre and A. Zlatkis, eds.). Elsevier, Amsterdam, 1979.
7. M. J. E. Golay, in *Gas Chromatography 1957* (East Lansing Symposium) (V. J. Coates, H. J. Noebels, and I. S. Fagerson, eds.), pp. 1–13. Academic Press, New York, 1958; see also *Gas Chromatography 1958* (Amsterdam Symposium) (D. H. Desty, ed.), pp. 139–143. Butterworth, London, 1958.
8. D. H. Desty, J. N. Haresnip, and B. H. F. Whyman, *Anal. Chem.* **32**:302 (1960).
9. R. Daneneau and E. Zerrener, *J. High Res. Chromatogr.* **2**:351 (1979).
10. W. Jennings, in *Glass Capillary Gas Chromatography in Clinical Analysis* (H. Jaeger, ed.), Chapter 1. Dekker, New York, 1985.
11. W. Jennings, *Gas Chromatography with Glass Capillary Columns'* 2nd ed. Academic Press, New York, 1980.
12. T. Shibamoto, in *Application of Glass Capillary Gas Chromatography* (W. Jennings, ed.), p. 455. Dekker, New York, 1981.
13. W. Jennings, *Comparisons of Fused Silica and Other Glass Columns for Gas Chromatography*, Huethig, Heidelberg, 1981.

14. K. Yabumoto, personal communication (1982).

15. J. C. Giddings, *Dynamics of Chromatography*, Part 1. Dekker, New York, 1965.

16. A. B. Littlewood, *Gas Chromatography: Principles, Techniques, and Applications*, 2nd ed. Academic Press, New York, 1970.

17. B. L. Karger, L. R. Snyder, and C. Horvath, *An Introduction to Separation Science*, Wiley, New York, 1973.

18. C. Horvath and W. R. Melander, in *Chromatography* (E. Heftmann, ed.), Part A. Elsevier, Amsterdam, 1983.

19. M. L. Lee, F. Yang, and K. D. Bartle, *Open Tubular Column Gas Chromatography: Theory and Practice*, Wiley (Interscience), New York, 1984.

20. International Union of Pure and Applied Chemistry, Commission on Analytical Nomenclature, Recommendations on Nomenclature for Chromatography, *Pure Appl. Chem.* **37**:445 (1974).

21. American Society for Testing and Materials (ASTM), Committee E-19, "Standard Recommended Practice for Gas Chromatography Terms and Relationships," ASTM E355-77. ASTM, Philadelphia, Pennsylvania, 1983.

22. L. Butler and S. J. Hawkes, *J. Chromatogr. Sci.* **10**:518 (1972).

23. J. M. Kong and S. J. Hawkes, *J. Chromatogr. Sci.* **14**:279 (1976).

24. W. Millen and S. J. Hawkes, *J. Chromatogr. Sci.* **15**:148 (1977).

25. T. Tait, *J. High Resol. Gas Chromatogr.* **17**:191 (1994).

26. J. J. van Deemter, F. J. Zuiderweg, and A. Klinkenberg, *Chem. Sci.* p. 5 (1956).

CHAPTER 2

THE OPEN TUBULAR COLUMN

2.1 General Considerations

The brief comparisons of packed and open tubular columns in Chapter 1 indicate that the latter offer several advantages, including possibilities for improved separations, shorter analysis times, and higher sensitivities. Elimination of the packing materials and the substitution of high-purity fused silica for less inert column materials also makes it possible to construct analytical systems that can generate data of greater quantitative validity [1, 2]. In spite of these advantages, packed columns continued to dominate the analytical field for many years. This retarded conversion to open tubular columns can be attributed to reasons that include the following:

1. The patent on open tubular columns restricted their commercial development because licenses were generally denied and the patent holder put little effort into developments and improvements. As a result, developments and improvements in column technology came largely from individual laboratories and institutions that constructed columns ostensibly for their own use. Column manufacture is an extremely labor-intensive process and few users could justify activity in this area. Hence, most industrial (and many non industrial) users continued to employ packed columns.

2. Instrument- and operator-related problems were initially encountered more frequently with open tubular columns because they are operated at much lower gas flow. As a result, extracolumn contributions, both instrumental and operator-related, become much more important in these low-flow open tubular systems.

Much of the instrumentation necessary had to be user-constructed or user-adapted. Injector and detector volumes had to be reduced and dead volumes (inadequately swept portions of the flow path) had to be eliminated wherever possible. Some chromatographers met these new challenges smoothly. Others preferred to stay with the less challenging (i.e., higher flow volume) world of the packed column. It was largely in response to the above that column manufacturers later directed attention to larger diameter (hence higher flow) open tubular columns. Larger-diameter columns tolerated larger carrier gas flows, and many of these special instrumental modifications could be avoided.

3. Many in the do-it-yourself community that pioneered these developments used stainless steel capillary tubing for their open tubular columns, but such tubing remains expensive and resulting columns usually show a high degree of activity. Segments of the petroleum industry that dealt primarily with hydrocarbon mixtures were less concerned with activity problems. These workers were able to employ stainless steel capillaries and proceeded to make important developments in inlets and detectors. Desty's machine [3] eventually gave us an inexpensive route to long lengths of glass capillary and added great impetus to the development of the modern-day open tubular column.

4. Suitable gas chromatographic instruments were not readily available. To retain a competitive position and remain commercially viable, instrument manufacturers designed their products with a view toward accommodating the largest number of users, and commercial chromatographs of this period were designed for packed columns. Investigators employing capillaries had to adapt packed column instruments to their special requirements. As the proportion of customers adapting their instruments to capillary columns continued to increase, manufacturers began offering options which conferred capillary capability to the packed column instrument; many such conversions were relatively crude. At this stage of development, the state of the art in instrumental design and instrumental modification resided not with the instrument manufacturer but with individual researchers.

5. Highly volatile or very low-molecular-weight solutes (i.e., low-k solutes such as methane, ethylene, ethane) could sometimes be separated on packed columns, but posed a formidable challenge to early open tubular columns. The number of theoretical plates required to separate low-k solutes varies inversely with the magnitude of k [Eqs. (1.22) and (1.23)]. The gas volume of the packed column is small, and gas holdup times (t_M) are shorter in packed than in capillary columns. Hence, solute retention factors (t_R'/t_M ratios) are usually larger in packed than in open tubular columns. The larger retention factors obtained on packed columns reduced the value of N_{req} for extremely low-k solutes to a degree that more than compensated for the lower plate numbers generated by the packed column. This is only true for very low-k solutes. As k becomes larger, the ratio $[(k+1)/k]^2$ approaches unity, and the above generalizations are no longer valid. Much of the advantage that the packed column had for the separation of low-k (or highly

volatile) solutes has now been eroded by the advent of low-β open tubular columns (i.e., open tubular columns with "super thick" films of cross linked, surface-bonded stationary phase). A second distinction is that an adsorptive packing could be used in packed columns designed for the analysis of fixed gases or very low molecular weight solutes. In such circumstances, solute retention is based on adsorption (gas–solid chromatography) rather than solution (gas–liquid chromatography), and solute retention factors become much larger. This causes a significant decrease in the $[(k + 1)/k]^2$ multiplier of Eq. (1.22), decreasing N_{req}. Such adsorptive packings are also available in porous layer open tubular (PLOT) and support-coated open tubular (SCOT) columns.

6. In times past, chromatographers referred to the column as "the heart of the chromatographic system." The overall performance of the system and the quality of the results obtained were most commonly dictated by the quality of the column. Column limitations could usually be attributed to the restricted powers of separation, which were usually related to (1) the limited theoretical plate numbers of the column and (2) its catalytic and adsorptive properties. As a result, both inadequacies in instrument design and imperfections in operator technique were masked. Much higher-quality open tubular columns are readily available today, and this places increased demands on the instrumental design and on operator skill and technique. With few exceptions, today's columns have improved to a point where operator technique and/or instrumental design are more often limiting than is column quality. Analysts (and instrument manufacturers) accept this challenge with varying degrees of enthusiasm.

7. From these considerations, it is understandable that open tubular column chromatography was long reserved to those willing (and able) to devote significant investments of labor, time, and instrumental modification to their analytical endeavors. Most industrial scientists are instead under pressure to produce results, and so their analytical endeavors must frequently employ equipment that can be purchased and operated per se.

Certainly these factors influenced those engaged in analytical methods development. The applicability of a method would be significantly reduced if it required (1) modification of an instrument and/or (2) construction of a suitable column. As one example, most of the earlier gas chromatographic procedures developed by regulatory agencies such as the Environmental Protection Agency (EPA) were necessarily based on packed column analyses. Most such procedures and methods have since been redesigned and now employ open tubular columns.

2.2   The Tubing

A major development that enabled many scientists to enter the field of open tubular gas chromatography occurred in 1960, when Desty *et al.* [3] suggested an elegantly simple machine capable of drawing long lengths of coiled capillary tubing from conventional (i.e., soda lime, borosilicate) glasses. Operating temperatures varied with the type of glass being handled but were approximately 600°C for both the softening oven and the bending tube. With such equipment, capillary tubing was produced at a rate of approximately 1–2 cm/min. Heavier-walled tubing was more common (~0.25 mm ID and 0.75 mm OD), but the advantage of increased flexibility conferred by thinner walls (~0.25 mm ID, 0.5 mm OD) was noted (e.g., [4]).

This machine made it feasible for many investigators to produce glass capillary tubing and, from that tubing, columns. Material costs of such columns were quite low, but labor costs were extremely high. Most of the developmental work in conventional glass capillary columns occurred in government and university laboratories, where labor accounted for an appreciable portion of a relatively fixed budget, permitting labor costs to be ignored. A number of such "centers of excellence," which devoted a major portion of their efforts to improving the glass capillary column and the instrumental hardware associated with it, came into being. In terms of theoretical plate numbers, column efficiencies were often quite high, but aggression toward some solutes could be demonstrated. The most critical points, however, were the fragility of those glass capillary columns and the manipulation required to prepare column ends for installation, factors that precluded their more general acceptance for routine analytical applications. Fused silica capillaries overcame these limitations and have essentially displaced soft glasses in capillary columns.

As a prelude to our considerations of column activity and column deactivation, it is probably wise to begin with the cautionary note that although not all activity problems are due to the column, it is most commonly faulted. Sample components can be abstracted by the container, the syringe, and the syringe needle (*vide infra*). Active sites, that lead to the disappearance of some components and the tailing (i.e., reversible adsorption) of others, often occur in the inlet and in portions of the detector (i.e., the FID flame jet). Diagnosis and correction of these problems are discussed in later chapters. Our attention here will be directed to the column per se.

Column activity can be evidenced as (1) total subtraction of a solute, (2) partial disappearance of a solute, or (3) skewed or tailing peaks. These may be engendered by catalytic activity or by adsorptive processes, which may be either reversible or irreversible. In coping with these problems, our goal is the production of a truly inert surface, fully wettable by the stationary phase, whose inertness will endure even under continuous high-temperature use. The response of the surface to the deactivating treatment will be influenced by the composition of the surface and by its previous history. Although fused silica is much more uniform in terms of composition, it does exhibit batch-to-batch variations, especially in the abundance of free silanols, hydrogen-bonded silanols, siloxane bridges, and their degree of strain. New untreated tubing also contains contaminants and atmospheric residues, some of which were present in the fused silica blank and some of which were generated from other materials during the high-temperature drawing process. Fused silica capillary is drawn at much higher temperatures than Desty's machine could handle (e.g., 2300°C). New tubing should be given a careful cleaning, followed by inert-gas flushing. The degree of surface hydroxylation on virgin fused silica tubing is usually less than that of conventional glass and varies considerably between lots. In accordance with the considerations discussed below, the tubing is usually given an acidic rinse to hydroxylate the surface and dried under conditions that will ensure the removal of free water and still preserve a maximum number of free silanols, avoiding the creation of strained siloxane bridges.

2.3   Sources of Activity

Conventional Siliceous Glasses

From the combined results of a great many investigators, many of which have been detailed elsewhere [4], a better understanding of the activity displayed by glass columns toward some solutes has finally emerged. The siliceous glasses are based on a lattice composed of alternating silicon and oxygen atoms (Fig. 2.1 [1]). The conventional siliceous glasses, however, also contain appreciable quantities of other substances [1, 5, 6], some of which occur fortuitously (e.g., metal oxides) while others are deliberately added during manufacture (e.g., boron). Metal ions (and boron) are characterized by the ability to interact with electron-donor solutes (e.g., many S-, N-, P-, and some O-containing compounds). In addition, the tetravalent silicon atoms exposed at the surface of the siliceous matrix satisfy some of their nonlattice bonds with —OH (silanol) groups, which can also serve as Lewis acid sites. These can interact with solutes to cause activity problems.

The combined contributions of many investigators ultimately led to methods for leaching these nonsiliceous materials (largely metal oxides and boron) from the surface of the glass capillary (see, e.g., [7–9]). However, only active sites that occur at the glass surface are available for removal by these leaching procedures. Potentially active sites buried in the bulk remain unaffected. Diffusion does occur in glass, but is normally very slow because of the high viscosity of the medium. As the glass temperature is increased, diffusivity increases [10]. Flame straightening of column ends (required for soft-glass columns but not for silica columns) accelerates the diffusion of ions buried within the matrix, some of which reach the surface and restore it to an active state. These various defects seriously limit the utility of conventional glasses for open tubular columns.

Quartz and Fused Silica Glasses

The average chromatographer faulted conventional glass capillaries on two points: (1) they were fragile and prone to breakage, and (2) they exhibited excessive

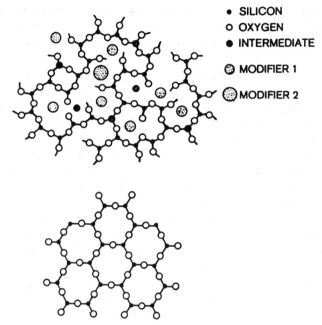

- SILICON
○ OXYGEN
● INTERMEDIATE
◉ MODIFIER 1
◉ MODIFIER 2

Fig. 2.1. Diagrammatic representations of (top) conventional siliceous glasses, and (bottom) fused silica. Silicon atoms, shown as solid black circles, are tetra valent. The additional bond extends upward or downward from these planar views. In the conventional glasses, the regularity of the —Si—O— matrix is disrupted by the inclusion of other atoms. The fourth silicon bond engages in only an occasional crosslink in conventional glasses, but the greater regularity of fused silica results in a highly crosslinked structure and a glass with higher softening and melting temperatures, and with immense tensile strength. Drawings adapted from [5], and reprinted with permission from [1].

activity; i.e., deactivation was a problem. Later studies (some of which are reviewed below) established meticulous procedures by which the surface of the conventional glass capillary could be rendered reasonably inert. Ideal chromatographic connections usually require straightened column ends and heating conventional glass to the point that it can be straightened has an adverse effect on the longevity of that deactivation treatment. In 1979, Dandeneau and Zerenner [11] investigated the chromatographic suitability of several glasses, including fused silica. It should be noted that although the words "quartz" and "silica" are sometimes used almost interchangeably (particularly in the older column literature), the two materials differ in origin and typical impurity levels. In capillary tubing both materials are in the form of silica glass rather than quartz, which is a crystalline material. "Fused quartz" is produced from naturally occurring quartz (principally from Brazil and Madagascar), and is available as bars, ingots, and tubing. It generally contains about 100 ppm (parts per million) of metallic oxides, dominated by those of aluminum and iron. Several refined grades of fused quartz are also commercially available. Refining generally consists of subjecting the finely ground quartz powder to acid leaching and remelting. In some grades, the metallic oxide content has been reduced to about 10 ppm.

The term "fused silica" is usually reserved for a synthetic product. Under "cleanroom" conditions, high-purity synthetic silicon tetrachloride is introduced as a fine spray into a hydrogen flame where the $SiCl_4$ decomposes and combines with water produced from hydrogen combustion to yield silicon dioxide, SiO_2, and HCl. The former is collected electrostatically and melted to produce fused silica. Synthetic fused silica can be produced with a metallic oxide content of less than 1 ppm. Both fused quartz and synthetic silica have been used in column manufacture. The demand for large quantities of high-quality silica from the fiberoptics industry has made this material available at modest prices. Essentially all commercial capillary columns are manufactured from synthetic silica today. Column manufacturers must also pay considerable attention to other variables, such as the level of free hydroxyl (silanol) in the quartz or fused silica and imperfections such as surface flaws that may occur in the tubing blanks.

Dandeneau and Zerenner [11] recognized that because its higher purity resulted in very large, highly crosslinked networks of silica, the tensile strength of this material was immense (refer to Fig. 2.1). This increased strength permitted the construction of exceedingly thin-walled columns (0.20 mm ID × 0.25 mm OD) that were inherently straight, extremely strong, and highly flexible. The highly crosslinked structure also necessitated higher working temperatures. The much more expensive blanks of these high-purity materials undergo softening somewhere above 2000°C. The lifetime of a conventional oven under these conditions would be relatively short. A flowing argon atmosphere is required, and, even then, vaporization of metal ions from the heated apparatus can cause serious problems when these ions are deposited onto the freshly drawn capillary. Consequently, state-of-the-art cleanrooms are typically used for drawing high-quality fused silica capillary. Even nitrogen is reactive at these temperatures and small air leaks into the glass furnace during the draw can seriously degrade the strength of the capillary as well as shorten the lifetime of the furnace. Fused silica and quartz (*vide infra*) also have very narrow ranges of suitable "working temperature." Another complication is that even water vapor and airborne dust can cause flaws in this high-purity tubing that lead to breakage. Because of these various considerations, it is desirable that drawing proceed as rapidly as possible once drawing conditions are idealized. Fused silica is typically drawn at rates that may approach 10 m/min. These restrictions make imperative the use of more sophisticated drawing machinery, operating under cleanroom conditions and based on fiberoptic technology. Figures 2.2 and 2.3 illustrate a state-of-the-art fused silica drawing tower.

These considerations lead to the conclusion that, compared to conventional glass capillary tubing, fused silica tubing is both material-intensive and labor-intensive;

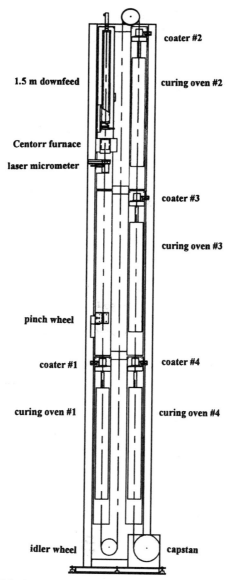

Fig. 2.2. Diagram of the front elevation of a state-of-the-art fused silica drawing tower. The fused silica blank enters the furnace at the upper right, is softened at *ca.* 2300°C, and the capillary is drawn through the laser micrometer to the pinch wheel. Furnace temperature, feed and draw rates are computer controlled, based on data supplied by the laser micrometer. A thin coating of polyimide deposited by the coater is then dried by the curing oven. The tubing passes over the idler wheel, returns to the top of the ten meter tower, receives three more exceedingly thin coatings of polyimide, and passes via the capstan to storage reels. *Courtesy of Roger Schirmer and Paul O'Conner, J&W Scientific, Inc.*

Fig. 2.3. Photograph showing a frontal view of the tower diagrammed in Fig. 2.2. All components detailed in Fig. 2.2 are duplicated on the rear of the tower, providing a second drawing unit that can be utilized in research without interfering with the production unit on the front. *Courtesy of Roger Schirmer and Paul O'Conner (shown at top right), J&W Scientific, Inc.*

it is an expensive product. Many of the "centers of excellence" that contributed so much to the understanding and development of capillary gas chromatography with glass capillary columns were effectively priced out of the market by the introduction of fused silica capillaries. To some degree, this situation has improved.

2.4   Structural Flaws

All glasses have surface flaws, and fused silica is no exception. Many such defects survive the drawing process to produce flaws in the drawn tubing. These flaws grow at a rate that is proportional to the stress on the tubing. When they become sufficiently large, the tubing breaks. To minimize these flaws, the fused silica blanks

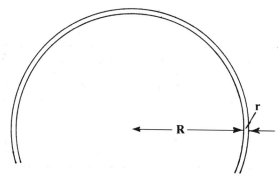

Fig. 2.4. Bending stress arises from the difference in the inner circumference $(2\pi R)$ and outer circumference $[2\pi((R + r)]$ of the bent capillary tubing. The inner circumference of the tube is compressed, and the outer circumference stretched, a situation that promotes breakage at the location of flaws in the glass.

used in column production are usually subjected to careful selection and then meticulously fire polished. Even so, some flaws persevere to produce weakened points in the drawn tubing. In addition, particulate matter (e.g., dust) that comes in contact with the tubing during these early phases of its manufacture will result in surface flaws. Drawing must take place under conditions of extreme cleanliness.

Flaws in the drawn tube are critical because the capillary is under high stress even in normal use. The inner and outer circumferences of the capillary are slightly different (Fig. 2.4) when it is coiled in the usual column configuration. The small differences in circumference combined with the unusually large Young's modulus of fused silica $[1 \times 10^7 \text{ lb/in}^2 \text{ (psi)}]$ results in very high tensile stress when the tubing is coiled. The configuration of the column, installed and connected, should offer the least possible strain. Tight bends will inevitably lead to eventual breakage. The magnitude of the stress increases rapidly with internal capillary diameter (Table 2.1 and Fig. 2.5), and these precautions can become even more important where larger-diameter tubing is connected to prepositioned gas sampling valves.

The typical fused silica capillary can withstand stress of at least 100,000 psi. Tensile strengths in excess of 1×10^6 psi are theoretically possible for perfect silica glass. Even very small flaws significantly degrade the strength of the capillary. The size of a critical flaw can be estimated by applying Griffith's theory [12, 13] to silica glass [14]. The largest flaw present in a silica capillary with a tensile strength of 100,000 psi is limited to approximately 5 nm in diameter. The largest flaw that can be present if a capillary is to withstand 36,000 psi—the approximate stress on a 0.53-μm-ID column wound in a 7-in coil—is only about 43 nm, far below the limits of detection by optical microscopy. The ready availability of capillary tubing with strengths suitable for chromatographic use is a remarkable accomplishment considering the extreme requirements imposed on the quality of the glass.

TABLE 2.1

Coiling Stress on a Capillary Column

Column coil diameter (inches)	Capillary internal diameter/external diameter (mm)		
	0.250/0.326	0.320/0.406	0.530/0.650
5.0	25.6	31.9	50.9
6.0	21.3	26.6	42.5
7.0	18.3	22.8	36.4
8.0	16.0	19.9	31.9

Coiling stress (in thousands of psi) on fused silica tubing of given ID and OD, at coil diameters ranging from 5 to 8 inches. At a given coil diameter, stress varies directly with the diameter (or radius) of the tubing. For tubing of a given ID/OD, stress varies indirectly with coiling diameter.

Michalske and Freiman [15] reported that the siliceous glasses undergo a time-related decrease in mechanical strength which is encouraged by static load and certain environmental factors. This structural weakening is due to a stress corrosion process consisting of a growth (usually slow) of preexisting surface flaws. Their data indicate that the velocity of flaw extension (i.e., crack growth) is related to the stress intensity and is encouraged by liquid water, moist gas, or ammonia. Stress frozen in the glass during drawing and the absence of grain boundaries to interrupt crack propagation make fused silica especially susceptible.

Conventional glass capillaries were forced through a curved tubular furnace (bending tube) to produce a continuous coil [4]. Interfacing the finished column to inlets and detectors usually required straightening the column ends. Fused silica is drawn as a long straight tube, which is then forced to a coiled configuration. The finished column is much easier to interface to inlets and detectors, but the coiled column is under stress. The degree of stress varies directly with the radius of the tubing and inversely with the coiling diameter (Fig. 2.5).

To inhibit these surface corrosion processes, the thin-walled flexible fused silica tubing is drawn through a reservoir of polyimide solution as it emerges from the drawing oven. The coating thus deposited on the outer tubing wall is imperfect and has pinholes. The freshly coated tubing is usually drawn through a tubular drying oven, recoated, and redried before the drawing rollers can contact the tubing. In some cases, it is then subjected to additional coating treatments [16]. The polyimide coating probably serves two functions: (1) it floods and fills existing flaws, which tends to discourage their continued growth; and (2) it seals the outer surface of the capillary with a waterproof barrier, limiting corrosive attack by (e.g.), water vapor. For reasons that are not completely understood, the inner surface of the tubing

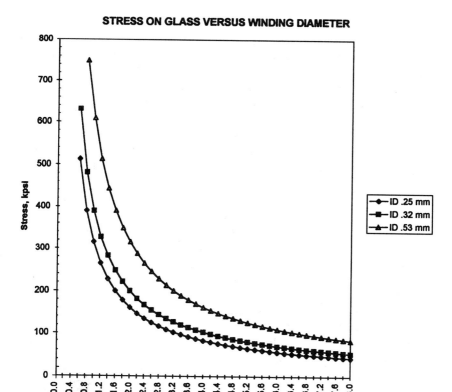

Fig. 2.5. Stress on glass tubing, in psi $\times 10^{-3}$, as a function of coiling diameter, for three different tubing diameters. See also Table 2.1. *Courtesy of Roger Schirmer, J&W Scientific, Inc.*

is much less susceptible to these corrosion problems, although acid leaching is known to decrease the strength of the capillary significantly. The interior surface of the "virgin" fused silica tubing is remarkably smooth [17], but is affected by the cleaning, hydroxylation, and deactivation processes to which it may be subjected in the column manufacturing process.

In attempts to detect flawed tubing so that it can be discarded as early as possible in the manufacturing process, most manufacturers first subject the tubing to dynamic and "proof" testing. The flaws that survive this test and the column manufacturing processes are usually minor and cause no problems. On rare occasions, user-sustained breakage can be traced to flaws that should have been revealed by the above procedures, but breakage in the hands of the user more commonly results from abrasion or scratching of the outer wall. This violates the integrity of the polyimide film and permits corrosion at that point.

While the polyimide coating has excellent mechanical resistance and good temperature tolerance, it also has a tendency to render the column opaque. When first applied, the better-quality material has a light golden color and is semitransparent. At temperatures in excess of 300°C, it becomes darker; the darkening is encouraged by oxygen and is apparently influenced by trace impurities in the polyimide that vary from batch to batch. In most cases, we would prefer a transparent column. It not only makes possible the discernment of localized defects in the column but also can be advantageous in certain injection modes (see Chapter 3). Some developments in pyrocarbon coatings have been rumored for years, but have never come to fruition, and there are periodic reports of thermally stable transparent coatings (e.g., [18]). Aluminum and nickel coatings have been applied to fused silica tubing by vapor phase deposition, but the cost of the completed tubing is significantly higher.

It is also possible to apply an outer metal sheath over the column by drawing the uncoated fused silica tubing through a bath of molten aluminum [19]. There are some reports that breakage of the fused silica inside this sheath, which would be difficult to detect, is not uncommon. Molten aluminum has a very high surface energy and would be less effective at flooding and sealing existing flaws in the surface. The major utility of such columns would appear to be for extremely high-temperature use (e.g., 450–500°C), where polyimide coatings would be entirely unsatisfactory. The upper feasible temperatures for gas chromatography, however, are imposed not only by the outer coating on the column but also by the temperature-dependent degradation of stationary phases (see below), and sometimes by the thermal stability of the solutes themselves. For analysis of crude oils and other applications requiring extreme temperatures ($>400°C$), the new generation of stainless steel columns offers a better route. Although the principles of their deactivation processes are apparently different, several manufacturers now offer stainless steel open tubular columns. Some examples appear in Chapter 10.

The temperature-related stresses to which the column is subjected become much more severe at these extreme temperatures, and column lifetimes may be considerably shortened. Stationary phases are, in general, subject to oxidative degradations that are both oxygen- and temperature-dependent: oxygen levels so minute that they may be tolerable at 300°C can lead to abbreviated column lifetimes at extremely high temperatures (see Chapters 4 and 10). Because of their greater resistance to thermal and oxidative degradation, the silarylene–siloxane phases discussed in Chapter 4 offer advantages for these applications. Even so, some very low volatility solutes are probably better analyzed by liquid or supercritical fluid chromatography.

2.5 Flexible Columns of Conventional Glasses

Consideration should also be given to another aspect which has been a source of confusion for some chromatographers. Certainly the blanks for drawing either quartz or fused silica capillaries are much more expensive than blanks of conventional glass, and the former require much more expensive drawing machinery and much more exacting processes. Because of these constraints, quartz and fused silica capillary tubing are much more expensive than conventional glass capillary tubing. Conventional glasses can also be drawn to very thin-walled capillaries, and there are reports that by coating such tubing with polyimide, one can produce less expensive columns that have the other advantages of quartz and fused silica (e.g., [20]). Most authorities regard those claims as highly exaggerated. The use of various conventional glasses to construct thin-walled flexible columns is one of the many alternative routes that column manufacturers have exhaustively explored and then abandoned. The fact that such tubing is weaker, less flexible, and more subject to static fatigue might be tolerable, but there is a more serious defect: columns produced from that tubing cannot be well deactivated, because the starting column is so thin that it cannot tolerate the leaching step required in most conventional deactivation treatments. Leaching renders that very thin-walled tubing extremely fragile [21]. Consequently, deactivation treatments are limited to one of the archaic and thermally labile treatments for masking active sites. The net result is a less flexible, very fragile column that is less inert (indeed, it can be extremely active) and that must be restricted to lower operating temperatures [22]. Conventional glass capillaries have also been coated internally with an inner envelope of polyimide prior to coating with the stationary phase [23]. Again, the polyimide-coated surface requires masking prior to application of the stationary phase, and interest in such columns was short-lived.

2.6 Silanol Deactivation

The results originally reported by Dandeneau and Zerrener [11], supported by many subsequent reports, indicated that fused silica columns are in most cases far more inert than columns of conventional glasses. The activity of the siliceous glasses is due to both metal oxides and surface silanols (plus strained siloxane bridges that can revert to silanol groups) [1]. In conventional glass columns, both contribute to chromatographic defects, and both require attention. As discussed earlier, metal contaminants in the surface of conventional glass columns are generally removed by leaching, rinsing, and careful drying, before attention is directed to the surface silanols. Fused silica has essentially no metal contamination, and the problem of deactivation is reduced to one of dealing with surface silanols, but there are some complicating factors. Because of the difference in the temperatures required for their drawing, conventional glasses possess a number of strained siloxane bridges [1] that readily rehydrate to silanols, while fused silica is less hydroxylated. Fused silica blanks can be obtained with high or low hydroxyl content, and drawn tubing can be readily rehydroxylated by washing with aqueous acids (e.g., [24]).

The concentration of silanols on the surface of porous silica particles has been widely studied. The concentrations on fully hydrated silicas range from 6 to10 silanols per square namoter of surface or 10–17 μmol/m^2 [25]. The smooth surface inside a synthetic silica capillary may be closer to 4 silanols/nm^2 and on a dehydrated capillary as low as 0.2 silanols/nm^2 [26]. The art of preparing a high-quality capillary column takes advantage of these groups to provide sites for surface modification while leaving as few free silanols as possible to contribute to column activity.

Earlier methods for silanol deactivation have been reviewed elsewhere (e.g., [4, 27]). Our treatment here will be abbreviated. Because it had been used successfully to deactivate silanol groups on siliceous packings, silylation, primarily with the chlorosilanes, was one of the earliest deactivation procedures investigated for glass capillaries (e.g., [28]). However, used either solely for the deactivation of glass capillaries or jointly for their deactivation and surface modification (by bonding other functional groups to the surface), these procedures usually produced columns that were faulted for excessive activity and/or low efficiency. The first encouraging silylation results employed hexamethyldisilazane (HMDS) at 300°C for up to 20 hr [29]. Grob et al. [7] reasoned that many of the inconsistencies subsequently reported by a variety of authors could be corrected by first resorting to a leaching process, which served two purposes: (1) dissolution and removal of metal oxides from the glass surface (the aggressiveness of these active sites would not be lessened by silylation) and (2) modification of the silica structure so that it presented an "optimum number" of silanol groups (i.e., the surface should be fully hydroxylated). One critically important step, then, is that following leaching; the surface must now be dehydrated, but not dehydroxylated. In other words, water, which would reduce the efficacy of silylating reagents and also react with disilazanes to produce high concentrations of ammonia, must be removed. Hydrogen-bonded silanols, however, should remain as free silanols and not be dehydroxylated to form strained siloxane bridges.

Wright et al. [30] also explored the use of chlorosilanes for high-temperature deactivation. Schomburg et al. [9] demonstrated that in situ-produced thermal decomposition products of polysiloxane phases effectively neutralized active sites. Stark et al. [31] reasoned that these decomposition products must be siloxane fragments and introduced the concept of using cyclic trimers and tetramers of —Si(CH$_2$)$_2$—O— in high-temperature vapor phase deactivation. Superior results were achieved with octamethyltetracyclosiloxane, and hexamethyltricyclosiloxane was reported to be less effective. The idea was further developed by Blomberg et al. [32, 33]. Lee et al. [34] reported that at 420°C the former reagent (tetramer) blanketed the fused silica surface with a thin film of glassy polysiloxane resin, in which localized pools of resin were evident. They suggested that this polymeric resinous coating was essential to effective deactivation and that simple silylation of surface silanols, which would be expected to occur at more moderate temperatures, was insufficient. They also reported that deactivation treatments utilizing the chlorosilanes were generally inferior to those employing either the cyclic siloxanes or polysiloxanes.

2.7 Column Coating

Stationary phases are discussed from the standpoint of "polarity," functionality, diffusivity, and other factors in Chapters 4 and 5. At this point our concern is confined to generalities of the coating operation. The goal, of course, is to deposit on the inner periphery of the cleaned, deactivated tubing a thin, uniform film of stationary phase. The role of many interrelated factors and the several methods used for coating stationary phases have been reviewed elsewhere [4]. The static technique (with various minor modifications) is most widely used today.

The stationary phase may endure on the column wall because (1) of cohesive or wetting forces; (2) the stationary phase in contact with the surface may be bonded to the surface to produce a stable monolayer, wetted by the remainder of the stationary phase; (3) it may be crosslinked to produce a liner or envelope of stationary phase polymer within the column; or (4) it may be both crosslinked and surface-bonded. Crosslinking can produce a stationary phase that is nonextractable under normal use conditions. This can offer distinct advantages for use with either splitless or on-column injection, where the column is flooded with liquid. It is also advantageous where the column has become contaminated with soluble but nonvolatile or high-boiling residues (as in on-column injection), as it enables these to be washed from the column with a suitable solvent (see later chapters). Crosslinking is advantageous for "superthick" (i.e., $1.5–5-\mu m$) film columns, in which the envelope of stationary phase can otherwise collapse and occlude the column [36]. Crosslinked surface-bonded phases are also desirable for columns used with supercritical fluid and liquid chromatography.

Surface bonding has been employed to attach stationary phases to packings, but this approach has not proved useful in capillary chromatography. The earliest results probably date to Bossart [36] and the polybutadiene layer bonded by Grob [37]. Because they utilized raw, untreated glass surfaces, the latter efforts produced columns that, while interesting, had low thermal stability. Later developments have made it clear that surface preparation is an integral part of the bonding process. Further developments can be traced to attempts by Blomberg and Wannman [38] and by Grob [39] and to the initial discovery by Jenkins [40] (see Chapter 4). Additional efforts that contributed to these developments, consideration of which is beyond the scope of this book, appear in Refs. 41–45. Grob's book [46] will prove especially valuable to those who wish to manufacture their own columns.

Details of stationary phase properties and other facets of the bonding and crosslinking reactions are discussed later in Chapter 4. In general, the initiators used to trigger free-radical reactions resulting in C—C crosslinking can be divided into peroxides (e.g., [47–53]), azo compounds (e.g., [54, 55]), gamma radiation from the cobalt isotope ^{60}Co [56–58], accelerated electrons from a Van de Graaff generator [59], and ozone [60]. Because the products of their decomposition are volatile, dicumyl peroxide (e.g., [47, 49]) and azo compounds (e.g., [54]) are widely used. Ionizing radiations have been faulted for causing a loss of flexibility in fused silica columns [61].

A final note, based on experience gained from the commercial manufacture of hundreds of thousands of columns over some 25 years, seems in order. Progress in the preparation of high-efficiency columns of increased inertness has been excellent. Until quite recently, most of our progress was based on the accumulation of bits and pieces of information from a variety of different investigations. Evaluation of this evidence required the reconciliation of apparent contradictions and discounting of some claims that seemed exaggerated. Most of these reports have been concerned with results obtained on one to several columns, and those samples are often too small for the law of averages to become evident. Results obtained with larger lots sometimes indicate that many of the results reported would appear to reflect an exception rather than a rule. In recent years, a much larger percentage of our progress in surface pretreatments, deactivation, and stationary phase synthesis is attributable to the research and development efforts of commercial manufacturers. Unfortunately, much of this information is regarded as proprietary.

References

1. W. Jennings, *Comparisons of Fused Silica and Other Glass Columns for Gas Chromatography.* Huethig, Heidelberg, 1981.
2. W. Jennings, *J. High Res. Chromatogr.* 3:601 (1980).
3. D. H. Desty, J. N. Haresnip, and B. H. F. Whyman, *Anal. Chem.* 32:302 (1960).
4. W. Jennings, *Gas Chromatography with Glass Capillary Columns*, 2nd ed. Academic Press, New York, 1980.
5. J. R. Hutchins, III and R. V. Harrington, *Encyclopedia of Chemical Technology*, 2nd ed., Wiley, New York, 10:533 (1966).
6. G. W. Morey, *The Properties of Glass*, 2nd ed. Van Nostrand-Reinhold, Princeton, New Jersey, 1954.
7. K. Grob, G. Grob, and K. Grob, Jr., *J. High Res. Chromatogr.* 2:31 (1979).
8. G. Schomburg, H. Husmann, and H. Borowitzky, *Chromatographia* 12:651 (1979).
9. G. Schomburg, H. Hussmann, and H. Behlau, *Chromatographia* 13:321 (1980).
10. R. H. Doremus, *Glass Science.* Wiley, New York, 1973.
11. R. Dandeneau and E. Zerenner, *J. High Resol. Chromatogr.* 2:35 (1979).
12. A. A. Griffith, *Trans. Royal Soc.* 221:163 (1921). (London).
13. S. W. Freiman, "Fracture Mechanics of Glass," in *Glass Science and Technology*, Vol. 5, *Elasticity and Strength in Glasses* (D.R. Uhlmann and N. J. Kreidl, eds.), Academic Press, New York, 1980.

14. R. A. Flinn and P. K. Trojan, *Engineering Materials and Their Applications*, 2nd ed. Houghton-Mifflin, Boston, 1981, p. 541.

15. T. A. Michalske and S. W. Freiman, *Nature* (London) **295**:511 (1982).

16. S. R. Lipsky, W. J. McMurray, M. Hernandez, J. E. Purcell, and K. A. Billeb, *J. Chromatogr. Sci.* **18**:1 (1980).

17. R. Barberi, M. Giacondo, R. Bartolino, and P. Righetti, *Electrophoresis* **16**:1445 (1995).

18. C. DeLuca, personal communication (1983).

19. R. Dandeneu, personal communication (1982).

20. R. P. W. Scott, *Trends Anal. Chem.* (pers. ed.) **2**:1 (1983).

21. S. R. Lipsky, *J. High Res. Chromatogr.* **6**:452 (1983).

22. W. Jennings, *Am. Lab.* **16**(1):14 (1984).

23. J. Balla and M. Balint, *J. Chromatogr.* **299**:139 (1984).

24. M. W. Ogden and H. M. McNair, *J. High Res. Chromatogr.* **8**:326 (1985).

25. R. K. Iler, *The Chemistry of Silica*, Chapter 6 Wiley, New York, 1979.

26. B. W. Wright, P. A. Peaden, G. M. Booth, and M. L. Lee, *Chromatographia* **15**:584 (1982).

27. M. L. Lee, F. J. Yang, and K. D. Bartle, *Open Tubular Column Gas Chromatography: Theory and Practice*. Wiley (Interscience), New York, 1984.

28. M. Novotny and K. Tesarik, *Chromatographia* **1**:332 (1968).

29. T. Welsch, W. Engwald, and C. Klaucke, *Chromatographia* **10**:22 (1977).

30. B. W. Wright, M. L. Lee, S. W. Graham, L. V. Phillips, and D. M. Hercules, *J. Chromatogr.* **199**:355 (1980).

31. T. J. Stark, R. D. Dandeneau, and L. Mering, *Pittsburgh Conf. Anal. Chem. Appl. Spectrosc., 1980*, Abstract 002 (1980).

32. L. Blomberg, K. Markides, and T. Wannman, *J. High Res. Chromatogr.* **4**:527 (1980).

33. L. Blomberg, K. Markides, and T. Wannman, *Proc. 4th Int. Symp. Capillary Chromatogr. 1981*, p. 73 (1981).

34. M. L. Lee, R. C. Kong, C. L. Wooley, and J. S. Bradshaw, *J. Chromatogr. Sci.* **22**:136 (1984).

35. K. Grob and G. Grob, *J. High Res. Chromatogr.* **6**:133 (1983).

36. C. J. Bossart, *ISA Trans.* **7**:283 (1968).

37. K. Grob, *Helv. Chim. Acta* **51**:729 (1968).

38. L. Blomberg and T. Wannman, *J. Chromatogr.* **168**:81 (1979).

39. K. Grob, *Chromatographia* **10**:625 (1977).

40. R. G. Jenkins and R. H. Wohleb, *Pap., Expochem.* 80 (1980).

41. K. Grob and G. Grob, *J. Chromatogr.* **213**:211 (1981).

42. K. Grob and G. Grob, *J. High Res. Chromatogr.* **5**:13 (1982).

43. T. J. Stark and P. A. Larson, *J. Chromatogr. Sci.* **20**:341 (1982).

44. P. A. Peaden, B. M. Wright, and M. L. Lee, *Chromatographia* **15**:335 (1982).

45. L. Blomberg, J. Buijten, K. Markides, and T. Wannman, *J. Chromatogr.* **239**:51 (1982).

46. K. Grob, *Making and Manipulating Glass Capillary Columns*. Huethig, Heidelberg, 1986.

47. J. Buijten, L. Blomberg, K. Markides, and T. Wannman, *Chromatographia* **16**:183 (1983).

48. P. Sandra, M. Van Roelenbosch, 1. Temmerman, and M. Verzele, *Chromatographia* **16**:63 (1983).

49. K. Markides, L. Blomberg, J. Buijten, and T. Wannman, *J. Chromatogr.* **254**:53 (1983).

50. K. Grob, G. Grob, and K. Grob, Jr., *J. Chromatogr.* **211**:243 (1981).

51. P. Sandra, G. Redant, E. Schacht, and M. Verzele, *J. High Res. Chromatogr.* **4**:411 (1981).

52. L. Blomberg, J. Buijten, K. Markides, and T. Wannman, *J. High Res. Chromatogr.* **4**:578 (1981).

53. J. Buijten, L. Blomberg, K. Markides, and T. Wannman, *J. Chromatogr.* **268**:387 (1983).

54. B. E. Richter, J. C. Kuei, N. J. Park, S. J. Crowley, J. S. Bradshaw, and M. L. Lee, *J. High Res. Chromatogr.* **6**:371 (1983).

55. B. E. Richter, J. C. Kuei, J. S. Bradshaw, and M. L. Lee, *J. Chromatogr.* **279**:21 (1983).

56. G. Schomburg, H. Husmann, S. Ruthe, and M. Herraiz, *Chromatographia* **15**:599 (1981).

57. W. Bertsch, V. Pretorius, M. Pearce, J. C. Thompson, and N. Schnautz, *J. High Res. Chromatogr.* **5**:432 (1982).

58. J. A. Hubball, P. R. Dimauro, S. R. Smith, and E. F. Barry, *J. Chromatogr.* **302**:341 (1984).

59. K. Markides, L. Blomberg, J. Buijten, and T. Wannman, *J. Chromatogr.* **267**:29 (1983).

60. J. Buijten, L. Blomberg, S. Hoffmann, K. Markides, and T. Wannman, *J. Chromatogr.* **289**:143 (1984).

61. J. A. Hubball, P. Dimauro, E. F. Barry, and G. E. Chabot, *J. High Res. Chromatogr.* **6**:241 (1982).

CHAPTER 3

SAMPLE INJECTION

3.1 General Considerations

Component separation is affected both by the ratio of solute retention factors (α) and by the width of solute peaks (Chapter 1). Peak widths are influenced by, among other things, the efficiency of the injection process: the shorter the length of column occupied by the injected sample, the shorter the band as it begins the chromatographic process, the shorter the band as it completes the chromatographic process and is delivered to the detector, the sharper the peak generated by the detector, and the better the separation. One of the most critical functions of the injection process is to introduce the sample so that it occupies the shortest possible length of column.

Solute bands lengthen, inevitably and inexorably, during the chromatographic process. In the ideal case, the increase in length would result only from longitudinal diffusion in the mobile phase. In actuality, it is usually influenced by other factors, such as active sites that delay or prolong the passage of some but not all molecules of a given solute, or irregularities in the stationary phase coating. These phenomena increase the degree of randomness in the behavior of identical solute molecules and lead to longer bands. Longer bands produce broader peaks which have adverse effects on component separation.

In assisting the reader to more easily correlate our discussion with other presentations, it should be pointed out that the above phenomenon is often referred to as "band broadening," probably because the end result is broadened peaks. However, the transverse dimension of the band (i.e., bandwidth) is obviously limited by the column, and so it is the longitudinal band dimension that varies. Consequently, it is actually the length of the band that is under consideration.

It is also important that the qualitative and quantitative composition of the injected band accurately reflect that of the original sample and that it contain detectable amounts of each substance to be quantitatively determined. Factors influencing the degree to which this is accomplished include

1. Discriminatory effects during transfer of the sample from the syringe to the inlet (these are usually, but not always, of greater concern in heated vaporizing injectors)
2. Discriminatory effects during transfer of the sample from the inlet to the column (for one thing, inlets often abound in active sites)
3. The length of column initially occupied by the injected sample (all modes)
4. The rate and efficiency of sample vaporization in the inlet (vaporizing modes, i.e., split, splitless)
5. The speed of sample transport from the inlet to the column (split, splitless, PTV)
6. The completeness of sample transport from the inlet (all modes)
7. The homogeneity (or heterogeneity) of (a) the temperature and (b) the phase ratio (both real and "apparent," i.e., condensed liquid) throughout the entire length of the sample band, i.e., "distribution constant" and "phase ratio" effects (*vide infra*)

Many facets of the injection process have previously been considered in some detail [1], and our treatment here will be largely supplementary. Different procedures are usually employed for injections into large-diameter (i.e., greater than >0.35-mm-ID) columns, as contrasted with small-diameter (i.e., less than <0.35-mm-ID) open tubular columns. Some basic considerations apply to both.

3.2 Extrachromatographic Phenomena Influencing Band Length

Syringe Technique

Discriminatory effects occasioned by the syringe have been considered elsewhere [1, 3, 4]. Briefly, hot-needle fractionation occurs when heat is transferred from the hot injection port to the needle. The heated needle then fractionates the sample in the seemingly short time of the injection process. Placing a 1- or 2-μL air plug in the syringe before and after the sample will mitigate this effect somewhat because the air plug will push out most of the sample during injection. Filled-needle injections or plunger-in-needle syringes are almost always undesirable in the split or splitless capillary modes, the so-called hot-vaporization injection modes. Split mode requires a fast injection. Although manual injection is not fast enough to avoid hot-needle fractionation, very fast auto injectors are available that will virtually eliminate this type of mass discrimination (or "split") [2]. The needle dwell time in the injector during a fast manual injection is on the order of a second, while a fast autoinjector dwell time is under 500 msec. Splitless injection usually gives better results when the "hot needle" technique [1, 3, 4] is employed.

Under some conditions, other facets of the syringe technique are critically important to the chromatographic results obtained. This is particularly true with on-column injections, where the syringe needle deposits the sample inside the chromatographic column. Rapid depression of the plunger on a standard 10-μL syringe expels the sample with considerable velocity. In free air, the unrestrained ballistic path can distribute a spray pattern of material over a distance exceeding 1.5 m. With the needle housed inside a column, that distance would be much less, because collision of the spray with the column wall would impose limits on the length of column subjected to the spray pattern. The linear movement of the mobile phase would complicate these considerations. Chromatographic defects such as peak splitting and peak broadening can be generated by too rapid injection of sample volumes greater than $\sim 1.0\,\mu$L. Too slow an injection, on the other hand, may result in wetting of the outside of the needle, and capillarity can then draw sample into the concentric space between the needle and the column. The band may be smeared still further as the needle is withdrawn from the column. Very large-volume injection, on the order of 50 to hundreds of microliters, using one of a variety of solvent removal techniques, also requires a carefully controlled injection speed to avoid overloading the system's capacity for solvent removal. Such systems profit from having a microprocessor-controlled autoinjector to deliver the sample at the optimum rate.

Liquid-Induced Spreading

Even in the case of an "ideal" on-column injection, where a short, discrete band is deposited on the coated column, problems may develop. The injected band of mixed solutes and solvent is not stationary, but advances through the column under the impetus of the mobile phase, albeit at a lower velocity. As it advances, solutes diffuse from the moving solvent-solute band to the stationary phase, smearing sample over an extended length of column. This phenomenon has been termed "band broadening in space." If the injected plug should "bridge" the column, the velocity of the sample plug will be that of the mobile phase, and the band will be smeared over an even longer section of column. The term "band broadening in space" was used to distinguish this spatial phenomenon from "band broadening in time" [5, 6]. The latter can be envisioned as a temporal (time-dependent) band-lengthening process occasioned by delivery of the sample to the column slowly and over a period of time. Before and during transport of the latter portions of the sample, portions deposited earlier have already begun their chromatographic development. This can be caused by (among other things) slow volatilization of the sample in the inlet and/or an excessive inlet volume relative to the gas flow (split or splitless). Other aspects of these band-lengthening processes are considered later in this chapter.

3.3 Chromatographic Factors Influencing Band Length

Following deposition of the sample onto the column, the length of column occupied by the band can be further affected. These effects can have positive or negative effects on the chromatographic results finally obtained.

Equation (1.14) can be rewritten as

$$k = K_c/\beta = c_S V_S/c_M V_M \tag{3.1}$$

The velocity of a chromatographing band varies directly with the velocity of the mobile phase and indirectly with the proportion of the analysis time that the solute spends in stationary phase:

$$v = \bar{u}/(k + 1) \tag{3.2}$$

Hence, the velocity of a solute band can be expressed as

$$v = [\bar{u}(\beta/K_c)] + 1 \tag{3.3}$$

If the velocity is identical at every point throughout the band, the band moves as a unit whose length is affected only by longitudinal diffusion and by heterogeneities in the flow path.

Distribution Constant Effects

If a solute band chromatographs under conditions where the distribution constant K_c is not uniform throughout the band, the velocity is not uniform throughout the band, and the length of the band will be affected. The term "distribution constant effects" has been proposed for these processes [7–9]. The concept is exploited in "cryogenic focusing" and in "thermal focusing" [10, 11]. Recall that the distribution constant is a function of temperature. As the temperature decreases, K_c increases, because of the increase in c_S and decrease in c_M. Conversely, an increase in temperature obviously causes a decrease in K_c. In both of the above focusing steps the band progresses from a warmer section of column to a cooler section of column. If a flowing coolant is employed, the direction of flow must be counter to that of the carrier gas. Because $K_{c, front} > K_{c, rear}$, $v_{front} < v_{rear}$ and the band is shortened or focused. In "thermal focusing" the band shortening phenomena are further exploited in the subsequent heating step, as heat is applied from the rear of the band toward the front (i.e., in the same direction as the carrier flow). Again, $K_{c, rear} < K_{c, front}$, $v_{front} < v_{rear}$ preserving or even enhancing the focus. "Secondary cooling," as it was originally employed in on-column injections [12], normally resulted in lower separation efficiencies, which were in part attributable to a distribution constant defocus that resulted from the positive temperature gradient [13]. Note that temperature programming does not normally exercise these effects. Where the temperature is changed over the entire length of the band simultaneously, the c_S/c_M ratios are decreased, but those decreases occur uniformly and simultaneously over the entire band; the entire band is accelerated as a unit. Distribution constant effects have also been exploited in coupled columns and in multidimensional gas chromatography. By using the column of greater polarity (higher distribution constant) last, bands are shortened as they enter that column [14]. The retention gap (see below) also utilizes a distribution constant focus.

Phase Ratio Effects

If a chromatographing band encounters conditions where the phase ratio β experienced by one section of the band differs from that experienced by other sections of the band, localized band velocities are not uniform and the length of the band is affected. The term "phase ratio effects" has been proposed for this phenomenon [7–9]. One classic example is the "solvent effect" (e.g., [15–17]) (not to be confused with the reverse solvent effect [18]), in which a more volatile solvent is condensed on the column just prior to the time when the less volatile solutes move into that area. The condensed solvent changes the apparent phase ratio of the column, and the front of the solute bands encounters a lower phase ratio than does the rear of the solute bands. Hence $v_{front} < v_{rear}$, and the band is shortened or focused. The phenomenon can also be observed when a major solute is present in amounts so great that it effectively changes the "apparent phase ratio" of the column for solutes in the immediate vicinity of that major component [18]. This concept, too, has been used in multidimensional gas chromatography, by following a standard film column with an ultra-thick-film column to achieve band shortening without cold trapping [14]. These phenomena are exploited in a variety of ways, sometimes fortuitously and sometimes deliberately. The retention gap (see below) also employs a phase ratio focus.

3.4   Hot Vaporizing Injection Methods

Direct injection into large-diameter columns, split injection and splitless injection constitute the so-called hot-injector vaporization methods. These involve the injection of a sample usually by syringe into a hot injector. The role of the injector is to rapidly vaporize the sample so in can be swept onto the column by the passing carrier gas. The issues and concerns of the sample leaving the syringe as well as how the vaporized sample is focused on the head of the column have been discussed in the previous section. In the next three sections the events that take place inside the injector are discussed. In a series of recent papers [19–21], Grob has considered the subjects of heat transfer, the time required for heat transfer and the capacity of vapor produced from a given injected liquid sample as key factors in the injection process with hot vaporizing injectors.

Split Injection

This injection mode is usually restricted to capillary columns and has been discussed in detail elsewhere [1]. Briefly, the sample is introduced into a heated injector, where it undergoes rapid vaporization and thorough mixing of the volatilized components. Most of the gas flowing through the injector, together with the major portion of the volatilized sample, is vented to the atmosphere, while a minor fraction of each is directed to the column. A schematic representation of a typical split injector is shown in Fig. 3.1. Because the inlet end of the column is coupled to or lies within the heated injector and the remainder of the column is housed in the cooler oven, the sample chromatographs through a negative temperature gradient: $K_{c,\,front} > K_{c,\,rear}$, $v_{front} < v_{rear}$, and the band is shortened or focused. Note that at the detector end of the column, the reverse situation prevails, and the solute bands ascend a positive temperature ramp. Because of the velocity gradient in the column (mobile phase velocity is highest, and the band is moving most rapidly at the outlet end of the column), the effect is usually very slight, but it can be demonstrated. The split mode [and, to a degree, programmed temperature vaporization (PTV); see Section 3.5] is differentiated from the other modes of injection in that

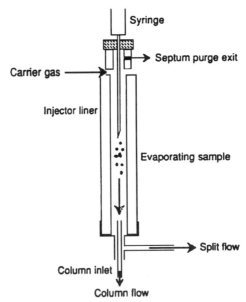

Fig. 3.1. A heated split injector during injection. The sample should be vaporized and homogeneous when it reaches the column opening. *Adapted from Ref. 19 with permission.*

it inherently provides a shortened injection band. The other modes of injection require some additional step(s) to achieve focusing of the injected band.

Split is a vaporizing injection mode, and it is desirable that the volume of the injection chamber be sufficient to accommodate the vaporized sample without engendering either a severe "pressure pulse" or "flashback" of the solvent (and solutes) into remote areas of the injector, against the face of the cooler septum, or back into the carrier gas line. These phenomena not only can affect split ratios and linearity but also can lead to tailing and "ghost peaks" on subsequent injections. At the same time, the volume of the injection chamber should not be so large that an inordinate length of time is required to flush the sample from the inlet. Capillary columns of "standard" dimensions are restricted to relatively low volumes of gas flow, yet the injection chamber must be flushed clean almost instantaneously to avoid the introduction of an elongated band of sample.

The apparent contradiction in these requirements is resolved by providing an inlet whose volume is large enough to accommodate the vaporized sample and that is swept by a relatively large gas volume. Only a small portion of that sample-laden gas is conducted to the column. The major portion is vented to atmosphere (i.e., "split") through some type of restriction. The primary function of a splitter is not merely to limit the sample size; far more important is the fact that the splitter permits a high flow rate through the injection chamber, so that *residual sample is flushed out of the inlet*. Carrier gas following the sample into the column is then *pure carrier and not exponentially diluted sample*.

Another aspect that should be considered here is that split injections require a very high rate of sample vaporization. If vaporization occurs only slowly, the band will be introduced into the column over an extended period of time, even if the vaporized sample is immediately transported to the column. The temperature of the injector should be sufficiently high to ensure "instant" vaporization of the solutes [especially true in isothermal runs and of (at least) the more volatile solutes in programmed runs; see below], but the *inlet temperature can be too high*. Not only does this subject the sample to an excessive and unnecessary thermal shock, but also the explosive vaporization of the solvent accompanying elevated inlet temperatures leads to solvent flashback and can generate excessive solvent tailing (splitters with a sufficient buffer volume are less prone to flashback problems). In programmed-temperature runs, higher-boiling solute peaks benefit more from a distribution constant focus and are broadened less by slower solute vaporization than are the lower-boiling solutes. As the solutes move from the heated inlet to the cooler column, the ratio K_{front}/K_{rear} is higher for higher boiling solutes. Vaporization within the inlet should not be so slow, or oven program rates so high, that the band begins the chromatographic process before all of that solute has been delivered to the column.

Because of this "instantaneous" vaporization requirement, split injectors are operated (for a given sample) at a higher temperature than is required by the other injection modes and may be unsatisfactory for thermally labile materials. Rapid vaporization is encouraged by a heat transfer surface of high thermal mass (e.g., glass bead packing), but such surfaces abound in active sites and may lead to rearrangement and degradation of some sample components. The removable glass (or quartz) liner should normally be subjected to deactivation. Untreated glass liners, glass beads, and glass wool expose the sample to both metal oxides and silanols. Substitution of quartz liners (and quartz wool) lessens the problem associated with metal ions, but silanols are still present (see Chapter 2). Methods of deactivating inlets are usually based on leaching and high-temperature silylation treatments derived from earlier observations on conventional glass capillaries; a typical procedure is detailed in Chapter 7.

The parameters for split injection, injector temperature, split ratio (which also influences the residence time in the heated injector), and sample solvent represent a balance and compromise between several variables that affect chromatographic performance. A higher gas velocity through the injector (i.e., a larger split ratio) usually improves the degree of sample mixing (which can affect linearity of the split) and the shapes of the resultant peaks. To accomplish sample vaporization, a finite amount of heat must be transferred to the sample. The amount of heat required will be influenced by the solvent, the size of the sample injected, the vapor pressures of the sample solutes, and the residence time in the injector (a function of the gas flows and the split ratio). Inadequate heat transfer from the injector to the injected sample can result in severe discrimination problems, often viewed as the Achilles heel of split injection. Although the theory of split injection is straightforward and is fairly consistent with the observed results, there is a significant amount of uncertainty and confusion about what happens to the solvent and analytes between the time when they leave the syringe and when they recondense at the head of the column. Grob has reported some interesting observations on the issues of heat transfer and residence time within the injector [19–21]. Table 3.1 shows the amount of heat required to vaporize 1 μL of a variety of common solvents used for sample introduction. The amounts of required heat become particularly significant when one considers the short amount of time available for heat transfer, and the heat contents and conductivities of the available heat sources: the heated syringe needle, heated carrier gas, heat content of any packing materials in the liner, the walls of the liner, and the base of the heated injector. Significant temperature drops, on the order of 30°, in the injector immediately following injection have been observed and reported.

The fidelity with which that portion of the sample split to the column reflects the actual sample composition is also of concern. A number of competent scientists have studied the effects of recognized variables on splitter performance and found that many interrelated factors play roles. In general, the splitter should achieve extremely rapid vaporization of the sample, which should then be thoroughly mixed and be subjected to expansion just prior to splitting. Although many of these inter-

TABLE 3.1

Energy Required for Evaporating 1 μL of Solvent and Heating It to 250°C

| | Energy consumed (in kcal) | | |
Solvent	Heating	Vaporization	Total
n-Hexane	89	80	169
Toluene	61	86	147
Diethyl ether	78	85	163
Chloroform	29	58	87
1-Propanol	77	166	243
Methanol	72	263	335
Water	220	539	759

Taken from Ref. 20 with permission.

relationships are still only partly understood, most would agree that the temperature and the degree of mixing are two factors that seem to exercise major effects on the linearity of a well-designed splitter. Even skilled chromatographers sometimes tend to use conditions that force discrimination with split injectors. Split injection subjects the sample to the most severe thermal shock of any of the common injection modes, and higher split ratios (which achieve better mixing, given the appropriate amount of turbulence in the liner) discard most of the sample. Those recognizing these points often have an understandable tendency to use both lower temperatures and lower split ratios. Inlet temperatures that are unnecessarily high can result in the degradation of thermally labile components, and the rapid vaporization of the sample can cause flashback, evidenced as solvent tailing. This can also affect quantitation (discussed later in this chapter and again in Chapter 10). Lower inlet temperatures can improve the chromatographic results, but if the temperature is not sufficiently high, higher-boiling solutes may generate broadened peaks. The split ratio governs the velocity of the injected sample through the mixing zone; too low a split ratio can lead to inadequate mixing and discrimination. To maintain the necessary velocity, the volume of the inlet liner should be reduced for those cases where very low split ratios are required. Properly performed, split injections are highly reproducible, and with careful calibration, splitters can be an excellent choice for quantitative work.

Splitless Injection

The splitless mode of injection was developed largely by the Grobs (e.g., [3–5, 19, 20]) as an alternative means of injecting samples into small-diameter open tubular columns. Again, major portions of this subject have previously been covered in

TABLE 3.2

Recommended Initial Temperatures for Splitless Injection

Solvent	Boiling point (°C)	Recommended initial temperature range (°C)
Methylene chloride	40	10–30
Chloroform	61	25–50
Carbon disulfide	46	10–35
Diethyl ether	35	10–25
Pentane	36	10–25
Hexane	69	40–60
Isooctane	99	70–90

Taken from Ref. 27 with permission.

detail [1]. Splitless injection is best suited to the analysis of trace levels of higher-boiling solutes that can be injected in a low-boiling solvent, under conditions where the solvent condenses on the column just before the solutes are transported into that region. Solutes move from an area where the "apparent" phase ratio (i.e., V_M/V_S) is high to an area where it is low, that is, the phase ratio ramp is negative. Referring to Eq. (3.3), the condensed solvent subjects the front of the solute band to a higher V_M/V_S ratio than is experienced by the rear of the band, $v_{front} < v_{rear}$, and the band is shortened or focused. The "solvent effect" is a manifestation of phase ratio focusing.

The solutes are generally presented in dilute solution in this mode of injection. The solvent must be low-boiling with respect to the solutes and high-boiling with respect to the column at the time of injection. Suggested initial oven temperatures for a variety of solvents used in splitless injection are presented in Table 3.2. These temperatures should provide a reasonable "solvent effect" focus. A reasonable starting temperature for solvents not listed is 10–30°C below the boiling point of the solvent [27]. The vaporization chamber has provision for the mobile phase to be introduced under conditions where either (1) the entire flow is through the inlet to the column or (2) the flow stream is divided, part of it entering the column as carrier gas and part backflushing the chamber through a purge valve to remove the last remnants of the injection. The injection is made with the purge valve closed and all flow passing first through the inlet and thence to the column. A splitless injector is schematically illustrated in Fig. 3.2. As with split injection, higher-boiling solutes thermally focus at the head of the column during splitless injection. To effectively cold-trap an analyte, the oven temperature should be 150°C cooler than the boiling point of the analyte [27].

The carrier gas velocity in the liner depends on the column flow and the internal diameter of the liner. A high carrier gas velocity in the liner leads to rapid and

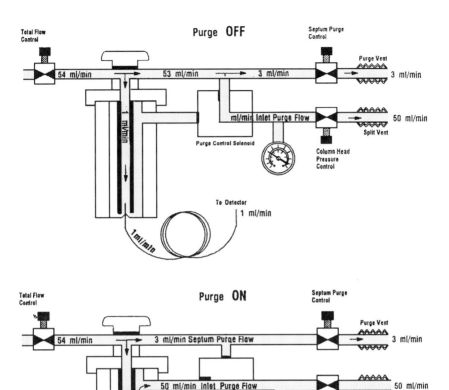

Fig. 3.2. A back-pressure regulated splitless injector during injection (purge off), and after sample transfer (purge on). Note that the total flow into and out of the injector remains constant. It is the way the carrier gas is routed through the liner and past the column inlet that changes. Taken from M. Klee, *GC Inlets—An Introduction*, Hewlett Packard Co., 1991 with permission.

efficient transfer of the sample onto the column, provided the temperature is sufficiently high to provide adequate heat transfer from the injector to the sample during this shorter exposure. Improved sample transfer is achieved in splitless injection using liners with smaller internal diameters (1.2 mm ID as opposed to 3.4 mm ID) [28]. Smaller-ID liners have also been observed to cause less sample degradation with sensitive analytes, presumably because they spend less time in the liner, which is the source of most of the activity in the system. Inasmuch as this is a vaporizing form of injection, the injection chamber must again have a sufficient volume to accommodate the vaporized mixture of solvent and solutes. An internal volume of ~ 1 cm^3 is probably average. Capillary columns of conventional dimensions can tolerate only restricted flows. Assuming a column flow of 2 cm^3/min, one volume of carrier gas would pass through the inlet every 30 sec. As an initial target, the purge function is activated after one and one-half volumes of gas have passed through (and flushed) the inlet. In our example, the first trial would activate the purge function at 45 sec, and the results would then be evaluated in terms of solute deliveries and solvent tailing. Purge activation times slightly shorter and slightly longer should also be tried to verify the most suitable delay. Shorter purge activation times tend to decrease both solvent tailing and delivery of solutes (especially higher-boiling solutes) to the column. Longer purge activation times increase both the degree of solvent tailing and delivery of solutes to the column. The existence of the purge step means that we are dealing with a contradiction in terminology: splitless injections are not "splitless"; they are splitless only in comparison to a split injection. Quantitation is usually least reliable with this injection mode.

For a given sample, the splitless injector can generally be operated at a lower temperature than would be required to introduce the same sample by split injection. In splitless injection solutes can volatilize somewhat more slowly than split injection because they spend more time in the injector, hence there is more time for heat transfer. The sample band (which is continuously conducted to the column until the time of purge activation) will be subjected to a phase ratio focus. In this injection mode the sample spends a longer time in the injector. As a result, active solutes (e.g., pesticides) have more time to interact with active sites and decomposed nonvolatile material present on the injector walls, base, and liner. These active sites can react catalytically in the hot injector to degrade chromatographic performance. For trace analysis (where splitless injection is most commonly used) routine maintenance and cleaning of the injector components is a critical factor for success. Pressure-pulsed injection can also reduce the decomposition of active solutes in splitless injection by reducing the time necessary to move the sample from the injector into the more inert environment offered by the column.

Regardless of the injection mode, hydrogen is the preferred carrier for capillary chromatography because it yields better overall separations and shorter analysis times [1, 4] (see also Chapter 5). With splitless injection, hydrogen has still another advantage. Because it is used at higher average linear carrier gas velocities, the flow through the inlet is higher and solutes are transferred from the inlet to the column more rapidly and more completely [24]. Columns coated with crosslinked (i.e., nonextractable) phases should always be used for splitless

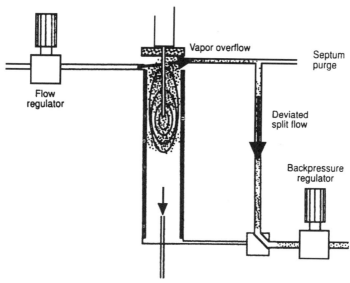

Fig. 3.3. Splitless injection when the needle is too short can result in flashback, sample loss and severe discrimination when the top of the injector overflows with vapor. *Taken from reference 28 with permission.*

injection. Noncrosslinked films undergo localized phase stripping, and columns exhibit shortened lifetimes due to the repeated on-column solvent condensation.

Grob has argued that splitless injections yield better results when the needle is 1–1.5 cm from the head of the column during the injection, and that the column end and needle tip should both be positioned low in the injector where the injector temperature is highest [28]. The top of the injector chamber is often cooler than the temperature set point or the center and lower portions of the liner. If the sample is injected too high in the liner and/or too far from the column end, losses through the septum purge vent as well as poor heat transfer can be substantial. In the backpressure-regulated system of gas supply, the split flow does not actually stop during splitless injection, but is instead diverted through the septum purge. This is illustrated in Fig. 3.3.

Computer-Controlled Pneumatics and Their Influence on Injection Efficiency

Computer control of carrier gas flow during the chromatographic separation has become a popular feature on modern gas chromatographs. The goal of these systems is to manipulate pressure at the head of the column to regulate the carrier gas velocity during the analysis in order to provide the optimum separation and reduce run times. These features are known by various names such as "Electronic

Pressure Control" (EPC), or "Programmable Pressure Control" (PPC). Generically termed "electronic pneumatic control" (EPC), or "computer-controlled pneumatics" [29], these systems, because their impact is delivered through the injector, can have a pronounced effect on both split and splitless injection. Note that two of these use the acronym "EPC." We will confine that usage to the more widely used term "Electronic Pneumatic Control."

In splitless injection the pressure in the injector is dramatically raised (or pulsed to a higher level) during the time of injection with the goal of forcing more of the vaporized sample onto the head of the column. The pressure pulse increases the flow onto the column to facilitate this. The increased pressure also reduces the expansion volume of the injected sample as it is vaporized thus reducing the migration of sample vapors into cooler parts of the injection system and lessening the possibilities of band broadening from slow sample transfer and ghost peaks on subsequent injections. Table 3.3 compares the expansion volumes as a function

TABLE 3.3

Approximate Gas Volumes for Common Solvents

Solvent	Inlet pressure (psig)[a]	μL vapor/1 μL solvent temperature ($^\circ$C)	
		200	300
Isooctane	10	140	170
	30	77	94
	60	46	56
Hexane	10	177	214
	30	98	118
	60	58	71
Toluene	10	217	263
	30	120	145
	60	72	87
Methylene chloride	10	361	437
	30	199	241
	60	119	144
Methanol	10	571	691
	30	315	382
	60	189	229
Water	10	1279	1550
	30	707	856
	60	423	513

[a] 10 psig is 69 kPa, 30 psig is 207 kPa and 60 psig is 414 kPa.
Adapted from S. Stafford, ed., *Electronic Pressure Control in Chromatography*, 1993, Hewlett-Packard Co.

of pressure as well as temperature. The volumes at 30 psig (lb/in^2 gauge) are almost half those observed at 10 psig. Following injection and an experimentally determined delay for the most efficient sample transfer, the inlet pressure is rapidly reduced to the pressure required for a normal chromatographic separation.

Pulsed-pressure splitless injection can offer several advantages over conventional backpressure-regulated inlet systems. These include less discrimination of early and late eluting solutes [30] and decreased decomposition of active solutes as a result of the sample spending less time in the injector [31]. However, there are some drawbacks to pressure-pulsed splitless injections that include band broadening of early eluting solutes. The band broadening is the result of the early eluting solutes spending the majority of their time in the column at the dramatically increased linear velocity that results from the pressure pulse. These compounds are not only broadened but their retention times are greatly reduced. The early eluting components never get a chance to refocus on the column. This broadening could probably be reduced by using a lower initial temperature or increasing the film thickness [30], namely, manipulating K_c or β to make k larger. Some sample discrimination of the more volatile components has also been observed and is attributed to loss through the top of the injector in the large-septum purge flow that is present during the splitless sampling period. This can be particularly bad for C_7 or smaller components [29] with open-ended splitless liners. The degree of solute loss depends on the injection volume and the septum purge flow rate during the sampling period. Mass discrimination in a forward-regulated system can be minimized by setting a low septum purge flow [32].

Injection into Large-Diameter Open Tubular Columns

In conventional (low-flow) capillary columns, the inlet may require modification so that a high gas flow can rapidly flush the vaporized sample from the injection chamber in spite of the restricted flow through the column. A major advantage of the larger-bore (e.g., 0.53-mm-ID) open tubular columns is that the standard packed column injector can be used without any special provisions for stream splitting. The carrier gas flow rate through these columns can be the same as that used in packed columns. As a result, the standard injector is rapidly cleansed. However, other injection-related problems can be encountered.

Figures 3.4–3.6 illustrate that direct vaporizing injections into these large diameter open tubular columns can create problems when used with the normal straight-bore inlet liner. Larger injections, or slower (more nearly optimal) carrier gas flows, often result in severe tailing of the solvent and more volatile solutes and discrimination against some higher-boiling solutes [8]. These problems are attributable to flashback. The severity of flashback increases with increasing injection size and with decreasing flow through the inlet, and is exacerbated as the differential between the temperature of the injector and the boiling point of the

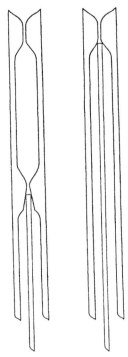

Fig. 3.4. Improved inlet liners designed for large diameter (0.45–0.53 mm OD) columns. Left, liner for direct flash injection, designed to seal with the square-cut column end. Carrier gas enters the liner from the top, and attains extreme velocity as it passes through the annular space between the syringe needle and the liner restriction. Good results can be obtained with liquid injections as large as 8–10 μL. Right, liner for hot on-column injections. This is a vaporizing mode of injection and should not be confused with cold on-column injection. Injection size, injection speed, and needle residence time following injection must all be correlated and vary with the internal volume. Adapted from the J. Chromatogr. Sci. **24**:34 (1986) and printed with permission of Preston Publications, Inc.

solvent (and solutes) became greater. Where higher-boiling solutes are forced against the face of the cooler septum, some of those entrained solutes condense and are lost to the analysis. These considerations led to the design of a new injector liner for the large-diameter open tubular column [22].

Figure 3.4 shows two types of liner recommended for use with large-diameter open tubular columns. That shown on the right is for flash-vaporized injections and is preferred in the vast majority of cases. The annular space between the inner wall of the upper constriction and the syringe needle is very slight, which forces the entering carrier gas to a very high velocity, greatly lessening the possibility of flashback. The end of the column should be cut square and seated snugly in the ground taper, which forms the lower restriction. Vaporization of

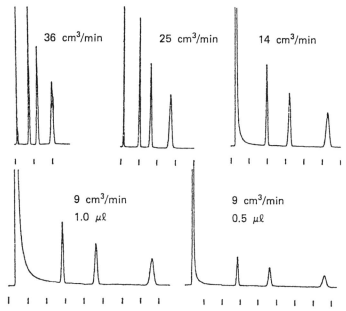

Fig. 3.5. Solvent tailing as a function of the carrier gas flow rate, using a "normal" injector liner and a Megabore^{TM} (column. C_{11-13} hydrocarbons in hexane, injector temperature 250°C. Top chromatograms, 1.0 μL injections, at (left) 36 mL/min; (center) 25 mL/min; (right) 14 mL/min. Bottom chromatograms (all at 9 mL/min): (right) 1 μL injection; (right) 0.5 μL injection. Where the volume of the vaporized injection fills or just slightly overfills the volume of the vaporization chamber within the liner, higher flows more effectively transport the sample to the column (first four chromatograms). At lower flows, the sample size must be decreased to avoid problems with "normal" liners (bottom chromatograms). Adapted from J. Chromatogr. Sci. **24**:34 (1986) and reprinted with permission of Preston Publications, Inc.

the sample still results in a pressure pulse, but as long as the syringe needle is in position, the column offers the path of least resistance. Instead of flashback, the sample is forced forward into the column. The syringe needle should be left in position until the pressure pulse is dissipated by flow into the column. The required delay amounts to approximately 1 sec per microliter injected. Solvent tailing provides a useful indication of too early syringe needle withdrawal [22].

The other liner illustrated in Fig. 3.4 is used for on-column injections into the large-diameter open tubular columns, as the syringe needle is guided directly inside the column. Some practitioners pack the longer portion with glass wool, Chromosorb, or similar materials and use it in the inverted position to protect the column during the injection of "dirty" samples. Nonvolatile materials then tend to remain in the inlet, which is periodically removed for cleaning or replacement.

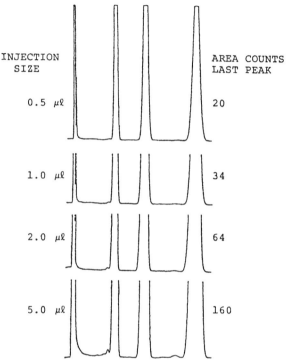

Fig. 3.6. Effect of injection size on the quantitative delivery of solutes, using a "normal" liner with the Megabore™ column. Helium carrier @ 30 mL/min. Other conditions same as Fig. 3.5 except for injection size. The severity of "flashback" correlates with the size of the injection, other conditions constant. Flashback forces vaporized sample (solvent plus solutes) to overflow the vaporization area. Components in that sample then condense on cooler surfaces (e.g., the septum face). The severity of discrimination correlates roughly with solute boiling point. Adapted from J. Chromatogr. Sci. **24**:34 (1986) and reprinted with permission of Preston Publications, Inc.

3.5   Programmed Temperature Vaporizing (PTV) Injector

This specialized form of injector can be used in split, splitless, or direct mode and employs an initial inlet temperature that is below the boiling point of all components, including any solvent [33–36]. After the sample is injected, the temperature of the inlet is programmed to increase at a high rate. Each component is vaporized and moved through the split point to the column as it gains sufficient volatility from the continually increasing temperature. Hence solutes should be exposed to less thermal shock. Excellent linearities up to the C_{28} hydrocarbon have been reported. Samples can be injected neat (i.e., without solvent); where solvent is present, it vaporizes less violently and the volume of the injection chamber can be more restricted. The application of PTV injection in the split mode is discussed below.

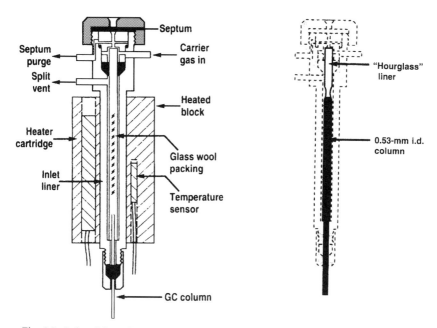

Fig. 3.7. Left, a Schematic representation of a PTV injector configured for split/splitless injection. The cooling coils were omitted for clarity. Right, the same injector modified for on-column PTV injection. The glass insert helps guide the needle directly into the column. *Taken from, J. Hinshaw, LC–GC, 10:748 (1992)* with permission.

A schematic of a generic PTV split/splitless injector is shown in Fig. 3.7 (left), which illustrates the similarity to a standard split/splitless injector. Differences occur in the size of the vaporization portion of the inlet. Liner volumes are smaller in PTV to permit more rapid heating and cooling. Neither does PTV demand as large a thermal mass as "flash" vaporizing inlets. PTV also requires provision for cooling, both for initial sample trapping and for postrun cooling (this can save considerable time in preparation for the next analysis), and for adjustable rates of heating that can be extremely fast. The injector temperature is usually externally controlled and programmable from heating rates comparable to the oven program, to ballistic heating at rates as high as 16°C per second. Some inlets allow for a hold at a slightly elevated temperature for the quick and selective evaporation of the solvent. This feature is helpful in some methods of large-volume injection and is discussed later.

Minor modifications, including auxiliary cooling and a liner designed to align the syringe needle with the column, permit on-column injections with the PTV (Fig. 3.7, right). In this mode the split vent is turned off and a 0.53-mm-ID column (or, preferably, a retention gap) is inserted into the injector. This permits a standard 0.47-mm-OD (28 gauge) syringe needle to be guided into the previously aligned column. Besides accepting the standard needle the retention gap will also collect any nonvolatile residues before they enter the analytical column. If the injection volume exceeds $2\,\mu L$, the retention gap will also reduce the band lengthening caused by condensed solvent flooding the head of the column.

With PTV, injection generally takes place at temperatures that are too low for fractionation within the hot needle, and this type of mass discrimination is rarely a problem. In addition, other adverse effects of hot vaporization injection (e.g., high mass discrimination, flashback, and labile sample degradation) can be minimized or avoided. The sample components are not exposed to a hot needle, nor are they subjected to heated injector components until they are warmed to their vaporization point. Each solute enters the column at a temperature close to its boiling point, and so avoids exposure to the higher temperatures that the injector reaches after their departure. Avoiding exposure to temperatures well above the solute boiling point can be especially meaningful in the case of labile analytes, which essentially "crack" at the elevated temperature of a flash vaporization injector. As illustrated in Table 3.3, the expanded gas phase volume of some solvents easily exceeds the volume of the injector. In a hot vaporization injector, this leads to flashback. Heated vapor is forced out of the injector and can be lost to septum purge or condensation in cooler portions of the system. This usually results in discrimination, and/or "ghost peaks" in subsequent injections. In PTV injection the rapid expansion is avoided, the solvent evaporates more slowly, and the problem is eliminated. PTV injection usually exhibits much lower discrimination toward high-boiling components, and quantitative performance is generally much better than that achieved with hot splitless injection. While losses of more volatile components are sometimes observed, results are usually superior with low volatility solutes such as steroids, polynuclear aromatic hydrocarbons (PAHs), and polychlorinated biphenyls (PCBs). PTV has yielded good results even with thermolabile pesticides, provided the injector inserts contained no glass wool or other packings [37].

PTV injectors have been successfully applied to a variety of analytical problems. Applications to high-boiling solutes, including the simulated distillation of crude oil fractions, triglyceride analysis in food products, nonionic surfactant analysis, and polymer additives, has been especially valuable. Many analysts find the PTV easier to operate, and report that their results are better than those they achieve with conventional on-column injection [38].

Although PTV is a great improvement over a conventional split/splitless injector, it is not without faults. Because vaporization is less rapid, sample transfer from the injector to the column can be slow. This means an additional focusing mechanism (e.g., thermal, retention gap or solvent) may be necessary. For this reason, PTV is not recommended for isothermal applications. Another concern is that a cool inlet will trap and concentrate impurities that may be present in the carrier gas or other parts of the inlet system. These impurities can later generate extraneous peaks when the inlet is finally heated.

3.6   On-Column Injection

On-column injection as it is known today also owes its origin largely to the efforts of the Grobs (e.g., [39, 40]; see also [41]). This technique holds its greatest utility for two types of samples: (1) those containing thermally labile solutes, and (2) those containing very high-boiling solutes. In the vaporizing modes of injection (split, splitless, and "PTV"), the sample is vaporized in the heated inlet, transported to a cooler column, and finally chromatographed. Although this sequence of events safeguards the column in that nonvolatile residues remain in the injector and the column is exposed only to materials that can be volatilized, thermally labile solutes may decompose. In on-column injection, the sample is deposited from the syringe directly on the column without undergoing vaporization. As the temperature of the column increases, solute vapor pressures rise and the rates of solute transport through the column increase (i.e., each solute experiences a decrease in k due to the decrease in c_S/c_M associated with increasing temperature). In this case, however, any nonvolatile or high-boiling solutes in the sample have also been deposited on the column. These deposits lead to severe "bleed" and baseline problems and to a rapid deterioration of column efficiency. High-temperature "baking" to remove the volatile residues in such deposits places an unnecessary stress on the column and leads to shortened column life (Chapter 6). Because they can be washed with a variety of solvents, columns coated with nonextractable crosslinked phases are preferred for on-column injection. A cautionary note seems advisable—if higher-boiling and/or nonvolatile residues are to be washed from the column, this process must be undertaken while those substances are soluble. Exposure to high temperatures for extended periods may render those deposits insoluble.

Originally, both the on-column injector and the column oven were cooled to temperatures below the boiling point of the solvent/solutes. For the analysis of higher-boiling solutes, it was then necessary to heat the oven to a point where those solutes exhibited suitable chromatographic properties. Following this, it was necessary to cool the entire assembly before the next analysis could be performed. This was inconvenient even with temperature programming and essentially impossible for higher-temperature isothermal operation. Galli *et al.* [12] suggested that "secondary cooling" offered an improvement; the column was housed in the heated oven and, during injection, cool air was discharged tangentially around the inlet end of the column. This cooled a short section of the column during the injection phase. Following injection, the secondary cooling was discontinued, and the column tracked the oven temperature.

Early results with on-column injection were sometimes disappointing. There were many reports of lowered separation efficiencies, poor quantitation, malformed peaks, and peak splitting. Knauss *et al.* [42] suggested that secondary cooling was responsible for some of these problems: the injected sample plug extended beyond the cooled zone, so that the front portion of the plug began the chromatographic process while only the latter portion was delayed by cold trapping. This, the authors suggested, might explain "peak splitting," an occurrence that was all too common in this mode of injection. Many workers also noted that component resolution usually suffered in on-column injections and that this was exacerbated by secondary cooling. The secondary cooling-engendered loss in separation efficiency has been rationalized [13, 43]. With split or splitless injections, band-shortening mechanisms come into play, as discussed above. With on-column injection, not only are these absent, but the use of secondary cooling requires the sample to advance over a positive temperature ramp; the c_S/c_M ratio is smaller at the band front than at the band rear, $v_{front} > v_{rear}$, and peaks become broader. Two complementary routes were quickly taken to correct this problem: (1) the inlet end of the column was housed in a nozzle that restricted diffusion of the cooling air so as to cool a longer segment of the column, and (2) the "retention gap" [44] was suggested as a means of reconcentrating bands and achieving band shortening. As it is most frequently employed, the retention gap is a section of uncoated, deactivated tubing attached to the inlet end of the column (see below). The concept is extremely useful in splitless and on-column injections, and by using large-diameter tubing for the retention gap, automated injectors can be used in on-column injection. Figure 3.8 illustrates (top) the phenomenon of peak splitting from cool on-column injection and (bottom) its correction as a result of using a retention gap.

Jenkins [45] suggested that the extreme flexibility and low thermal mass of the fused silica column permitted an elegantly simple approach to cold on-column injections, regardless of the oven temperature; in which a short section of the column is retracted from the oven and immediately cools to ambient temperature. The injection is made into that cooled portion of the column, which is then reinserted into the heated oven. Abruptly "stepping" the column to the higher temperature avoids the band lengthening associated with the positive temperature ramp [11].

With any on-column (and, to a lesser degree, splitless) injection technique, long "smeared" solute bands can lead to serious complications. The susceptibility of the sample to this physical smearing is governed by a number of sometimes interrelated factors, including the size of the sample injected (larger samples accentuate the problem), the "wettability" of the column by the solutes/solvent (influenced by the deactivation treatment, the type and thickness of stationary phase coating, and the "polarities" of the solutes, solvent, and stationary phase), how long the liquid plug endures (the boiling points of the solvent and solutes, the temperature of the column at the point of injection, and the subsequent temperature profile), and the

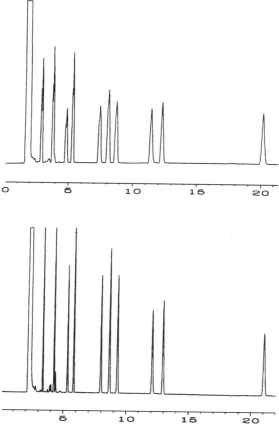

Fig. 3.8. Top, the higher volatilities of low-k solutes can result in split, broadened peaks as they reach the detector. Split peaks can lead to misassignments. Bottom, the addition of a 3 meter retention gap to the same system provides a focusing mechanism, eliminating peak splitting, and generating sharper peaks.

velocity of the carrier gas. Various aspects of these band-lengthening phenomena and methods of reconcentrating those bands have been considered in some detail (e.g., [5, 15–18, 46, 47]).

Band lengthening due to solvent movement can be reduced by making the injection under conditions that encourage quiescent evaporation of the solvent immediately beyond the point of injection, leaving the solutes cold-trapped on the column [43]. Another approach is a "stop flow" injection, that is, interrupting the flow of carrier gas during the injection [48–50]. These approaches can also benefit from incorporation of the retention gap concept. But here again the retention gap must be "wettable" by the sample solvent or long enough to hold an especially long band. Aqueous samples are very problematical in this regard and retention gaps often need to be long enough and subjected to a more polar deactivation treatment or a very thin coating of a polar stationary phase such as DB-1701 (14% cyanopropyl methylsiloxane) [51] to provide a wettable surface with some resistance to chemical attack. Unfortunately aqueous samples and reactive components in extracts can quickly degrade these deactivation treatments.

Retention Gap Focusing

As discussed above, prefacing the coated separation column with a section of uncoated, but deactivated, tubing achieves several desirable effects:

1. Nonvolatile residues tend to accumulate in the "guard column" and limit contamination of the separation column.

2. By using a precolumn of sufficient diameter, standard-sized needles can be utilized in on-column injections into small-diameter columns (which also permits automating such injections).

3. The abrupt decrease in the phase ratio effectively shortens the band beginning the chromatographic process.

4. The abrupt increase in the distribution constant also exercises a focusing effect.

One-meter retention gaps are usually sufficient to achieve the focusing benefits, but analysts with particularly dirty samples sometimes use 5-m lengths. This is because contamination usually occurs well within the first .05 m of the column so the longer retention gap will allow for 10 quick "clips" prior to the more involved procedure of changing the connector to the analytical column. Retention gaps can easily be inspected for contamination by holding the column up to a strong light. Sample residues and contamination will readily be visible in the gap. Another factor in the determination of retention gap length is important to analysts injecting large samples, on the order of 2–5 μL. Here the volume of the gap must be able to accommodate the sample's solvent volume without allowing it to bridge the column and be blown for several meters by the carrier gas. Such bridging spreads the initial band well into the analytical column and makes a narrow initial band impossible. Wider-bore retention gaps are often used with narrower-bore analytical columns with no adverse affects. The wider-bore is especially advantageous to on-column injection with standard needles and the injection process can easily be automated.

On the other hand, the coating process itself has a pacifying action. Very thinly coated columns often exhibit excessive activity, and uncoated precolumns can be extremely active [52]. Figure 3.9 shows (top) severe activity contributed by a 1-m retention gap composed of raw (untreated) fused silica tubing, as contrasted with (bottom) the results obtained with a section of deactivated fused silica tubing. Octamethyltetracyclosiloxane, diphenyltetramethylsiloxane, and

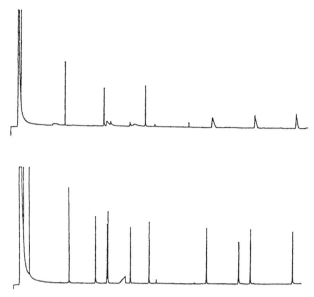

Fig. 3.9. Influence of active sites in the retention gap. Both chromatograms utilized cold on-column injection of the Grob test mixture through a one meter fused silica retention gap to a 30 m × 0.32 mm fused silica column coated with DB-1. Raw fused silica tubing was used as the retention gap in the top chromatogram, while the tubing used as retention gap in the bottom chromatogram was first deactivated with octamethylcyclotetrasiloxane. The asymmetry in peak number 4 (2-ethylhexanoic acid) of the lower chromatogram testifies to a polarity mismatch between this solute and the dimethylpolysiloxane stationary phase; it does not indicate activity.

tetraphenyldimethylsiloxane have all given excellent results when employed as vapor phase deactivating agents on fused silica tubing [52]. Commercially available retention gap tubing is usually deactivated in this manner to give a methyl or phenyl methyl silane surface. These modify the surface energy and provide a wettable surface. Otherwise the sample can break into several bands and/or spread through the length of the column. However, such surfaces are not very satisfactory for the injection of methanol of other polar solutions. Better injection results can be obtained with these solutions by first rinsing the retention gap (prior to connection and installation) with an extremely dilute solution of Carbowax 20M or OV-1701 (e.g., 5–10 mg/mL in methylene chloride) [51]. It should be emphasized that these films should be extremely thin, or the focusing benefits of the retention gap will be lost.

3.7 Large-Volume Injection

The injection of large volumes, ranging from 10 to 1000 μL, is beginning to become more than an academic curiosity. Practical gas chromatographers are also beginning to explore the utility and benefits of these techniques. The movement to large-volume injections is being driven by several important benefits. The first is the obvious improvement in sensitivity. With a detector capable of measuring 10 pg, a 1-μL injection will allow the measurement of a 10 ppb (parts per billion) solution of the solute, and a 200-μL injection into the same detector will allow the measurement of a 50 ppt (parts per trilllion) solution of the same solute. A second benefit to large volume injection is a savings of time and effort in the preparation of samples for GC analysis. Most schemes for sample preparation prior to trace analysis involve the concentration of a large volume of extraction solvent, usually by the Kadurna-danish method. The elimination of this step not only decreases the sample preparation time but also improves the accuracy of the procedure by eliminating losses associated with the concentration step. Finally the development of automated systems that combine sample preparation with GC analysis is moving rapidly. Large-volume injection will fill a role in facilitating the transfer of sample to the instrument for analysis.

The major technical challenge associated with large-volume injection is removing the solvent without losing the analytes of interest. To select the best method for large-volume injection, several things need to be considered: how much solvent must be injected, the nature of the solvent, and the volatility and polarity of the analytes of interest. The common concerns with all methods developed for large volume injection are [53]

1. The injection process should introduce the smallest possible amount of solvent to the analytical column.

2. To protect some detectors, solvent vapor is best vented either from the injector or some other vent, usually between the retention gap or precolumn and the analytical column.

3. The solvent venting process, time, flow, and temperature, needs to be optimized to minimize the loss of the most volatile analytes of interest.

PTV Injection with Solvent Split

PTV injectors can be used for the injection of large sample volumes either by multiple injections [54] or solvent venting in the cold split mode. For a multiple-injection method, large volumes are injected in several smaller fractions (up to eight injections) with the solvent evaporating at low temperatures, and exiting through the open split vent, between each injection. Once a sufficient quantity of trace analytes is built up, the injector temperature is rapidly increased to elute the solutes onto the column. For a hexane solution of n-alkanes, discrimination was observed up to C_{16} [54]. No discrimination of the higher boiling hydrocarbons was observed.

A method of cold split injection with solvent venting has been described by Vogt and coworkers [55]. This allowed the injection of up to 250 μL into a cold

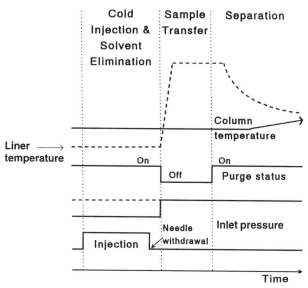

Fig. 3.10. Parameters to be optimized for a large volume PTV injection in the solvent split mode. *Adapted from C. Cramers and H. Jannsen with permission.*

liner which was filled with glass wool. During injection the solvent was vented through an open split vent. When most of the solvent was removed the split vent was closed and the injector was rapidly heated in order to vaporize the analytes onto the column. For the reproducible injection of large sample volumes in the solvent split mode the following parameters need to be optimized and controlled: glass liner design, liner packing, sample solvent, speed of sample transfer into the injector, solvent evaporation temperature and venting time, split flow rate, injector heating rate and final temperature. These steps are outlined in Fig. 3.10. The advantages and disadvantages of large-diameter liners (>2 mm ID) for large volume injection have recently been described [56]. The larger liners, packed with glass wool, will allow for the rapid injection of up to 150 μL. Optimization of injection speed is unimportant, and no special syringe devices are required. For injection volumes greater than 150 μL, the rate of injection must be controlled. In this case the smaller-ID liners are advantageous in that the efficiency of transfer of the sample from the liner to the column is increased and the time is reduced. For sensitive analytes the amount of decomposition is greatly reduced with the smaller liner.

An advantage of this technique is that discrimination of high-boiling components is virtually eliminated and quantitative performance is much better than hot splitless injection. While losses of volatile components are observed, the effective analysis of low-volatility solutes such as steroids, PAHs, and PCBs has been reported. The analysis of thermolabile pesticides has been effected, but only when using inserts without glass wool or packing.

Vapor Overflow Technique

The vapor overflow technique is an extension of the PTV method for large-volume injection that differs most notably in the use of the septum purge instead of the split vent for solvent removal and the injection takes place into a hot injector. The vapor overflow technique for the injection of large-volumes was originally described for the PTV splitless injector but has also been applied to conventional vaporizing splitless injections [57]. The technique is robust and effective. As illustrated in Fig. 3.11, the entire sample is quickly injected directly into the hot (270–370°C) injector. The injector contains a large bore liner packed with Tenax. The split vent is closed during the injection period while the septum purge is open (150 mL/min). Driven by the rapid expansion, vapor is pushed up and exits the system via the septum purge vent. The packing bed that supplies the heat of vaporization of the exiting solvent becomes cooler, and the low- and medium-volatility solutes are trapped on the packing [58]. However, volatile solutes are vaporized and lost with the solvent. Once the solvent evaporation is acceptably complete the septum flow

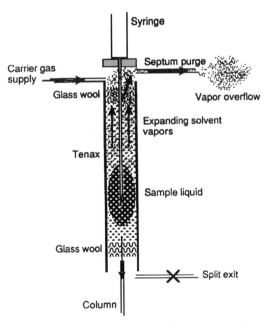

Fig. 3.11. A large volume injection with the vapor overflow technique. The heat of vaporization of the sample solvent cools the liner packing enough to help trap the higher boiling analytes of interest. *Reproduced from [59] with permission.*

is closed and as the packing regains high temperature, the trapped solutes are desorbed onto the column. A refocusing mechanism must be applied, but the column temperature should be maintained at a level that prevents column flooding by condensed solvent [59]. The method has also been applied to large-volume aqueous samples by employing an organic cosolvent to wet the Tenax packing [60, 61].

On-Column Large-Volume Injection

This method involves injection of the entire sample onto the column, a retention gap is attached to the front of the column, whose volume is sufficiently large to accept the entire sample. The solvent is then vaporized and stripped from the analytes of interest [62, 63]. The sample must "wet" the surface of the retention gap and provide a thin film. Beading or bridging samples will be blown over the entire length of the retention gap, and their reconcentration prior to the separation is difficult at best. The length of the retention gap is determined by the method of solvent removal and the size of the sample. If the sample is under 100 μL, it can be accommodated by a retention gap of reasonable length, such as a 15-meter length of 0.32-mm-ID deactivated tubing [64]. Larger samples can be accommodated by a larger retention gap, but it is accepted practice to use the same retention gap and control the rate of sample injection such that it corresponds to the rate of evaporation from the retention gap. The solvent is evaporated by the flowing carrier gas from the rear of the band to the front as the more volatile solutes are concentrated at the rear of the shrinking flooded zone (see Fig. 3.12).

Inlets have also been developed for "partial concurrent solvent evaporation." In this method up to 90% of the solvent injected is evaporated during injection [53]. The solvent is vented through a three-way valve located between the retention gap and the analytical column. This diverts the majority of the solvent out of the analytical system, and permits a much faster analysis. Concurrent solvent evaporation is facilitated by an oven temperature that approaches the pressure corrected boiling point of the column. Attaining this temperature, however, would create a flashback, where the vapor and some analytes would be driven back into the injector and the pneumatics. The thin solvent film of the flooded zone retains the analytes. Hence the solvent *should not be completely vented* before it reaches the separation column. Once the injection is complete and the flooded zone evaporated, the solvent vent is closed and the oven program started. There is no retention in the pre-column, and analytes migrate to the head of the column where they are focused by the abrupt decrease in β. A short length of coated precolumn is sometimes used between the retention gap and the solvent vapor vent in order to improve the recovery of the more volatile solutes (Fig. 3.13, top). There are a variety of applications for this type of large volume injection. In a modification of the above method some workers have developed a separate heating system for the

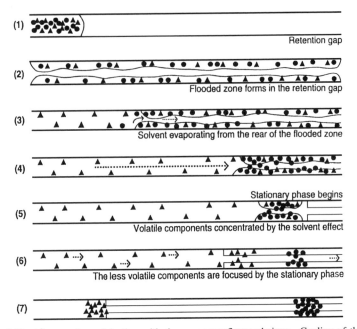

Fig. 3.12. A large volume injection with the vapor overflow technique. Cooling of the heated packing in the liner by the evaporating sample solvent causes higher boiling analytes to be trapped on the packing. As the packing regains heat, the higher boiling analytes begin the chromatographic process. See text for discussion. From [58]. Reprinted with permission.

pre-column and successfully analyzed a variety of samples including polynuclear aromatics from water samples (Fig. 3.13, bottom) [65–67].

Large-volume injection by the on-column method and the use of a retention gap has several advantages over the other methods of large volume injection. These include the capability of injecting milliliters of sample [68], and generating accurate quantitative results. (By virtue of the solvent effect, this avoids discrimination toward volatile solutes and gives enhanced recoveries for sensitive and thermolabile analytes.) On-column methods suffer when the samples are dirty (the retention gap is quickly contaminated), or with aqueous samples that are unable to wet the retention gap surface. Commercial instrumentation is available that is specifically designed for on-column large volume injection, and that features microprocessor-controlled optimization of injection rate, temperature, and pressure, based on the specific solvent being used [69].

LC–GC

In 1980, Majors coupled an LC to a GC via an autosampler [70]. The utility of this coupling hinges on the superiority of HPLC separations as a sample cleanup

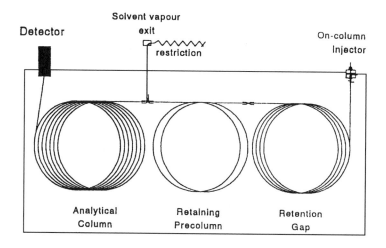

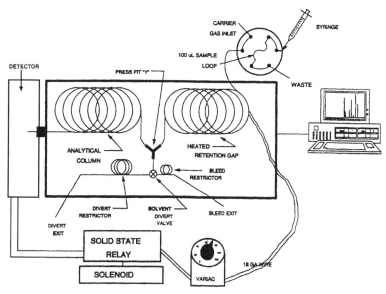

Fig. 3.13. Instrumental configuration for on-column large volume injection (top). On column large volume injection with an externally heated retention gap (bottom). *Adapted from Ref. 69 and Ref. 67 with permission.*

tool. HPLC fractions should provide much cleaner fractions than other methods of sample preparation, specific fractions can be selected for GC analysis, and the non-destructive detection methods employed in LC facilitate method development. The major challenge of coupling GC with HPLC lies in the removal of large volumes of solvent while providing a focused injection band. Several methods have been devised to accomplish solvent removal, and some methods have been automated. The application of a deactivated precolumn with partially concurrent solvent evaporation, as described above for standard large-volume injection, is effective and illustrated schematically in Fig. 3.14. The subject has been thoroughly reviewed [71–73] and is still a fertile research area. Applications of coupled HPLC-GC to group hydrocarbon separations [73], heroin metabolites in urine [74], and citrus oil characterization [75] have been published. Figure 3.15 illustrates the automated analysis of sterol dehydration products in a refined edible oil using a loop-type transfer method [28].

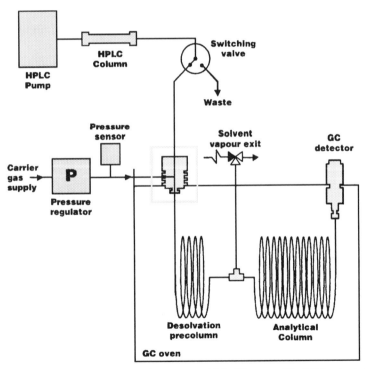

Fig. 3.14. Schematic representation of an automated LC–GC system. The LC fraction is injected by a modified on-column large volume injection method. *Taken from [75] with permission.*

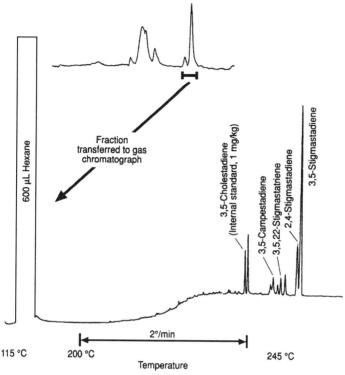

Fig. 3.15. The automated analysis of sterol dehydration products from a refined edible oil sample by LC–GC. The top chromatogram is from the LC and the lower chromatogram is from the injected GC fraction. Reproduced from [28] with permission.

3.8 Purge-and-Trap Sampling

Purge and trap sampling is used to prepare and introduce samples into the gas chromatograph, and is considered briefly in this chapter. The methodology has been reviewed [76], and is an effective method for sampling and analyzing low levels of volatile organic compounds from matrices such as drinking water, waste water, soil, and sludge. Used in conjunction with a mass spectrometer or selective detectors in series (e.g., PID and ELCD), it is the preferred method for evaluation of water purity in the United States and is stipulated in several US EPA methods. One of these methods, EPA 524.2 Revision 4, allows for the quantitation of 82 analytes at a detection limit of 0.1 ppb in a single analysis.

A schematic representation of a purge-and-trap concentrator/gas chromatograph is shown in Fig. 3.16. Briefly, the sample is purged for a specified time and temperature with a purge gas, usually helium, and the volatile analytes are absorbed in the trap (Fig. 3.16, top). The trap is an absorbant material such as Tenax, silica

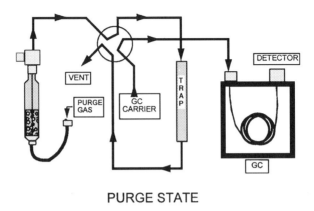

PURGE STATE

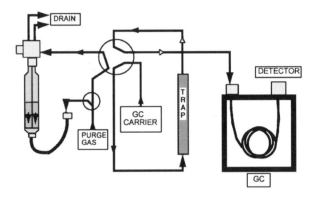

DESORPTION & DRAIN

Fig. 3.16. Schematic representation of a purge and trap concentrator in the (top) purge, and (bottom) desorb modes. The six port valve is automated.

gel, charcoal, or any one of several "synthetic charcoals" or molecular sieves or a mixture of these. After a specified trapping time, the valve is switched and the trap is rapidly heated at a specified rate to desorb the volatile analytes that are carried onto the column by the carrier gas (Fig. 3.16, bottom). The variables of time, temperature, and flow rate need to be optimized for calibration of the system. The large volume of gas required to quantitatively desorb the sample and conduct the

Fig. 3.17. The application of purge and trap with a 0.25 mm I.D. column to the analysis drinking water: (top) without, and (center) with cryogenic cooling at the head of the column. The improvement of the peak shape in the initial analytes is readily apparent. The method for reconcentrating at the head of the column is schematically illustrated in the bottom figure. *Taken from [77] with permission.*

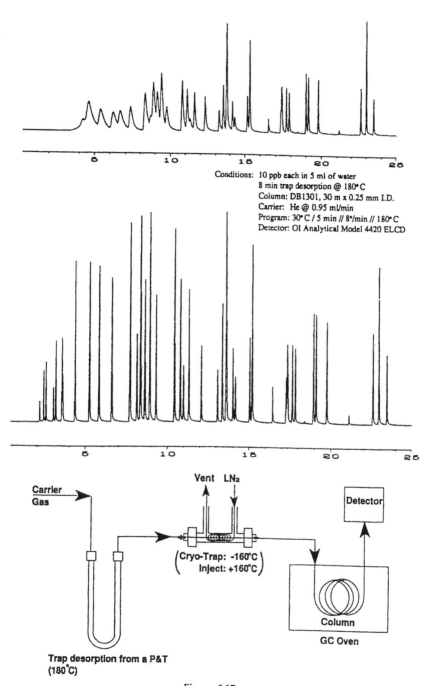

Conditions: 10 ppb each in 5 ml of water
8 min trap desorption @ 180° C
Column: DB1301, 30 m x 0.25 mm I.D.
Carrier: He @ 0.95 ml/min
Program: 30° C / 5 min // 8°/min // 180° C
Detector: OI Analytical Model 4420 ELCD

Figure 3.17.

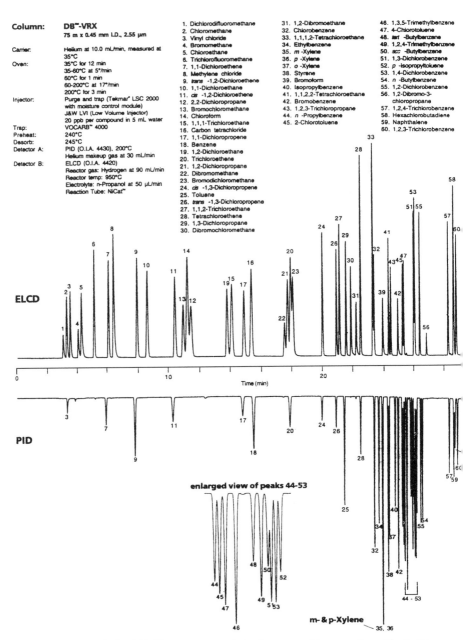

Column: DB™-VRX
75 m × 0.45 mm I.D., 2.55 μm

Carrier: Helium at 10.0 mL/min, measured at 35°C

Oven: 35°C for 12 min
35-60°C at 5°/min
60°C for 1 min
60-200°C at 17°/min
200°C for 3 min

Injector: Purge and trap (Tekmar® LSC 2000 with moisture control module)
J&W LVI (Low Volume Injector)
20 ppb per compound in 5 mL water

Trap: VOCARB™ 4000
Preheat: 240°C
Desorb: 245°C
Detector A: PID (O.I.A. 4430), 200°C
Helium makeup gas at 30 mL/min
Detector B: ELCD (O.I.A. 4420)
Reactor gas: Hydrogen at 90 mL/min
Reactor temp: 950°C
Electrolyte: n-Propanol at 50 μL/min
Reaction Tube: NiCat™

1. Dichlorodifluoromethane
2. Chloromethane
3. Vinyl chloride
4. Bromomethane
5. Chloroethane
6. Trichlorofluoromethane
7. 1,1-Dichloroethene
8. Methylene chloride
9. trans -1,2-Dichloroethene
10. 1,1-Dichloroethane
11. cis -1,2-Dichloroethene
12. 2,2-Dichloropropane
13. Bromochloromethane
14. Chloroform
15. 1,1,1-Trichloroethane
16. Carbon tetrachloride
17. 1,1-Dichloropropene
18. Benzene
19. 1,2-Dichloroethane
20. Trichloroethene
21. 1,2-Dichloropropane
22. Dibromomethane
23. Bromodichloromethane
24. cis -1,3-Dichloropropene
25. Toluene
26. trans -1,3-Dichloropropene
27. 1,1,2-Trichloroethane
28. Tetrachloroethene
29. 1,3-Dichloropropane
30. Dibromochloromethane

31. 1,2-Dibromoethane
32. Chlorobenzene
33. 1,1,1,2-Tetrachloroethane
34. Ethylbenzene
35. m -Xylene
36. p -Xylene
37. o -Xylene
38. Styrene
39. Bromoform
40. Isopropylbenzene
41. 1,1,2,2-Tetrachloroethane
42. Bromobenzene
43. 1,2,3-Trichloropropane
44. n -Propylbenzene
45. 2-Chlorotoluene

46. 1,3,5-Trimethylbenzene
47. 4-Chlorotoluene
48. tert -Butylbenzene
49. 1,2,4-Trimethylbenzene
50. sec -Butylbenzene
51. 1,3-Dichlorobenzene
52. p -Isopropyltoluene
53. 1,4-Dichlorobenzene
54. n -Butylbenzene
55. 1,2-Dichlorobenzene
56. 1,2-Dibromo-3-chloropropane
57. 1,2,4-Trichlorobenzene
58. Hexachlorobutadiene
59. Naphthalene
60. 1,2,3-Trichlorobenzene

Fig. 3.18. The analysis of drinking water contaminates by purge and trap sampling *without cryogenic focusing*. The combination of a thick stationary phase film and a stationary phase that increases the K_c of the more volatile analytes (1–6) provide shortened bands and sharper peaks.

analytes to the column can have a major impact on the quality of the subsequent gas chromatographic separation. The trap can be viewed as a small packed column, and analyte desorption is best accomplished at gas flows of 30–40 mL/min, volumes that are totally incompatible with open tubular columns. Hence, desorption is usually accomplished by using lower flows for longer times. This has the result of lengthening the injected band, which must then be refocused. If the focusing step is omitted, the long starting band generates broad peaks, with adverse effects on both separation and sensitivity. It is not unusual for several of the low-k solutes (e.g., chloromethane, chloroethane, vinyl chloride) to broaden to such a degree that they emerge only below the limits of detection (Fig. 3.17, top). For 0.25- or 0.32-mm-ID columns, the solute bands can be sharpened, benefiting both separation and sensitivity, by cryogenic focusing (Fig. 3.17, center). To conserve the expensive coolant, focusing should be applied to only a small section at the head of the column rather than cooling the GC oven (Fig. 3.17, bottom). The separation can also be accomplished on 0.53-mm- or 0.45-mm-ID columns without cryogenic cooling. Early efforts hoped to accomplish ambient temperature refocusing by using thicker film columns of extraordinary length (105-m). The resolution of the lower-k solutes improved, but it was later demonstrated that the improvement was due not to the column length per se, but instead to the pressure drop generated by the vary long column. Similar results can be achieved by placing a restrictor on the end of a much shorter column (Section 6.5). Subsequent work utilized shorter (75-m) columns, which can accomplish the same goals in much shorter times (Fig. 3.18).

The use of a split injection interface has been demonstrated to help compensate for the flow mismatch between the trap and the separation column. Some workers have explored the design and application of microtraps to improve the flow match between the trap and column [77].

3.9 Selecting the Proper Injection Mode

Selection of the proper injection mode is, to some degree, a tradeoff between the "separation needs" and the nature of the specific sample. The "separation needs" might be viewed as the column and chromatographic conditions required to successfully analyze the sample: adequate resolution, capacity, and analysis time. However, many analysts have little latitude in this regard; only one or two instrumental/column configurations may be available, and the injection mode decision is then predicated by the sample and the nature of the analytes of interest. The sample based considerations include, sample concentration, the presence of nonvolatile contaminants, and the stability (thermal and otherwise) and volatility range of the analytes of interest. For one reason or another, some analytical methods specify a particular injection mode. In other cases, the sample-based flow chart shown in Fig. 3.19 can be viewed as a general guide to aid in injection mode selection.

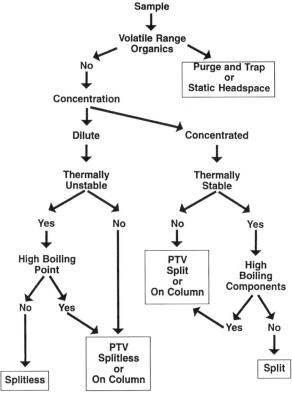

Fig. 3.19. A sample-based flow chart for selecting the injection mode. As an example, if the sample did not contain volatile range organics, was reasonably concentrated, thermally stable, and lacked high boiling components, the injection mode of choice is split.

Capillary Columns

With smaller-diameter open tubular columns, gas flow volumes are much more restricted and capillary injectors are required. Several factors should be considered in deciding which injection mode is most suitable for a given sample. That decision is usually based on the nature and composition of the sample and the goals of the investigation. Preliminary sample workup or preparation may be necessary in some cases. Many such procedures are based on isolation or concentration steps that employ cold-trapping, distillation, extraction, and/or adsorption. There is often no way to avoid these steps, but it must be remembered that both quantitative and qualitative changes in sample composition can result from these treatments [11].

Direct headspace injection frequently offers the most direct and elegantly simple approach to sample introduction. A cold trap can be employed to exercise a distribution constant focus on vapor sample injections as large as 1 mL, which can be made directly into a fused silica column using a macrosyringe fitted with a fused silica needle [7, 78]. Not only does the increased inertness of this sampling system make it possible to detect some of the more reactive solutes (including sulfur- and nitrogen-containing compounds), but dilution of the sample by carrier gas is minimized by direct injection into the column rather than into an injection chamber, whence additional carrier gas further dilutes the sample while conducting it to the column. These cold-trapped solute (or sample) bands are not stationary. Unless their vapor pressures are zero, some mass of those materials enters the mobile phase and must be chromatographically transported through the column. Normally their vapor pressures in the trapping area are so low (and their distribution constant so small) that the steady-state concentration in the mobile phase is below the limits of detection. The amount of a solute lost to detection is usually minuscule, but the conclusion that sharp peaks testify to quantitative cold trapping is obviously invalid. With solutes of higher vapor pressure (lower boiling solutes or higher trapping temperatures), losses would be higher, and in cases where the mobile phase concentration of a solute is above the limits of detection, breakthrough becomes evident. Components that negotiate the trap eventually ascend a positive temperature gradient as they leave the trap and enter the uncooled column. This subjects them to a distribution constant defocus; $v_{front} > v_{rear}$, and the shapes of those peaks are badly degraded [7, 78]. On-column vapor phase injections hold particular appeal where the volatiles of interest have sufficient volatility, or where their volatilities can be increased, such as by increasing the temperature of the sample. Such injections introduce relatively long bands to the column, and some means of achieving band focusing (as discussed earlier) should be employed prior to the chromatographic process. This subject is considered again in Chapter 8.

The inlet splitter is probably the simplest of the specialized injectors used with capillary columns, and when it is properly employed split injections can be quite satisfactory. The sample should be liquid (or in solution), and the solutes should have reasonable thermal stabilities. Further, their concentrations in the sample should be approximately 0.01–10%. If the volatilities of the solutes fall within the approximate range of the C_6–C_{19} hydrocarbons, it is usually possible to select operational conditions (temperature, flow path, internal volume, split ratio, etc.) that yield good linearity. The range of solute volatilities over which good linearity is achieved can sometimes be extended by employing programmed temperature vaporization or by varying some of the above parameters.

Where the solutes exist as a dilute solution of higher-boiling solutes, splitless may be the preferred injection mode. Solute volatilities should approximate those of the C_{10}–C_{26} hydrocarbons. While detectable amounts of solutes beyond this range will be delivered to the column, discrimination occurs. The severity of discrimination varies with solute volatility and other factors. Hydrogen is almost always the preferred carrier gas in open tubular columns (see Chapter 5), but it offers some additional benefits in splitless injection. Hydrogen yields higher optimum linear velocities, which translate to larger volumetric flows. These larger flow volumes are more efficient at transporting the sample from the inlet to the column [24].

On-column injection holds its greatest utility for thermally labile solutes that tend to degrade or rearrange in the vaporizing modes of injection and for higher-boiling solutes that resist (complete) vaporization in split or splitless modes.

Some on-column inlets are designed to utilize fused silica needles in sample introduction [7, 78]. These offer an extremely inert system, which permits the analysis of even low levels of active solutes. Some recent developments in specialized techniques of on-column injection show promise of further extending our analytical capabilities [8]. Samples that are heavily contaminated will quickly deteriorate the column in on-column injection. For these samples PTV on-column injection provides most of the benefits while leaving the nonvolatile materials in the injector liner.

Small-Diameter Capillary Columns (Microbore Columns)

As the internal diameter of the capillary column drops below 0.20 mm the difficulties faced in injection increase. For 0.05–0.10-mm-ID columns (whose popularity is increasing among those seeking faster analyses), the complications of higher optimum velocities and higher inlet pressures challenge both the equipment and the analyst. Because of the comparatively large interior volumes of conventional injectors, the split mode of injection is mandated [79], and split ratios of 1 : 500–1 : 5000 are required. At the present state of the art, this is the only feasible way to shorten the sample band delivered to these small-diameter low-flow columns. Attention should also be directed to the boiling point range of the analytes because sample residence time in the vaporization area is so short that transfer of adequate heat to higher boiling-constituents may be a problem. Discrimination toward high boilers can be severe under these conditions. To obtain *very* fast analysis times, on the order of a few seconds, special injectors are required. These include the fluid-gate-type [80] and cold-trapping [81] injectors. These areas are discussed in greater detail in Chapter 8.

Large-Diameter Open Tubular Columns

The larger-diameter open tubular columns, 0.53 and 0.75 mm ID, that can be directly substituted for packed columns can tolerate much larger gas flow volumes. As a result, normal-size injections into these columns usually pose no special problems. They are generally attached directly to the standard packed column injector by virtue of a reducing adapter. As explored above, performance can be improved by employing injection liners specifically designed for these columns. The direct flash vaporizing injector, in which the injected sample is rapidly vaporized and then conducted to the column (Fig. 3.4), is usually preferred for most applications. For thermally labile or high-boiling solutes, the on-column liner is usually advantageous. As mentioned above, if the latter is packed with an appropriate material and installed in an inverted position, it is possible to confine injection residues to the inlet.

References

1. W. Jennings, *Gas Chromatography with Glass Capillary Columns*, 2nd ed. Academic Press, New York, 1980.
2. J. Hinshaw, *LC–GC* **9**:338 (1991).
3. K. Grob, Jr., *J. High Res. Chromatogr.* **2**:15 (1979).
4. K. Grob and G. Grob, *J. High Res. Chromatogr.* **2**:109 (1979).
5. K. Grob, Jr., *J. Chromatogr.* **213**:3 (1981).
6. K. Grob, Jr., *Anal. Proc.* (London) **19**:233 (1982).
7. W. Jennings, in *Sample Introduction in Capillary Gas Chromatography* (P. Sandra, ed.), p. 23. Huethig, Heidelberg, 1985.
8. M. F. Mehran and W. Jennings, *J. Chromatogr. Sci.* **24**:34 (1986).
9. M. F. Mehran, W. J. Cooper, M. Mehran, and W. Jennings, *J. Chromatogr. Sci.* **24**:142 (1986).
10. J. A. Rijks, J. Drozd, and J. Novak, in *Advances in Chromatography* (A. Zlatkis, ed.), p. 195. Elsevier, Amsterdam, 1979.
11. W. Jennings and A. Rapp, *Sample Preparation for Gas Chromatography*, Huethig, Heidelberg, 1983.
12. M. Galli, S. Trestianu, and K. Grob, Jr., *Proc. 3rd Int. Symp. Capillary Chromatogr.*, p. 149 (1979).
13. W. Jennings and G. Takeoka, *Chromatographia* **15**:575 (1982).
14. W. Jennings, *J. Chromatogr. Sci.* **22**:129 (1984).
15. K. Grob and G. Grob, *Chromatographia* **5**:3 (1972).
16. K. Grob and G. Grob, *J. Chromatogr.* **94**:53 (1974).
17. K. Grob and K. Grob, Jr., *J. High Res. Chromatogr.* **1**:275 (1978).
18. R. J. Miller and W. Jennings, *J. High Res. Chromatogr.* **2**:72 (1979).
19. K. Grob, *J. High Res. Chromatogr.* **15**:190 (1992).
20. K. Grob and M. DeMartin, *J. High Res. Chromatogr.* **15**:335 (1992).
21. K. Grob and M. DeMartin, *J. High Res. Chromatogr.* **15**:399 (1992).
22. R. R. Freeman, E. J. Guthrie, and R. Lautamo, personal communication (1985).
23. K. Grob, and G. Grob, *J. Chromatogr. Sci.* **7**:587 (1969).
24. K. Grob, Jr. and A. Romann, *J. Chromatogr.* **214**:118 (1981).
25. G. Schomburg, H. Behlau, R. Dielmann, F. Weeke, and H. Husmann, *J. Chromatogr.* **142**:87 (1977).
26. G. Schomburg, H. Husmann, and R. Rittmann, *J. Chromatogr.* **204**:85 (1981).
27. R. Freeman, *High Resolution Gas Chromatography*, 2nd ed. Hewlett-Packard Co., 1981.
28. K. Grob, *Anal. Chem.* **66**:1017A (1994).
29. J. Hinshaw, personal communication (1996).
30. P. Wiley, R. Phillips, K. Klein, M. Thompson, and B. Hermann, *J. High Res. Chromatogr.* **14**:649 (1991).
31. P. Wiley, K. Klein, M. Thompson, and B. Hermann, *J. High Res. Chromatogr.* **15**:763 (1992).
32. J. Hinshaw, *J. High Res. Chromatogr.* **16**:247 (1993).
33. F. Poy, S. Visani, and F. Terrosi, *J. Chromatogr.* **217**:81 (1981).

34. F. Poy, *Chromatographia* **16**:343 (1982).
35. G. Schomburg, H. Husmann, F. Schultz, G. Teller, and M. Bender, *Proc. 5th Int. Symp. Capillary Chromatogr.*, p. 280 (1983).
36. F. Poy and L. Cobelli, *J. Chromatogr.* **279**:689 (1983).
37. H. Mol, P. Hendriks, H. Janssen, C. Cramers, and U. Brinkman, *J. High Res. Chromatogr.* **18**:124 (1995).
38. H. Van Lieshout, H. Janssen, and C. Cramers, *Am. Lab.*, **27**:38 (Aug) (1995).
39. K. Grob and K. Grob, Jr., *J. Chromatogr.* **151**:311 (1978).
40. K. Grob, *J. High Res. Chromatogr.* **1**:263 (1978).
41. G. Schomburg, R. Diehlmann, H. Husmann, and F. Weeke, *Chromatographia* **10**:383 (1977).
42. K. Knauss, J. Fullemann, and M. P. Turner, *J. High Res. Chromatogr.* **4**:681 (1981).
43. R. Jenkins and W. Jennings, *J. High Res. Chromatogr.* **6**:228 (1983).
44. K. Grob, Jr. and R. Mueller, *J. Chromatogr.* **244**:185 (1982).
45. R. Jenkins, personal communication (1982).
46. K. Grob, Jr., *J. Chromatogr.* **251**:235 (1982).
47. K. Grob, Jr., *J. Chromatogr.* **253**:17 (1982).
48. F. Hougen, personal communication (1981).
49. E. Bayer and G. H. Liu, *J. Chromatogr.* **256**:201 (1983).
50. F. Pacholec and C. F. Poole, *Chromatographia* **18**:234 (1984).
51. G. Rene van der Hoff, P. Van Zoonen, and K. Grob, *J. High Res. Chromatogr.* **17**:37 (1994)
52. W. Jennings and G. Takeoka, in *Neu Ulmer Gesprache 1985* (H. Jaeger, ed.). Huethig, Heidelberg, 1987.
53. J. Vreuls, H. Mol, J. Jagesar, R. Swen, R. Hessels, and U. Brinkman, *Proc. 16th Int. Symp. Capillary Chromatogr.*, Riva del Garda, Huethig, Heidelberg, p. 1181 (1994).
54. L. Lendero, E. Gerstel, and J. Gerstel, *Proc. 8th Int. Symp. Capillary Chromatogr.*, Riva del Garda, Huethig, Heidelberg, p. 342 (1987).
55. W. Vogt, K. Jacob, and H. Obweber, *J. Chromatogr.* **174**:437 (1974).
56. H. Mol, H. Janssen, C. Cramers, and U. Brinkman, *Proc. 16th Int. Symp. Capillary Chromatogr.*, Riva del Garda, Huethig, Heidelberg, p. 1107 (1994).
57. K. Grob, *J. High Res. Chromatogr,* **13**:540 (1990).
58. K. Grob and S. Brem, *J. High Res. Chromatogr.* **15**:715 (1992).
59. K. Grob, S. Brem, and D. Frohlich, *J. High Res. Chromatogr.* **15**:659 (1992).
60. K. Grob and D. Frohlich, *J. High Res. Chromatogr.* **16**:224 (1993).
61. K. Grob, S. Brem, and D. Frohlich, *J. High Res. Chromatogr.* **17**:792 (1994).
62. K. Grob, G. Karrer, and M. Riekkola, *J. Chromatogr.* **333**:129 (1985).
63. K. Grob, H. Neukom, and M. Riekkola, *J. High Res. Chromatogr.* **7**:319 (1984).
64. K. Grob and B. Schilling, *J. High Res. Chromatogr.* **7**:531 (1984).
65. P. Morabito, J. Hiller, and T. McCabe, *J. High Res. Chromatogr.* **12**:347 (1989).
66. P. Morabito, J. Hiller, and T. McCabe, *J. High Res. Chromatogr.* **16**:5 (1993).
67. P. Morabito, J. Hiller, T. McCabe, and D. Zakett, *J. High Res. Chromatogr.* **16**:90 (1993).
68. K. Grob, personal communication with the author (1996).
69. F. Munari, P. Colombo, P. Magni, G. Zilioli, S. Trestianu, and K. Grob, *J. Microcolumn Sep.* **7**:403 (1995).
70. R. Majors, *J. Chromatogr. Sci.* **18**:571 (1980).
71. K. Grob, *J. Chromatogr. A*, **703**:265 (1995).
72. K. Grob, *On-Line Coupled LC–GC*, Huethig, Heidelberg, (1991).
73. A. Trisciani and F. Munari, *J. High Res. Chromatogr.* **17**:452 (1994).
74. F. Munari and K. Grob, *J. High Res. Chromatogr.* **11**:172 (1988).
75. F. Munari, G. Dugo, and A. Cotroneo, *J. High Res. Chromatogr.* **8**:601 (1985).

76. S. M. Abeel, A. K. Vickers, and D. Decker, *J. Chromatographic Sci.* **32**:328 (1994).
77. R. Snelling, paper, *Pittsburgh Conf. Anal. Chem. Appl. Spectrosc., 1996*, Chicago; also OI Analytical Application Note 08750396.
78. G. Takeoka and W. Jennings, *J. Chromatogr. Sci.* **22**:177 (1984).
79. G. Gaspar, *J. Chromatogr.* **556**:331 (1991).
80. G. Gaspar, R, Aninino, C. Vidal-Madjar, and G. Guiochon, *Anal. Chem.* **50**:1512 (1978).
81. A. Van Es, J. Janssen, C. Cramers, and J. Rijks, *J. High Res. Chromatogr.* **11**:852 (1988).

CHAPTER 4

THE STATIONARY PHASE

4.1   General Considerations

Gas chromatographic activities were formerly dominated by packed columns, which are restricted to relatively low numbers of theoretical plates. Separation efficiencies in these "low-N" columns are very much dependent on solute α values, i.e., stationary phase selectivity [see Eq. (1.23) and the accompanying discussion]. As a result, a large number of stationary phases were utilized in packed columns. Only a few of the many stationary phases that were developed over the years have been discarded; most of them are still available as packed column phases. Many of these are unsuitable for use in open tubular columns, because they rapidly lose viscosity at higher temperatures. This causes "beading" on the surface, which is disastrous to column efficiency and column lifetime. Gum-type stationary phases that retain their viscous characteristics at higher temperatures maintain their status as thin films and give columns of much longer lifetime. While this limits the materials that can be used as stationary phases in open tubular columns, the much higher theoretical plate numbers available usually translate to superior separations, even with a less selective stationary phase. Modern analysts usually regard the vast majority of the more than 200 stationary phases listed by the supply houses catering to the packed column chromatographer as obsolete. The less selective stationary phases offer several benefits: they are usually the "less polar" phases, and columns coated with those phases exhibit higher efficiencies, longer lifetimes, lower bleed rates, and shorter analysis times; the latter normally result in higher sensitivities. On the basis of individual tests of many thousands of columns, column efficiencies, in terms of theoretical plates per meter of column length, usually decrease in the following order:

Polydimethylsiloxane = 5% diphenyl–95% dimethylpolysiloxane > 12–14% phenylcyanopropyl–86–88% dimethylpolysiloxane > 24–28% phenylcyanopropyl–72–76% dimethylpolysiloxane = 25% trifluoropropyl–75% dimethylpolysiloxane > 50% phenyl–50% dimethylpolysiloxane > 50% phenylcyanopropyl–50% dimethylpolysiloxane = polyethylene glycol.

Efficiencies are highest for columns coated with nonpolar stationary phases, and in general decrease with increasing polarity of the phase. In general, nonpolar stationary phases also give columns of longer lifetime, and should be used whenever possible. There are some separations that do require both high numbers of theoretical plates and greater selectivity; selected examples are discussed later.

4.2 Stationary Phase Polarity and Selectivity

The terms "selectivity" and "polarity" are often misused in gas chromatography and sometimes (incorrectly) used interchangeably. A polar substance (stationary phase or analyte) possesses a permanent dipole. Polar stationary phases are those containing functional groups such as —CN, —CO, —C—O, and —OH. Most hydrocarbon-type stationary phases (e.g., squalane and the Apiezon greases) are apolar, but their characteristics restrict their use to packed columns. They are not normally suitable for use in open tubular columns. The low-polarity stationary phases 100% polydimethylsiloxane and 95% dimethyl–5% diphenylpolysiloxanes are extremely useful and presently enjoy the broadest application. As the percentage of diphenylsiloxane subunits increases with respect to the methylsiloxane, the stationary phase becomes more polar. Phases containing 35–50% diphenyl substitution are considered "midpolarity" phases. Greater polarity can be conferred by substituting cyanopropyl (or cyanoethyl) for up to 50–100% of the methyl groups [1]. Phases of extreme polarity [e.g., 1,2,3,4-tetrakis(2-cyanoethoxy)butane, 1,2,3-tris(2-cyanoethoxy)propane] are available in packed columns, but yield unstable films on open tubular columns. Polyester-type stationary phases are considered the most polar of the commercially available phases that are usable in open tubular columns.

It is important to recognize that in gas chromatography, solutes elute in an order mandated by their "net" vapor pressures, where the net vapor pressure of a solute might be viewed as the latent vapor pressure, reduced by the sum of all interactions between that solute and the stationary phase under a given set of chromatographic conditions. The strengths of the individual solute–stationary phase interactions are very much a function of the solubility of that solute in that stationary phase. These forces attain their full potential only if the solute is fully soluble in the stationary phase. In a Grob test mixture on a dimethylsiloxane column, octanol (bp 194.5°C) invariably elutes *before* nonanal (bp 192°C). These are relatively polar solutes, and neither is fully soluble in this apolar phase. Both solutes are eluting "too early," but the exclusion effect is greater for the more polar eight-carbon-chain alcohol than for the less polar nine-carbon-chain aldehyde (Fig. 4.1).

Both the stationary phase and the solutes play roles in "selectivity." A stationary phase that exhibits high selectivity for one type of solute may exhibit intermediate or low selectivity for solutes with different functional groups. A generality

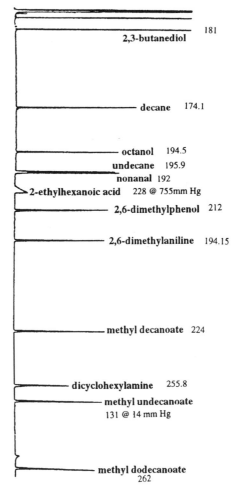

Fig. 4.1. A Grob test mixture on a dimethylpolysiloxane column. Numbers at the top of each peak denote the boiling point of that solute, at atmospheric pressure unless otherwise noted. Note discrepancies in boiling points and elution order, as well as the shape of the 2-ethyl hexanoic acid peak. See text for discussion.

(to which there are exceptions) is that the larger the solute K_c values are in a stationary phase, the more selectivity that stationary phase displays toward those solutes. Apolar hydrocarbon solutes would exhibit larger values of K_c in an apolar stationary phase than in a polar phase. Polar solutes would exhibit larger values of K_c in a more polar stationary phase than in an apolar stationary phase. Selectivity for hydrocarbon solutes would be higher in the apolar and lower in the polar stationary phase, whereas selectivity for more polar solutes would be greater in a more polar stationary phase. If a given solute is chromatographed under a particular set of conditions on columns coated with different stationary phases, the solute retention factor will be largest for the stationary phase that has the higher degree of interaction with that solute. Three types of solute–stationary phase interactions are of concern in gas chromatography: dispersion, dipole, and proton sharing. Most of the following discussion hinges on work directed by Hawkes [2] and contributions of Stark *et al.* [3].

Dispersion Interaction

The electrons of an atom or molecule oscillate through several different and distinct positions, each of which is characterized by a particular pattern of electrical asymmetry. Each of those short-lived electronic configurations causes the atom (or molecule) to display an overall instantaneous dipole, which leads to polarization of adjacent atoms or molecules and generates attractive forces between those atoms or molecules (similar to London–van der Waals forces of attraction). These intermolecular attractions constitute the major part of the total attractive force between a hydrocarbon solute and the stationary phase, and they are significant for any dissolved solute molecules and the surrounding stationary phase solvent. There have been attempts to relate the potential for dispersion interaction to the refractive index [4]. Burns and Hawkes [2] used a modification of this approach in calculating the dispersion indices shown in Table 4.1.

TABLE 4.1

Stationary Phase Interaction Indices

Stationary phase	Dispersion index	Dipole index	Basicity index	Acidity index
Dimethylsiloxane	9	0	0	0
Phenylmethylsiloxane	11.6	0	0	1
3-Cyanopropylsiloxane	10.4	11	3	0
Trifluoropropylsiloxane	8.6	3	0	1
Polyethylene glycol	8.6	8	4	0

Data from Burns and Hawkes [2].

Dipole Interactions

When both the stationary phase and the solute possess permanent dipole moments, the alignment of the two dipoles can result in a strong interaction between the stationary phase and the solute. In some cases the close proximity of a strong dipole in the stationary phase can generate an induced dipole in the solute (and vice versa). Such dipole interactions usually involve individual functional groups of the solute and of the stationary phase. Burns and Hawkes [2] used benzene and hexane to calculate the relative strengths of permanent dipoles in the stationary phase. They reasoned that the difference in the adjusted retentions of hexane and benzene on a given stationary phase could be attributed to an induced dipole (benzene)– permanent dipole (stationary phase) interaction, provided dispersion effects were taken into account. "Polarity" was earlier defined as the presence of a permanent dipole; hence the magnitude of the "dipole index value" in Table 4.1 can be considered a measure of the relative polarity of a given stationary phase. Permanent dipole-to-permanent dipole interactions are usually stronger than permanent dipole–induced dipole interactions, but both are considered "dipole interactions."

Base–Acid Interactions

With hydroxyl-containing solutes that exhibit similar dipole and dispersion potentials, the formation of hydrogen bonds with the stationary phase is an important separation mechanism. The ability of a stationary phase to participate in hydrogen bonding is usually a measure of its Bronsted basicity. The hydroxyl group can, however, also behave as a base and is retained by Lewis and Bronsted acids. Because relatively few of the more commonly used stationary phases exhibit acidity, the increased retention of an alcohol (as compared to another solute exhibiting similar dispersion and dipole interactions) has been used as an index of stationary phase basicity [2]. These values are also shown in Table 4.1.

Gas chromatography is essentially a volatility phenomenon, in which the rate at which a given solute traverses the column varies directly with the average velocity of the mobile phase and indirectly with the proportion of the time the molecules of that solute spend in the stationary phase:

$$v = \bar{u}/(k + 1) \qquad (4.1)$$

Both solute retention factors k and solute vapor pressures are governed largely by (1) the temperature and (2) the degree of interaction between the solute and the stationary phase. When dispersive forces are the dominant form of solute– stationary phase interaction, both solute elution orders and the relative degree of solute separation can be predicted on the basis of solute boiling points, because these then dictate the solute vapor pressures.

The "retentiveness" of a stationary phase for a given solute under a particular set of conditions is evidenced by the retention factor k; the more retentive the phase (toward that solute), the larger the k. The ratio of two solute retention factors (k_2/k_1) determines their separation factor, and (assuming constant peak width) the larger the separation factor of two solutes, the greater their resolution [Eq. (1.23)].

There are several disadvantages to increased stationary phase retention. A larger K_c usually translates to lower sensitivity; the larger K_c, the lower the solute concentration in mobile phase and the longer the time required to transport that solute from the end of the column to the detector (at constant carrier flow). This results in broader and less intense peaks. From Eq. (1.14), as K_c increases, the solute retention factor must also increase, and longer analysis times will be required. These effects can be countered by increasing the column temperature at the time of detection (which would lower the K_c). For higher-molecular-weight solutes or those engaging in a higher degree of solute–stationary phase interaction, the higher required temperature may place increased stress on the stationary phase and lead to shorter column life. Active solutes that tend to produce tailing peaks generally exhibit greater peak distortion with increased retention. There have been several attempts to establish systematic rationales for stationary phase selection (e.g., Rohrschneider or McReynolds constants), but the process is in most cases still based on empirical judgments. Where improved solute separation is the goal, a stationary phase that would be expected to engage in a greater degree of interaction with those particular solutes is employed. On the other hand, a less retentive phase would be used to encourage the elution of extremely high-boiling solutes or those with more reactive functional groups. It is also possible to select stationary phase mixtures or to synthesize new stationary phases that maximize the separation factors of a given group of solutes to yield optimized separations. This latter approach is considered in later chapters.

4.3 Polysiloxane Stationary Phases: General Comments

The polysiloxane phases have been the subject of an updating by Blomberg [1] and a thorough review by Haken [5]. They are normally viewed as constituting the most "abuse-tolerant" group of stationary phases. However, the nature of their substituent groups and their homogeneity and purity strongly influence behavior under stress; they are not indestructible. A poorly deactivated glass or fused silica surface can cause phase deterioration, as can the free silanol groups that are normally present in the polymers themselves. In the better stationary phase preparations, free silanols have been converted to inactive groups, a process that for the terminal silanols is termed "end-capping." Even minute traces of the catalysts used in the preparation of the polysiloxane polymers will lead to phase deterioration. All of these problems are exacerbated by higher temperatures (see Chapter 10). High-quality columns demand stationary phases of the highest purity, homogeneity, and batch-to-batch reproducibility.

4.4 Dimethyl Siloxane Stationary Phases

As discussed in Section 4.1, polymer viscosity, especially at higher temperatures, is critical to column longevity. The viscosity of the polydimethylsiloxanes is influenced only slightly by temperature. This has been attributed to their structure. In the absence of dilutents such as solvents, a linear polysiloxane molecule forms a coiled helix in which the siloxane bonds are shielded by orientation toward the axis and from which the alkyl groups project outward [6].

With the polydimethylsiloxanes, a temperature increase brings about two opposing effects: (1) it tends to increase the mean intermolecular distance, and (2) it expands the helices and thus tends to decrease the distance. Hence the tendency to increase the mean intermolecular distance is countered by the expansion of the helix, and temperature has little effect on the viscosity of the polydimethylsiloxanes. The substitution of functional groups such as phenyl or cyanopropyl (see below) disrupts the helical structure, and the viscosities of these siloxanes are more temperature-sensitive.

The dimethylsiloxane polymers have several properties that are desirable in stationary phases. They are thermally stable, their viscosity is affected only slightly by temperature, they are readily crosslinked and can be covalently bonded to the support surface, and they have good wetting properties. In addition, columns made from these polymers generally have much lower levels of column bleed than do those utilizing, e.g., polyethylene glycol or Apiezon-type phases. In the polydimethylsiloxanes, the R positions are occupied by methyl groups. To impart greater "functionality," other groups may be substituted for some of the methyl groups. Table 4.2 lists the approximate compositions of some of the more popular siloxane phases.

The trivial name "silicones" is still used in some segments of industry for the polysiloxanes, and has an historical basis. In 1901, Kipping and Lloyd [7] obtained oily and gluelike products from syntheses that were directed toward producing silicon-containing analogs of organic ketones. Although they later perceived that these actually exhibited an —Si—O—Si— structure, they continued to use the word "silicones" for these products. Today, the trivial name "silicone oils" is often used for linear polydimethylsiloxane fluids (e.g., SF-96, OV-101, SP-2100); "silicone gums" for higher-molecular-weight linear polydimethyl siloxanes (e.g., OV-1, SE-30); "silicone rubber" usually but not always denotes a crosslinked polysiloxane gum (e.g., DB-1, Ultra-1) [5]. The wetting characteristics of these methyl siloxanes are such that they form films of the highest uniformity on glass and fused silica. Columns produced with these stationary phases typically exhibit very high efficiencies.

One of the major limitations of the polydimethylsiloxane stationary phase is its limited selectivity. The interaction between solutes and a dimethylsiloxane phase is limited largely to dispersion forces. Solute elution order is based largely on solute vapor pressures, and in polydimethylsiloxane, extradispersion forces (e.g.,

TABLE 4.2

Composition of Selected Stationary Phases

Phase	$CH_3{}^a$ (%)	Phenyl (%)	$CNPr^b$ (%)	Other
SE-30, OV-101, DB-1	100	0	0	0
SE-54, DB-5	94	5	0	1% vinyl
OV-17, DB-17	50	50	0	0
OV-210, DB-210	50	0	0	50% TFP^c
OV-225, DB-225	50	25	25	0
DB-1301	94	1	1	0
OV-1701, DB-1701	88	6	6	0
SP-2330, DB-2330	25	0	75	0

[a] In the bonded (DB) polysiloxane phases, the percent CH_3 figure is nominal; a few of the bonds indicated as methyl exist as crosslinks or surface bonds.

[b] Cyanopropyl.

[c] Trifluoropropyl.

hydrogen bonding, dipole interactions) play little or no role. Solutes that cannot be sufficiently differentiated on the basis of their dispersive interactions require the use of stationary phases with which they can engage in other types of interactions.

4.5   Other Siloxane Stationary Phases

Polydimethylsiloxane is relatively apolar, but polarity and selectivity can be manipulated, within limits, by substituting other functional groups (e.g., phenyl, vinyl, cyanopropyl, or trifluoropropyl) for some of the methyls. The 5% phenyl–95% methylpolysiloxane has endured for many years as one of the most widely used stationary phases. The substitution of more "polar entities" can produce polymers that not only can engage in dispersive interactions, but also are capable of different degrees of dipole and/or acid–base interactions. These will display increased selectivity toward particular solutes. Such substitutions will also affect thermal stability of the polymer [8]. Referring back to the schematic structure just preceding Section 4.4, substitution of an electronegative group for R would decrease the strength of the Si—C bond and increase that of the Si—O—Si bond. Conversely, substitution of an electron donor would strengthen the Si—C bond and weaken the Si—O—Si bond. This would lead one to predict that substituting phenyl for much of the methyl would yield a higher-temperature polysiloxane phase [8]. Although it is true that the thermal stability of the polymer itself is improved, that greater stability is often of only limited utility to the chromatographer. The surface of the fused silica tubing requires other special treatments to encourage wetting by this high-phenyl polymer, and the thermal limits of those treatments may now dictate the upper temperature limit of the column, as reflected in Table 4.2.

Cyano groups are also strongly electron-attracting, and their substitution on the polysiloxane chain results in weakening the Si—C bond. This is especially true if the cyano group is in the α position. A γ-substituted cyanosiloxane reportedly has essentially the same oxidative thermal stability as a methylsiloxane [1, 8].

4.6   Aryl-Substituted Siloxanes

In the polysiloxane structures detailed above, the backbone is composed of polarizable siloxy units. At temperatures of 200–250°C, that backbone displays a thermodynamic predilection toward ionic rearrangements that yield a mixture of lower-molecular-weight cyclic siloxanes. Sveda [9] was the first to increase the resistance to thermal and oxidative degradation by inserting the phenyl *into* the polysiloxane chain as an *aryl* inclusion, producing a *silarylene–siloxane*. These polymers possess increased resistance to thermal and oxidative degradations. They also display slightly different selectivities than their alkyl–substituted siloxane counterparts. Since the early 1990s, several column manufacturers began offering proprietary "low-bleed" columns containing *poly(tetramethyl-1,4-silphenylenesiloxane)* stationary phases.

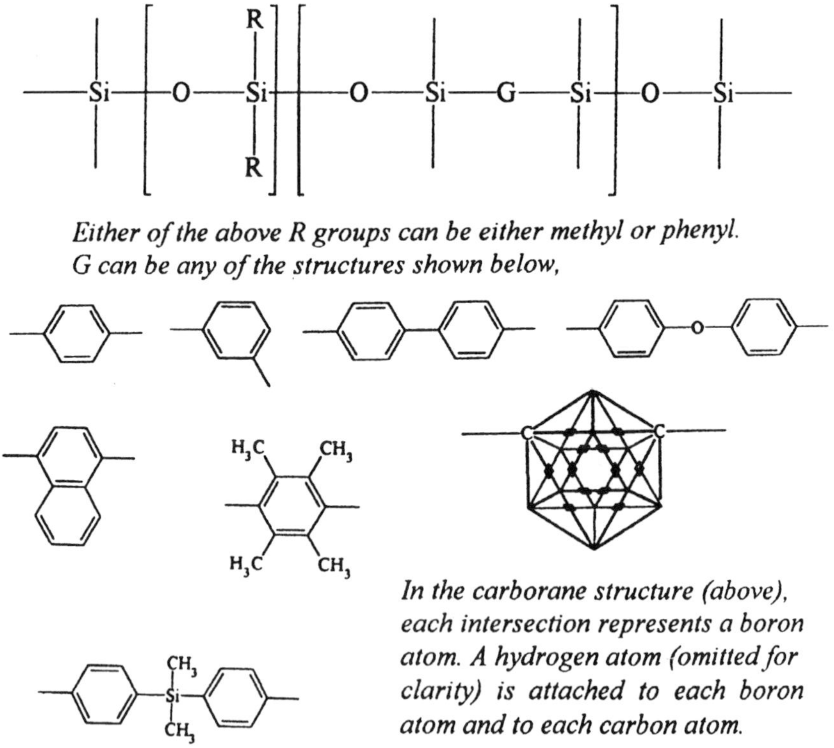

Either of the above R groups can be either methyl or phenyl.
G can be any of the structures shown below,

In the carborane structure (above), each intersection represents a boron atom. A hydrogen atom (omitted for clarity) is attached to each boron atom and to each carbon atom.

Fig. 4.2. Groups suggested as inclusions in the polysiloxane chain to effect "stiffening" [10].

Based largely on the nature of the degradation products formed, routes have been proposed for the thermal, oxidative, and silanol-triggered degradation routes for siloxane polymers. The commonly termed "backbiting" reaction involves a terminal silanol attacking the polysiloxane chain to generate fragments that recombine to a variety of cyclic siloxanes. Again, these are usually dominated by the trimer and the tetramer. As these more thermodynamically favored products are split off, a new terminal silanol is left on the chain. In that sense, the reaction is self-sustaining until the phase is exhausted. This reaction is facilitated by flexibility in the siloxane backbone, permitting closer proximity of two normally separated points in the chain. Hence, this particular degradative reaction might be inhibited by inserting into the polysiloxane chain groups that would make the chain more rigid and restrict its flexibility. Routes to higher-temperature and more resistant siloxane polymers have been reviewed by Dvornic and Lenz [10]. Figure 4.2, from that reference, shows some of what might be termed "chain-stiffening groups." These concepts have been utilized to produce a whole new series of improved stationary phases.

In some cases, the goal of the column manufacturer was an improved phase that could be substituted for an existing product. Figure 4.3 illustrates results

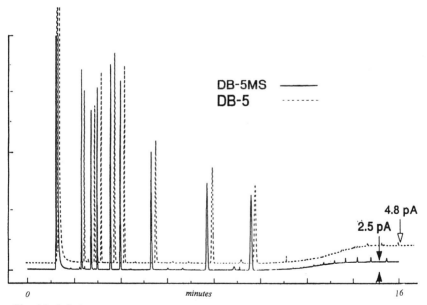

Fig. 4.3. Polarity test mixture on a 5%-phenyl-95%-dimethylpolysiloxane (DB-5, dotted line) versus a lower bleed silarylene-siloxane poly-(tetra-methyl-1,4-silphenylenesiloxane) (DB-5MS, solid line) that was carefully "tailored" to produce "equivalent" selectivity. Isothermal @ 150°C, ramped to 320°C. Note difference in bleed levels at the higher temperature. In order of elution, 1,6-hexanediol; 4-chlorophenol; methyl nonanoate; 4-propylaniline; tridecane; 1-undecanol; acenaphthylene; pentadecane.

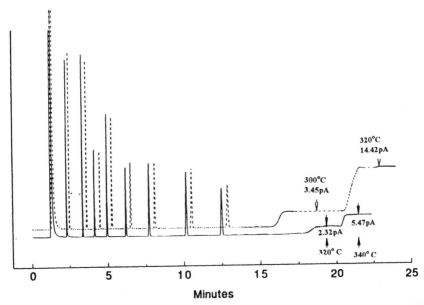

Fig. 4.4. Test mixture chromatograms generated on (dotted line) a standard 35%-phenyl-65%-dimethylpolysiloxane versus (solid line) a more thermally stable silarylene-siloxane stationary phase (DB-35MS) that was "tailored" to produce, as nearly as possible, a duplicate selectivity. Isothermal @ 125°C, and ramped to 320°C (standard 35% phenyl column) and 340°C (DB-35MS). Note differences in bleed at higher temperature. In order of elution, benzyl amine, tridecane, 4-chlorophenol, tetradecane, 1-undecanol, pentadecane, biphenyl, hexadecane. The retention indices for biphenyl were 1556.9 and 1556.4, respectively. It is also worth noting that bleed levels for the latter are significantly lower, even at higher temperatures.

generated on a 5% diphenyl–95% dimethylpolysiloxane, versus a lower-bleed silarylene–siloxane poly(tetramethyl-1,4-silphenylenesiloxane) that was carefully "tailored" to mimic the selectivity of the original phase.

Figure 4.4 shows chromatograms generated on (dotted line) a standard 35% phenyl–65% dimethylpolysiloxane versus (solid line) a more thermally stable silarylene–siloxane stationary phase that was "tailored" to produce, as nearly as possible, a duplicate selectivity. The retention indices for biphenyl were 1556.9 and 1556.4, respectively. It is also worth noting that bleed levels for the latter are significantly lower, even at higher temperatures.

In other cases, the goal is to design a new improved stationary phase and accept the fact that its selectivity will be unique; indeed, this can be advantageous. DB-XLB is an example of a silarylene–siloxane polymer where the silarylene groups were chosen with a view to "stiffening" the chain, as discussed earlier. The manufacturer lists upper temperature limits of 340°C isothermal, and 360°C on programming; these are fairly conservative limits. Short segments of a DB-XLB

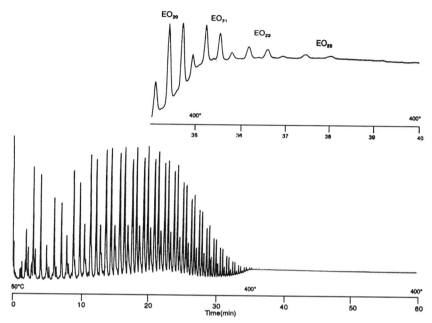

Fig. 4.5. A 3 m segment of a 0.25 mm DB-XLB column programmed to 400°C and held. The molecular weight of the last discernable solute eluting at 38 min in the expanded printout is 1238. From reference [11].

column have been used on programmed analyses to 400°C (Fig. 4.5 [11]), but this is hardly conducive to long life. The structure of the polymer inhibits some classic routes of thermal degradation, and in this case, produced a phase possessing unique selectivity. Some examples of how these unique selectivities have been exploited, including the resolution of critical PCB isomers, are explored in Chapter 9.

4.7 Bonded, Crosslinked, and/or Immobilized Stationary Phases

"Bonded" phases, as the term is usually used in open tubular columns, resulted from the commercial discovery that conventional glass capillary columns coated with SE-54 by a high-temperature, high-pressure process [12] exhibited greatly extended lifetimes. It was deduced that the vinyl moiety of the SE-54 stationary phase reacted under these very stringent conditions to produce a crosslinked stationary phase. That discovery was exploited by incorporating into other prepolymeric oligimers a small amount of vinyl. Both high temperatures and peroxide additions were used to initiate free radicals, and "bonded phase" columns became commercially available [13]. Grob *et al.* [14] reported their first findings on "immobilization" of the OV-61 stationary phase at almost the same time, indicating that they had been working independently along the same lines. That publication was soon followed by the "immobilization" of OV-1701 and a "crosslinked surface-bonded" OV-1701 [15]. From this point in time, developments occurred rapidly. While the practicing analyst continues to reap benefits from these activities, their consideration is beyond the scope of this book. The interested reader is referred to further publications of K. Grob, G. Grob, and K. Grob, Jr., and to those from groups headed by L. Blomberg, G. Schomburg, M. L. Lee, and S. Lipsky, as well as many others.

Initially, the term "bonded" elicited considerable controversy over its meaning and usage, resulting in some degree of semantic confusion. Some equated the wetting forces holding deposited films with covalently bonded films. Others debated whether these modified stationary phases were surface-bonded or crosslinked, and there were suggestions that the distinction was perhaps unimportant and that all such stationary phases should be termed "immobilized." There is, in fact, a difference; some phases are merely crosslinked, and others are both crosslinked and bonded to the surface by means of covalent bonds [9, 13]. Under normal circumstances the distinction is probably unimportant to the average gas chromatographer. It is of greater importance to those using splitless and on-column injection, and it can become painfully clear to those using such columns for open tubular liquid chromatography or for supercritical fluid chromatography. Columns that are coated with stationary phases that are both crosslinked and surface-bonded exhibit longer lifetimes in these applications [16].

Current procedures for column preparation usually involve pretreatments of the interior surface of the fused silica tubing. Some of the surface preparation steps are designed to render the fused silica surface silanols amenable to stationary phase bonding, are considered proprietary, and fall outside the scope of this book. Peroxides may be added to the coating solution, and static coating procedures [17, 18] are generally employed to deposit a film of vinyl-containing stationary phase oligimers on the interior wall of the tubing. Heating causes peroxide decomposition, which yields free radicals and initiates crosslinking (and, if the surface has been properly prepared, surface bonding) when the column is heated. The preferred peroxide is generally dicumyl peroxide, which decomposes to form volatile products that are dissipated during the heating step [9]. The use of ozone to initiate crosslinking in both nonpolar and medium-polar siloxane stationary phases has been suggested [18], and attention has also been directed to the use of azo compounds [19, 20]. A methyl-2-phenylethylpolysiloxane was reported to undergo crosslinking without addition of a free-radical initiator [21]. In this respect, it is similar to the SE-54, which really started all of this activity. Ionizing radiation has also been used as a free-radical initiator for the polysiloxanes [22–25], but such treatments can have detrimental effects on the flexibility and integrity of the protective outer polyimide coating; imperfections in that coating can lead to column breakage [26].

There have also been several recent publications on the synthesis of other polysiloxane stationary phases [27–30] and on methods of characterizing stationary phases by liquid chromatography [31], by supercritical fluid chromatography [32], and by gas chromatography of their hydrolytic products [33].

4.8 Polyethylene Glycol Stationary Phases

Probably because its higher average molecular weight permits its use at higher temperatures, Carbowax 20M has been the polyethylene glycol most widely used in gas chromatography. Structurally, it can be represented as

$$HO-CH_2-CH_2-[O-CH_2-CH_2]_n-OH$$

Where as the average molecular weight is 20,000, the range is not specified and wide batch-to-batch variations occur. One of its major disadvantages is oxygen lability, especially at higher temperatures. In common with the normal polysiloxane phases, even traces of oxygen are detrimental at higher temperatures, but with the polyethylene glycol phases, even with less oxygen and lower temperatures, the penalty is higher. Reducing the oxygen concentration in the carrier gas can help prolong the lifetime of columns coated with polyethylene glycol phases. Decomposition products of polyethylene glycols include acetaldehyde and acetic acid (see Section 10.4).

Other disadvantages for which Carbowax 20M is faulted include its solubility in water and low-molecular-weight alcohols and its relatively high low-temperature limit; it solidifies to a waxy solid at 50–60°C. The first of these sometimes poses limitations on the types of sample that can be used, especially with splitless and on-column modes of injection.

The high low-temperature limit can also be troublesome. In Chapter 2, the influence of solute diffusivity in the stationary phase was explored. With stationary phases of normal diffusivity ($D_S \geq 10^{-7}$ cm^2/sec [34, 35]), the limitations imposed by the D_S term are minimal on open tubular columns until d_f exceeds about 0.4 μm. At larger values of d_f or smaller values of D_S, chromatography suffers. If a stationary phase that is normally liquid is operated under conditions where it becomes a solid, D_S increases greatly. The lower operating temperature limit of a stationary phase is normally dictated by this provision (Fig. 4.6 [36]).

There has also been progress in bonding polyethylene glycol stationary phases. Methods that have been reported include coating leached Pyrex glass with Carbowax 20M and then recoating with a second solution containing the same stationary phase plus dicumyl peroxide [37]. The use of dicumyl peroxide to bond Carbowax 20M to an adsorbed layer of graphitized carbon has been described [38]. The columns reportedly tolerate "several" water injections, but the phase is extractable with dichloromethane. Some preliminary results have also been reported by Morandi *et al.* [39].

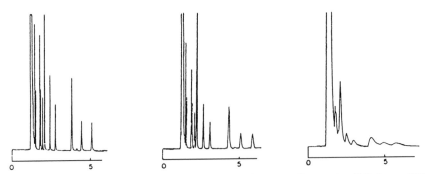

Fig. 4.6. Isothermal chromatograms of a polarity test mixture on a "conventional" Carbowax 20M fused silica capillary column at (from left to right): 55°C, 50°C, and 45°C. The anomalous results at lower temperatures reflect severe lowering of stationary phase diffusivity due to phase transition. *Adapted from Journal of Chromatographic Science, 22:129 (1984) with permission of Preston Publications. Inc.*

Although several commercial companies offer columns coated with bonded forms of polyethylene glycol, no details of the procedures used in their manufacture have been released. It is apparent that different methods are employed, because columns from different suppliers differ in their resistance to extractability with solvents such as water or methanol, and in their minimum and maximum operating temperature limits. With most stationary phases, users are most influenced by the latter, usually selecting that column with the highest upper operating temperature limit. With the polar stationary phases, however, the lower operating temperature limit is often more critical than the upper operating temperature limit. Those who find themselves limited by the upper temperature limit would often do better to reduce retentiveness by selecting a shorter- and/or higher-phase-ratio (thinner-film, larger-diameter) column. This would tend to accomplish elutions at lower temperatures where separation factors are usually higher. More retentive columns require higher elution temperatures, which usually have a negative impact on separation factors.

To analysts concerned with the separation of low-molecular-weight polar solutes (e.g., methanol, acetone, acetaldehyde), the minimum operating temperature of the phase becomes the crucial factor. With conventional Carbowax 20M phases where the temperature must be elevated to 50–60°C in order to maintain an acceptable diffusivity, highly volatile solutes flash through without separating. A polyethylene glycol phase whose temperature can be lowered below the boiling points of those volatile solutes without suffering a disastrous loss in diffusivity readily achieves separation of these lower-molecular-weight solutes (Fig. 4.7).

"Extended range" versions of polyethylene glycol phases are available from several manufacturers. In general, these are based on a trademarked crosslinkable

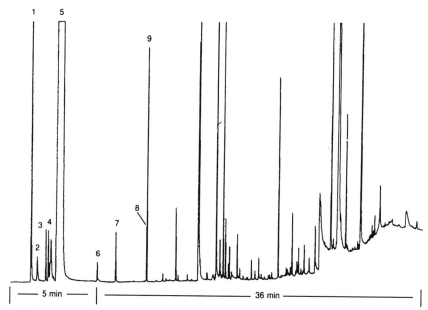

Fig. 4.7. Split injection of a white wine on "DB-WAX," a polyethylene glycol stationary phase capable of lower temperature operation. Peak identification; 1) acetaldehyde; 2) methyl acetate; 3) ethyl acetate; 4) methanol; 5) ethanol; 6) 1-propanol; 7) *iso*butanol; 8) 2-methyl-1-butanol; 9) 3-methyl-1-butanol. Column, 30 m × 0.25 mm, d_f 0.25 μm; 35°C (5 min) to 230°C @ 6°C/min. *Reproduced by permission of J&W Scientific, Inc.*

polyethylene glycol-type material [40]. Columns containing this material have been reported to exhibit minimum operating temperatures of ~35°C, and upper operating temperatures in the range of 240–280°C. Unfortunately, shifts of retention indices are not uncommon, and may be exacerbated by exposure to higher temperatures (Table 4.3 [41]).

TABLE 4.3

Retention Indices Stability Tests on 30 m × 0.25 mm Columns, 0.25-μm Bonded Innophase

Compound	1st 270° cure[a]	1st rinse[b]	2nd 270° cure[a]	2nd rinse[b]	3rd 270° cure[a]
Aniline	1776.47	1774.60	1766.99	1764.87	1760.78
2-chlorophenol	1866.83	1865.10	1856.77	1854.60	1850.19
1-undecanol	1874.14	1870.72	1866.65	1862.88	1861.41

[a]Cured at 270°C under normal hydrogen flow for 12 hrs.
[b]Rinsed with 10 mL CH$_2$Cl$_2$.
Data generated by Roy Lautamo [41].

The repeatability and reliability of retention indices are especially important to those in the flavor and fragrance industries who maintain vast libraries of these data as determined on polydimethylsiloxane (e.g., SE-30, OV-101, DB-1) and on a polyethylene glycol phase.

4.9   Enantiomer Separations

In many cases, the biological effects of two members of an enantiomorphic pair are entirely different. Enantiomers often exhibit differences in degree of interaction with a biological receptor, in route and/or mechanism of biological transport, and in the manner in which they are metabolized. Enantiomer differentiation can be critically important where the materials in question interact with biological systems, including drugs and pharmaceuticals, insect pheromones, and aroma and flavor compounds. In addition, enantiomer separations have applications to forensic analysis, to the dating of some fossil materials, and (in some cases) in the detection of synthetics added to or used in the formulation of a "natural" product. Table 4.4 [42] lists several compounds whose enantiomers elicit different physiological responses.

A solute chromatographs at a velocity that can be represented as $v = \bar{u}/(k+1)$ (Chapter 3). Since $K_c = \beta k$, k is a function of K_c, which is affected not only by the normal physical equilibria (e.g., dispersion forces) but also by any interplay between the solute and the stationary phase (or, to our occasional distress, the solid support). It is really this chemical interplay that determines the "selectivity" of a given column for a given solute.

The differentiation of diastereomers and enantiomers is important to the following considerations. Diastereomers are not enantiomers, and their chromatographic separation usually poses no special problem. Because they differ structurally, their physical and/or chemical properties must differ during the chromatographic process [42]. Enantiomers, however, are identical in all properties except chirality and exhibit identical physical properties in any isotropic medium. As emphasized by Lochmueller [44], the solute retention factor is a function of the solute vapor pressure and the activity coefficient, and because the separation factor of two solutes is the ratio of their retention factors:

$$\alpha = k_{(2)}/k_{(1)} = y_{(1)} p_{(1)}^{\circ} / y_{(2)} p_{(2)}^{\circ} \tag{4.2}$$

where y represents the activity coefficient and p° the vapor pressure of the solute in question. In the case of diastereomers, vapor pressures differ and separation is possible on achiral stationary phases. The vapor pressures of enantiomers are identical, and to achieve their separation, they must be chromatographed under conditions where their activity coefficients differ. This can occur only in a chiral stationary phase. For this reason, the stationary phases considered to this point

TABLE 4.4

Enantiomer Specific Characteristics of Selected Chiral Compounds

Compound	Characteristics
R-(−)-Sodium Glutamate	no synergistic action[a]
S-(+)-	flavor enhancing[a]
D-(−)-Ascorbic Acid	no activity[b]
L-(+)-	antiscorbutic[b]
R-(+)-Penicillamine	extreme toxicity[c]
S-(−)-	antiarthritic[c]
R-(+)-Thalidomide	soporific; no secondary effects[c]
S-(−)-	teratogenic to one strain of mice and Natal rats[c]
R-(−)-Asparagine	sweet taste[c]
S-(+)-	bitter taste[c]
P-(+)-Gossypol	no activity[d]
M-(−)-	antispermatogenic[d]
(+)-Androst-4-,16-diene-3-one	sweaty, urine-like (sexual hormone of boars)[e]
(−)-	odorless[e]
R-(−)-Carvone	spearmint aroma[f]
S-(+)-	caraway seed aroma[f]
R-(+)-Limonene	orange-like aroma[g]
S-(−)-	terpene-like aroma[g]
$5R,6S,8R$-(+)-Nootketone	grapefruit aroma[h]
$5R,6S,8R$-(−)-	woody, spicy aroma[h]
R-(−)-4-Methyl-3-heptanone	blocks the effect of S (below), *Popillia japonica* alarm pheromone[i]
S-(+)-	alarm pheromone, *Popillia japonica*[i]
$1R,3R,4S$-(−)-Menthol	sweet, fresh, minty: cooling effect[k]
$1S,3S,4R$-(+)-	herbaceous; less mint-like, less cooling[k]

[a] J. Solms, *Chimia* **21**:169 (1967); [b] T. C. Daniels, in Wilson and Griswold: *Textbook of Organic Medicinal and Pharmaceutical Chemistry*, R. F. Doerge, Ed.; 8th ed., Lipincott: Philadelphia (1982); [c] D. Enders and W. Hoffmann, *Chem. Uns. Zeit* **18**:140 (1985); [d] V.R. Meyer, *Pharm. Uns. Zeit* **18**:140 (1989); [e] V. Prelog, L. Ruzicka, P. Meister, and P. Wieland, *Helv. Chim. Acta* **28**:618 (1945); [f] T. J. Leitereg, D. G. Guadagni, J. Harris, T. R. Mon, and R. Teranishi, *Nature* **230**:455 (1971); [g] L. Friedman and J. G. Miller, *Science* **172**:1044 (1971); [h] H. G. Haring, F. Rijkens, H. Boelens, and A. van der Gen, *J. Agric. Food Chem.* **20**:1018 (1972); [i] R. G. Riley and R. M. Silverstein, *Tetrahedron* **30**:1171 (1974); [j] J. H. Tumlinson, in *Chemical Ecology: Odour Communication in Animals*, F. J. Ritter, Ed., Elsevier, Amsterdam (1979) p. 1301; [k] R. Emberger and R. Hopp, in *Topics in Flavour Research*, R. N. Berger, S. Nitz and P. Schreier, Eds., Eichner, Marzling (1985). Table adapted from *Analysis of Chiral Organic Molecules: Methodology and Applications*, P. Schreier, A. Bernreuther, and M. Huffer, de Gruyter, Berlin, New York, 1995.

are generally unable to distinguish between the enantiomers of optically active compounds. There are two ways in which such compounds can be separated: (1) if one enantiomer of an optically active derivatizing reagent can be obtained in a high degree of purity, the resultant diastereomers (which are no longer enantiomers) can be separated on a standard stationary phase [43, 44]; (2) alternatively, the enantiomers or the enantiomeric derivatives can sometimes be separated directly on a chiral stationary phase [45, 46].

The first route has been faulted for several disadvantages: (1) the derivatizing reagent must be optically active, which greatly limits the choices, (2) the reagent must be optically pure, (3) systematic errors arise due to differences in the reaction kinetics of the two enantiomers, and (4) racemization may occur at one of the asymmetric carbons during derivatization.

A chiral phase, on the other hand, can be used directly for chiral separations of suitably volatile materials. Less volatile materials can be treated with any common volatility-enhancing derivatizing reagent, followed by direct separation of the reactants. Where the hydrogen-bonded associations between the chiral phase and the two enantiomers are sufficiently different, separation is achieved [46]. A number of chiral phases have been synthesized and investigated, and most of those efforts have been well reviewed (e.g., [47, 48]). Our earlier chiral phases included dipeptide and derivatized amino acid phases. One of the more widely used phases was Chirasil-Val, whose useful temperature range has been listed as 70–240°C [48]. The maximum operating temperatures of such phases were limited both by the vapor pressures of the materials themselves and an unfortunate tendency to racemize and lose chiral specificity at higher temperatures [49]. Work on bonded polymeric chiral phases opened other possibilities. An excellent discussion of these and more recent efforts appears in Ref. 42.

Cyclodextrin-based phases have led to a major renaissance in chiral separations. These chiral toroidal-shaped molecules are composed of six or more D(+)-glucose units bonded through α-(1,4)-glycosidic linkages, with the glucose units in chair conformation. Cyclodextrins have been coated directly on the column in their pure state [49, 50], and as a solution of the cyclodextrin in a polysiloxane phase of suitable polarity [50, 51]. Immobilization has also been accomplished by bonding the cyclodextrins to polysiloxanes [52, 53]. The selectivity of the stationary phase and the concentration of cyclodextrin both influence the degree of enantiomer separation realized.

4.10 Other Special-Selectivity Stationary Phases

Stationary phases capable of certain specific types of interaction with specific types of solute are occasionally used to perform specialized separations. The separation of enantiomers by chiral stationary phases described above is one such example that has significance to most biologically oriented analysts. Some other selective stationary phases have been reported, but their applicability is less general, and our treatment of these will in most cases be limited to a few key references.

Unsaturated and aromatic hydrocarbons have been separated on stationary phases that exercised their selectivity partially by formation of a π-charge complex. Typical of these electron donor stationary phases are dialkyltetrachlorophthalates and 2,4,7-trinitro-9-fluorenone [54–59].

Liquid crystal stationary phases have been used for the separation of some iso-meric solutes having structural rigidity, such as certain polychlorinated biphenyls, substituted benzenes, polynuclear aromatic hydrocarbons, and steroids [58–60]. Within a specific temperature range, the properties of a liquid crystal phase are intermediate between those of a liquid and those of a crystalline solid. Mechani-cally, it more nearly resembles a liquid, but a liquid with a high degree of order. As a result, it still has some of the anisotropic properties of the solid. A thermotropic liquid crystal is one in which the liquid crystal state is generated above the melt-ing point of the solid and exists through a discrete range up to some "clearing temperature," where it becomes an isotropic liquid.

The thermotropic liquid crystals are classified into nematic, cholesteric, and smectic types, depending on the liquid crystalline structure. Smectic phases have highly ordered layers of parallel rodlike molecules. The interlayer dimension is determined by the length of those molecules. Nematic phases maintain the parallel orientation, but do not exhibit the layered structure. In this case, the movement of liquid molecules restricted only by the parallel configuration. The cholesteric phase requires that the liquid crystal have a chiral center, resulting in a twisted nematic structure. Both the smectic and cholesteric types generally exhibit the liquid crystal structure only over a very narrow range of temperatures. The nematic type usually has a broader range of operating temperature and is dominant among the liquid crystals that have been used as stationary phases [58–60].

The retention characteristics of liquid crystal columns have been reported to vary unpredictably, and they are often characterized by lower theoretical plate numbers that can affect separation efficiencies [61–63]. Laub $et\ al.$ [64] reported that more efficient columns containing the liquid crystal N, N'-bis(p-butoxybenzylidene)-a,a'-bi-p-toluidine (BBBT) could be prepared by dissolving the liquid crystal in the 5% phenyl polysiloxane SE-52. Direct comparisons of their results are precluded by the massive shift of the retention factor of the test solute, triphenylene (6.92 in SE-52; 29.4 in BBBT), but the column efficiencies seemed low. Crude comparisons indicate that the theoretical plate numbers ranged from approximately 70% of that theoretically possible for the pure SE-52, to 16% for the pure BBBT, to 43% for 80% BBBT in SE-52. Markides $et\ al.$ [65] used aliphatic hydrocarbons as flexible spacers to couple mesomorphic moieties to a polysiloxane backbone. The result, a smectic biphenylcarboxylate ester–polysiloxane polymer, was capable of undergoing crosslinking and reportedly produced highly efficient columns whose smectic properties endured from 118 to 300°C [66].

4.11 Gas–Solid Adsorption Columns

Porous layer open tubular (PLOT) and support-coated open tubular (SCOT) columns contain a layer of adsorptive material affixed to the column wall. With adsorptive-type supports, the separation mechanism is gas–solid adsorption chro-matography rather than the gas–liquid partition chromatography with which we have dealt so far. Coated with materials such as Al_2O_3, molecular sieves, and synthetic polymers, these work very well for the analysis of light hydrocarbons at above-ambient temperature. Examples of such an analysis, as well as some unique separations requiring PLOT columns coated with silica gel or certain of the porous polymers, are discussed in later chapters.

References

1. L. Blomberg, *J. High Res. Chromatogr.* **5**:520 (1982).
2. W. Burns and S. J. Hawkes, *J. Chromatogr. Sci.* **15**:185 (1977).
3. T. J. Stark, P. A. Larson, and R. D. Dandeneau, *Proc. 5th Int. Symp. Capillary Chromatogr.*, p. 65 (1983).
4. R. A. Keller, B. L. Karger, and L. R. Snyder, in *Gas Chromatography 1970* (R. Stock and S. G. Perry, eds.), p. 125. Institute of Petroleum, London, 1971.
5. J. K. Haken, *J. Chromatogr.* **300**:1 (1984).
6. H. W. Fox, P. W. Taylor, and W. A. Zisman, *Ind. Eng. Chem.* **39**:1401 (1947).
7. F. S. Kipping and L. L. Lloyd, *J. Chem. Soc.* **79**:449 (1901).
8. W. Noll, *Chemistry and Technology of Silicones*, Academic Press, New York, 1968.
9. M. Sveda, U.S. Patent 2,561,429 (1951) and U.S. Patent 2,562,000 (1951).
10. P. R. Dvornic and R.W. Lenz, *High Temperature Siloxane Elastomers*, Huethig & Wepf, Basel, Heidelberg, New York, 1990.
11. W. Jennings and P. Stremple, paper, *Pittsburgh Conf. Anal. Chem. Appl. Spectrosc.*, Chicago, 1996.
12. W. Jennings, K. Yabumoto, and R. H. Wohleb, *J. Chromatogr. Sci.* **12**:344 (1974).
13. R. Jenkins (J & W Scientific, Inc.), personal communication (1979).
14. K. Grob, G. Grob, and K. Grob, Ir., *J. Chromatogr.* **213**:211 (1981).
15. K. Grob and G. Grob, *J. High Res. Chromatogr.* **5**:13 (1982).
16. R. Houcks (Suprex, Inc.), personal communication (1985).
17. K. Grob and G. Grob, *J. High Res. Chromatogr.* **6**:153 (1983).
18. J. Bouche and M. Verzele, *J. Gas Chromatogr.* **6**:501 (1968).
19. M. Giabbai, M. Schoults, and W. Bertsch, *J. High Res. Chromatogr.* **1**:277 (1978).
20. K. Grob, *J. High Res. Chromatogr.* **1**:93 (1978).
21. J. Buijten, L. Blomberg, S. Hoffmann, K. Markides, and T. Wannman, *J. Chromatogr.* **283**:341 (1984); ibid. **289**:143 (1984).
22. M. L. Lee, P. A. Peaden, and B. W. Wright, *Pittsburgh Conf. Anal. Chem. Appl. Spectrosc., 1982*, Abstract 298 (1982).
23. S. R. Springston, K. Melda, and M. V. Novotny, *J. Chromatogr.* **267**:395 (1983).
24. J. S. Bradshaw, S. J. Crowley, C. W. Harper, and M. L. Lee, *J. High Res. Chromatogr.* **7**:89 (1984).
25. G. Schomburg, H. Husmann, S. Ruthe, and M. Herraiz, *Chromatographia* **15**:599 (1982).
26. W. Bertsch, V. Pretorius, M. Pearce, J. C. Thompson, and N. G. Schnautz, *J. High Res. Chromatogr.* **5**:432 (1982).
27. Gy. Vigh and O. Etler, *J. High Res. Chromatogr.* **7**:620 (1984).
28. O. Etler and Gy. Vigh, *J. High Res. Chromatogr.* **7**:700 (1984).
29. J. A. Huball, P. DiMauro, E. F. Barry, and E. Chabot, *J. High Res. Chromatogr.* **6**:241 (1983).
30. J. C. Kuei, J. 1. Shelton, L. W. Castle, R. C. Kong, B. E. Richtet, J. S. Bradshaw, and M. L. Lee, *J. High Res. Chromatogr. Chromatogr.* **7**:13 (1984).
31. M. Ahnoff and L. Johansson, *Chromatographia* **19**:151 (1984).

32. J. S. Bradshaw, N. W. Adams, B. J. Tarbet, C. M. Schregenberger, B. S. Johnson, M. B. Andrus, K. E. Markides, and M. L. Lee, *Proc. 6th Int. Symp. Capillary Chromatogr.*, p. 132 (1985).
33. M. W. Ogden and H. M. McNair, *Proc. 6th Int. Symp. Capillary Chromatogr.*, p. 149 (1985).
34. L. Butler and S. Hawkes, *J. Chromatogr. Sci.* **10**:518 (1972).
35. J. M. Kong and S. J. Hawkes, *J. Chromatogr. Sci.* **14**:279 (1976).
36. G. Takeoka and W. Jennings, *J. Chromatogr. Sci.* **22**:177 (1984).
37. V. Martinez de la Gandara, J. Sanz, and I. Martinez-Castro, *J. High Res. Chromatogr.* **7**:44 (1984).
38. M. V. Russo, G. C. Goretti, and A. Liberti, *J. High Res. Chromatogr.* **8**:535 (1985).
39. F. Morandi, D. Andreazza, and L. Motta, *Proc. 6th Int. Symp. Capillary Chromatogr.*, p. 103 (1985).
40. Innophase Corporation, 95 Brownstone Ave., Portland, CT 06480.
41. R. Lautamo, personal communication (1996).
42. P. Schreier, A. Bernreuther, and M. Huffer, *Analysis of Chiral Organic Molecules: Methodology and Applications*, de Gruyter, Berlin, New York, 1995.
43. C. J. W. Brooks, M. T. Gilbert, and J. D. Gilbert, *Anal. Chem.* **45**:896 (1973).
44. C. H. Lochmueller, *Separ. Purif. Meth.* **8**:21 (1979).
45. W. A. Koenig, W. Rahn, and J. Eyem, *J. Chromatogr.* **133**:141 (1977).
46. E. Gil-Av, B. Feibush, and R. Charles-Sigler, in *Gas Chromatography* (A. B. Littlewood, ed.), p. 227. Institute of Petroleum, London, 1966.
47. W. Paar, J. Pleterski, C. Yang, and E. Bayer, *J. Chromatogr. Sci.* **9**:141 (1971).
48. H. Frank, G. J. Nicholson, and E. Bayer, *J. Chromatogr.* **146**:197 (1978).
49. W. A. Koenig, S. Lutz, P. Mischnick-Luebbecke, B. Brassat, and G. Wenz, *J. Chromatogr.* **441**:471 (1988).
50. D. W. Armstrong, W. Y. Li, and J. Pitha, *Anal. Chem.* **62**:214 (1990).
51. V. Schurig and H. P. Nowotny, *J. Chromatogr.* **441**:155 (1988).
52. P. Fischer, R. Aichholz, U. Boelz, M. Juza, and S. Krimmer, *Angew. Chem.* **102**:439 (1990).
53. V. Schurig, D. Schmalzing, U. Muehleck, M. Jung, M. Schleimer, P. Mussche, C. Duvekot, and J. C. Buyten, *J. High Res. Chromatogr.* **13**:730 (1990).
54. D. L. Meen, F. Morris, and J. H. Purnell, *J. Chromatogr. Sci.* **9**:281 (1971).
55. M. Ryba, *Chromatographia* **5**:23 (1984).
56. S. H. Langer, *Anal. Chem.* **44**:1915 (1972).
57. J. H. Purnell and O. P. Srivastava, *Anal. Chem.* **45**:1111 (1973).
58. J. H. Laub and P. L. Pecsok, *J. Chromatogr.* **113**:47 (1975).
59. C. L. de Ligny, *Adv. Chromatogr.* **14**:265 (1976).
60. H. Kelker, *Adv. Liq. Cryst.* **3**:237 (1978).
61. G. M. Janini, *Adv. Chromatogr.* **17**:231 (1979).
62. Z. Witkiewicz, *J. Chromatogr.* **251**:311 (1982).
63. J. Szulk and J. Witkiewicz, *J. Chromatogr.* **262**:141 (1983).
64. R. J. Laub, W. L. Roberts, and C. A. Smith, *J. High Res. Chromatogr.* **3**:335 (1980).
65. K. E. Markides, H. C. Chang, C. M. Schregenberger, B. J. Tarbet, J. S. Bradshaw, and M. L. Lee, *J. High Res. Chromatogr.* **8**:516 (1985).
66. K. E. Markides, M. Nishioka, B. J. Tarbet, J. S. Bradshaw, and M. L. Lee, *Anal. Chem.* **57**:1296 (1985).

CHAPTER 5

VARIABLES IN THE GAS CHROMATOGRAPHIC PROCESS

5.1 General Considerations

There are a great many "analytical variables" responsible for the chromatographic result obtained when solutes making up a particular sample are separated using capillary column gas chromatography. Some are under the control of the operator up to the time of sample injection and might be termed "operational parameters." These include the temperature of a isothermal analysis or the parameters of a temperature-programmed analysis, and the average linear velocity of the carrier gas. Other operational variables depend on the injection mode (i.e., the split ratio with split injection and the purge activation time with splitless injection). The choice of sample solvent can also influence the results obtained, especially with splitless and on-column injections. One aspect of this concerns the relative "polarities" of the solute, solvent, and stationary phase. In all cases, syringe and injection techniques can exercise a variety of effects.

There are other variables that, immediately preceding sample injection, are no longer under operator control. These normally include the choice of carrier gas, the choice of injection and detection modes, certain design features of the injector and the detector (including whether a "retention gap" was incorporated, the manner in which it was attached, and the method by which it was deactivated), and the column variables of length, diameter, stationary phase film thickness, and stationary phase type. This latter group might better be considered "design parameters."

Many of these variables have been discussed under headings found in other chapters. The utility of a particular injection mode for any given sample is covered primarily in Chapter 3, with some additional comments relative to specialized problems in Chapter 8. The influence of instrumental design features (e.g., excessive or unswept volumes in the flow path, suitability of the detection mode, speed and fidelity of the data-handling equipment) are considered in Chapter 7, and factors influencing the selection of the stationary phase are considered in Chapters 4 and 6.

The present chapter is concerned primarily with the proper choice of carrier gas and the effects that variables such as carrier gas velocity, column length, column diameter, and stationary phase film thickness might be expected to have on chromatographic results. In later chapters, this information will be used to rationalize decisions concerning the choice of stationary phase, column dimensions, instrument conversion and adaptation, special analytical considerations, and even specific applications. The latter portion of the present chapter will also consider how changes in these variables can affect solute elution order.

One of the decisions most frequently faced by the practicing chromatographer is the selection of carrier gas velocity. This is the first decision to be made after column installation, and it recurs whenever any of a number of other parameters are changed. This seemingly simple decision can have a profound effect on separation efficiency, analysis time, sensitivity, and even the extent to which thermally labile solutes survive the chromatographic process.

5.2 Volumetric Column Flow

The flow of carrier gas through an open tubular column is the direct consequence of a difference in pressure at the column inlet and column outlet. This pressure difference constitutes what is generally referred to as the "pressure drop across the column" and is typically established by applying carrier gas pressure to the head of the column (i.e., column inlet). The volume of carrier gas flowing through the column can be accurately estimated using the Hagen–Poiseulle equation for the viscous flow of ideal fluids in a closed channel (Eq. 5.1) where p_i and p_o are the inlet and outlet pressures, respectively in Pascals,

$$F_0 = \left[\left(p_i^2 - p_o^2\right)pr_c^4\right]/\left[16\eta(T)p_oL\right] \qquad (5.1)$$

where r is column radius in mm, $\eta(T)$ is dynamic viscosity in centipoise as a function of temperature (see Section 5.3), L is column length in cm, and F_0 is flow in/cubic centimeters per second at the column outlet. Commonly referred to as "volumetric flow," this is what many analysts believe they are measuring when they turn off their detector gases and attach a flow measuring device to the detector. This is not always true. To accurately determine column flow from physical measurements at the column outlet, it is important to account for differences between the conditions under which such flow occurs (e.g., column temperature) and the conditions under which the flow is being measured (normally at ambient temperature). Although a difference in pressure may also exist, errors attributable to this are less common since p_o is usually equivalent to atmospheric pressure (the most notable exception is mass-sensitive detectors operated under vacuum).

$$F_0 = \left\{ \left[(p_i^2 - p_o^2) \, pr_e^4 \right] / (16\eta p_o L) \right\} (T_a / T_c) \tag{5.2}$$

Equation (5.2) can be used to calculate volumetric column flows that more closely resemble measurements made at room temperature and atmospheric pressure. Where T_a and T_c are the absolute ambient temperature and absolute column temperature, respectively in K, Eq. (5.2) will yield values of reasonable accuracy in the vast majority of applications. To calculate volumetric flows that compare still more closely to those measured experimentally, it becomes necessary to employ one final correction to account for the nonideality of real carrier gases. This final correction factor assumes the form of an exponent on the (T_a / T_c) ratio presented above in Eq. (5.2). That the value of this exponent is essentially unity (1.005) only emphasizes the fact that Eq. (5.2) is already sufficiently accurate for use in most applications.

5.3   Carrier Gas Viscosity

Although the dynamic viscosity of carrier gas viscosity is essentially independent of pressure, it does vary with temperature. As temperature *increases*, so does carrier gas viscosity. As a result, what is typically included as η in Eqs. (5.1) and (5.2) is properly included here as $\eta(T)$. Some regard this behavior as inexplicable in that it is the opposite of what is typically encountered with liquids (most liquids appear to become less viscous as temperature increases). Nevertheless, it is logical that carrier gas viscosity does increase with temperature.

Consider an ensemble of gas molecules at an absolute temperature K, moving randomly about the volume to which they are confined, with an average kinetic energy equal to one half the product of their molecular mass and squared velocity. Since $0.5mv^2 = K$, these molecules possess an average molecular momentum equal to mv and move randomly about at an average speed equal to c (i.e., velocity without regard to direction). As they occasionally collide with each other, they exchange some measure of their individual momentum (i.e., molecules possessing greater than average momentum will impart some measure of that momentum to molecules possessing less than average momentum). Through such collisions, the momentum of individual molecules is redistributed or transported throughout the bulk material with an efficiency that is essentially a measure of the dynamic viscosity, η.

As the temperature of such a system increases, so does the average speed of the molecules making up that system. As this occurs, the frequency with which such molecules collide increases and so the efficiency with which momentum is transported throughout the system increases. This results in an increase in dynamic viscosity. By analogous reasoning, a decrease in temperature results in a decrease in dynamic viscosity.

Although similar collisions occur in liquids, intermolecular forces acting over the much smaller mean distances between molecules constituting liquids serve to reduce the elasticity of collisions responsible for the redistribution or transport of molecular momentum. Until the temperature of such systems becomes high enough to overcome the effect of such intermolecular forces, the dynamic viscosity of liquids typically decreases rather than increases as the temperature is raised.

Although one might expect that increasing the pressure of a gas at constant temperature and volume would cause it to behave more like a liquid, this is rarely observed very far from the critical conditions at which such gases undergo a bonafide gas–liquid phase transition. Even though the mean distance between molecules composing gaseous systems may be reduced through the application of pressure, any resulting increase in the frequency of intermolecular collisions in the absence of strong intermolecular forces serving to reduce the elasticity of such collisions merely means that the average distance over which a molecule travels before engaging in a momentum transferring collision is reduced. As a result, the redistribution of molecular momentum is actually rendered less efficient, because molecules tend to collide more frequently with other molecules having essentially the same momentum. In other words, a greater number of collisions are wasted and the anticipated change in dynamic viscosity with pressure remains largely negligible.

In gas chromatography, the influence of dynamic viscosity can be observed experimentally as an increase in the gas holdup time (time required for an unretained solute to traverse the column) if the column inlet pressure is held constant while the column temperature is increased. As the temperature increases, so does the dynamic viscosity of the mobile phase (i.e., the carrier gas becomes *less* fluid) and a greater head pressure is required to maintain the flow that was observed at lower temperatures (Fig. 5.1).

Sufficiently accurate carrier gas viscosities can be calculated using one of two methods. The first (and perhaps the most straightforward) has been presented by Ettre [1]. Slightly more accurate results can be obtained with the more rigorous method of Hawkes [2]. In the first method, a general relationship describing how carrier gas viscosity varies with temperature is established. This relationship is presented as

$$\eta(T) = A(T/273.15)^B \tag{5.3}$$

where T is absolute temperature in kelvins, and A and B are constants (Table 5.1), and $\eta(T)$ is dynamic viscosity in micropascals. In the second method, slightly different equations are used to describe the behavior of each different carrier

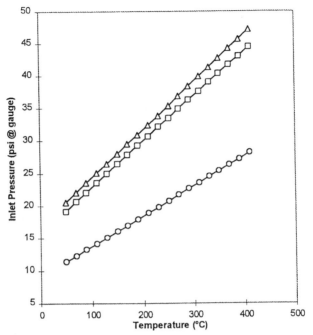

Fig. 5.1. Inlet pressure required to maintain a 1.0 mL/min volumetric flow of carrier gas though a 60 m × 0.25 mm × 1.0 μm column; helim (Δ); nitrogen (□); and hydrogen (○).

gas. Where T is absolute temperature in kelvins, the relationship for hydrogen is presented as [3].

$$\eta(T) = 0.1679(T)^{0.6967} \tag{5.4}$$

For helium, the relationship is somewhat more complicated, but is presented below as [4].

$$\eta(T) = 0.7840374(T)^{0.5} f / \Omega \tag{5.5}$$

where f and Ω have the following definitions [(Eqs. (5.6a–e) and (5.7a–d)], T is

TABLE 5.1

Carrier Gas Viscosity Constants for Use in Eq. 5.3

Carrier gas	A	B
Hydrogen	8.35	0.68
Helium	18.6	0.646
Nitrogen	16.59	0.725

absolute temperature in kelvins, $T^* = T/10.40$, $a = -0.126516$, $b = -1.230553$, $c = 2.171442$, and $\alpha = 13.65299 - \ln(T^*)$.

$$f = 1 + (3/196)(8E - 7)^2 \tag{5.6a}$$

$$E = 1 + [1/(4\Omega)] - [(2\Omega)/\alpha]$$
$$+ 0.00635209\alpha^2(-C_1 - C_2 - C_3) \tag{5.6b}$$

$$C_1 = [2a/\ln(T^*)^3] \tag{5.6c}$$

$$C_2 = [3b/\ln(T^*)^4] \tag{5.6d}$$

$$C_3 = [4c/\ln(T^*)^5] \tag{5.6e}$$

$$\Omega = 0.00635209\alpha^2 1.04 + C_4 + C_5 + C_6 \tag{5.7a}$$

$$C_4 = [a/\ln(T^*)^2] \tag{5.7b}$$

$$C_5 = [b/\ln(T^*)^3] \tag{5.7c}$$

$$C_6 = [c/\ln(T^*)^4] \tag{5.7d}$$

When this method is used to calculate the dynamic viscosity for helium between 25°C and 425°C, results within 0.5% of experimental values can be obtained [2]. If those results are corrected using Eq. (5.8), the corrected values are within 0.1% of experimental results [2]:

$$\eta(T) = \text{calculated}[0.995 + (T - 300)2.5 \times 10^{-5}] \tag{5.8}$$

For nitrogen, a similar approach must be taken [5]:

$$\eta(T) = 1.07113T^{0.5}/\Omega \tag{5.9}$$

Where Ω is defined by Eq. (5.10), and $T^* = 104.2$, this method can be used without correction between 25°C and 350°C.

$$\Omega(T) = \exp\{C_7 - C_8[\ln(T^*)] + C_9[\ln(T^*)^2]$$
$$+ C_{10}[\ln(T^*)^3] - C_{11}[\ln(T^*)^4]\}$$

$$C_7 = 0.431433$$

$$C_8 = 0.472647$$

$$C_9 = 0.0883435 \tag{5.10}$$

$$C_{10} = 0.0105232$$

$$C_{11} = 0.00516796$$

Both methods illustrate that different carrier gases not only exhibit significantly different intrinsic viscosities at any particular temperature, but they also respond differently to changes in temperature that cause their viscosities to vary. In some circumstances, this can be a matter of considerable practical significance. In applications requiring higher gas flows through long narrow columns (especially at high temperatures), the use of helium or nitrogen may prove impractical. Conversely, in applications requiring the use of fairly short and/or large-bore columns, the use of hydrogen may be precluded due to an inability to effectively control the application of such low head pressures (i.e., <1.0 psi).

5.4   Comparing Calculated to Experimental Volumetric Flows

When comparing calculated volumetric flows to those measured experimentally, there are several things to keep in mind. Volumetric flows bear a fourth-order dependence on column radius. As a result, large discrepancies between calculated and experimentally measured flows will be observed whenever the inside diameter is not known with some accuracy, or is not consistent throughout the length of the column. It is also important to recognize that variations in experimentally measured volumetric flows may be implicit to the method or methods by which such flows are measured. For example, measurements made with conventional bubble flowmeters must be corrected for both temperature and water vapor pressure prior to a direct comparison to flows calculated using equations presented in Section 5.2. Solid-state flowmeters do not require such correction, but should be calibrated against known standards prior to use.

5.5   Volumetric Column Flow and Average Linear Velocity

The distinction between volumetric column flow and the average linear velocity of carrier gas is important on several levels. Although the volumetric flow of carrier gas at the outlet of a column will remain constant if the column temperature, inlet pressure, and outlet pressure are also constant, the velocity of carrier gas will vary at various points along the length of the column. Because carrier gases are "compressible," the volume of the gas responds to changes in pressure that occur over the pressure drop established to cause flow through the column (see Section 5.1). If it were possible to divide the typical capillary column into infinitesimally small segments without influencing the behavior of the carrier gas flowing through each segment, it would be possible to observe that the pressure of gas in each segment conforms to the following relationship:

$$p(x) = p_0 \sqrt{\left(\frac{p_i}{p_o}\right)^2 - \left(\frac{x}{L}\right)\left[\left(\frac{p_i}{p_o}\right)^2 - 1\right]} \tag{5.11}$$

where p_i and p_o are column inlet and outlet pressure, respectively, L is column length, and x is distance from the inlet, $p(x)$ is pressure at distance x from the column inlet. The ratio (p_i/p_o) is sometimes referred to as the carrier gas

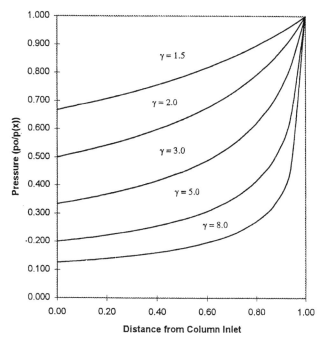

Fig. 5.2. Carrier gas pressure (reduced) as a function of the distance from the column inlet at various carrier gas compression ratios (γ).

compression ratio. Plots of Eq. (5.11) for several different compression ratios (Fig. 5.2) reveal that as the magnitude of the pressure drop increases, so does the compression ratio and so does the curvature of Eq. (5.11).

The plots presented in Fig. 5.2 have been generalized to emphasize the fact that column ID, choice of carrier gas (hydrogen, helium, nitrogen, etc.), and actual column length do not uniquely affect this behavior. In all cases, however, the compressibility of carrier gases results in an increase in the linear velocity of such media as they expand during flow from regions under higher pressure (nearer the inlet) to regions under lower pressure (nearer the outlet) (Fig. 5.3).

Figures 5.2 and 5.3 seem connected by the fact that each curve in Fig. 5.3 intercepts the y axis at the compression ratio (γ); this is not a coincidence. In fact, while Eq. (5.11) can be used to describe how carrier gas pressure varies with distance from the column inlet, with only minor modification it can also be used to describe how the linear velocity of carrier gas increases with distance from the column inlet:

$$u(x) = \frac{u_o}{\sqrt{\left(\frac{p_i}{p_o}\right)^2 - \left(\frac{x}{L}\right)\left[\left(\frac{p_i}{p_o}\right)^2 - 1\right]}} \tag{5.12}$$

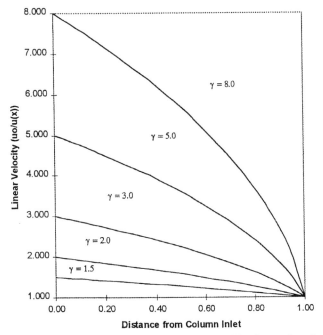

Fig. 5.3. Linear velocity of carrier gas (reduced) as a function of the distance from the column inlet at various carrier gas compression ratios (γ).

Figure 5.3 and Eq. (5.12) further emphasize that variation in the linear velocity of carrier gas with distance from the column inlet does not depend upon the choice of carrier gas, column ID, or actual column length. As a result, the relationship between the pressure and linear velocity of carrier gas as a function of distance along the column can be simply stated as

$$\frac{p(x)}{p_0} = \frac{u_0}{u(x)} \tag{5.13}$$

When factors affecting the volumetric flow of carrier gas are combined with the effect of carrier gas compressibility, it becomes possible to illustrate the behavior of such mobile phase media in a way that reveals the concerted influence of column dimensions, temperature, and choice of carrier gas in a single readily accessible experimental parameter known as the average linear velocity.

In practice, the average linear velocity of a carrier gas (u) can be determined directly from a measurement of the time (t_M) it takes an unretained solute to traverse the length (L) of the column:

$$\bar{u} = L/t_0 \tag{5.14}$$

where L is in cm and t_M is in seconds, $\bar{u}$ will be in centimeters per second and will equal the same average linear velocity calculated from accurately known values of volumetric flow (F_0), inlet pressure (p_i), outlet pressure (p_o), and column radius (r):

$$\bar{u} = F_0 j / \pi r_c^2 \tag{5.15}$$

where j is a factor that was first used by James and Martin to correct for carrier gas compressibility [6]:

$$j = \frac{3}{2} \frac{\left(\dfrac{p_i}{p_o}\right)^2 - 1}{\left(\dfrac{p_i}{p_o}\right)^3 - 1} \tag{5.16}$$

These same equations can be used to calculate reliable flows, column holdup times, and average linear velocities from any variety of temperatures, column dimensions, and carrier gas viscosities (Fig. 5.4).

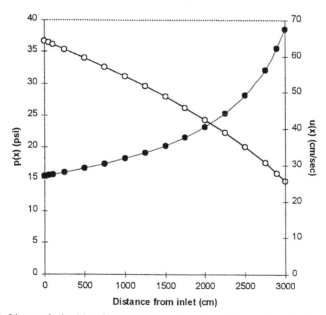

Fig. 5.4. Linear velocity (•) and pressure (○) as a function of distance from the inlet of a 30 m × 0.25 mm × 1.0 μm DB-1 column operated with helium carrier gas at a temperature of 175°C, an inlet pressure of 22 psig, and an outlet pressure of 14.690. Under these conditions, the carrier gas viscosity is 0.0026 cP, volumetric flow is 1.99 mL/min (measured as 1.30 mL/min at 20°C), column hold-up time is 1.378 min, and the average linear velocity of carrier gas is 36.34 cm/sec.

5.6   Regulation of Gas Flow and Gas Velocity

The carrier gas velocity is usually regulated either by controlling the carrier gas *pressure* at the column inlet or outlet, or by controlling the *flow* to maintain a constant *mass* of carrier gas flowing into the column. With "forward pressure regulation," the sensing point (and normally the column) is located downstream from the controlling valve. With "backpressure regulation," the sensing point (and normally the column) is located upstream from the control valve. In either case, if pressure at the sensing point deviates from the set pressure, mechanical or electronic feedback triggers a proportionate opening or closing of the control valve, and pressure (at the sensing point) returns to the set point. Fixed and/or variable restrictors (e.g., needle valves) are sometimes incorporated into the gas control system [7].

With a pressure-regulated system, the linear velocity (and the flow rate) of the carrier gas decrease with increasing temperature. With mass-flow-regulated systems, a constant mass of gas is introduced to the column, and the average linear velocity (and the flow rate) of the carrier increases as the higher temperature forces the constant mass of gas to expand to a larger volume.

A recent extension of pressure-regulated systems is electronic pneumatic controlled (EPC) gas chromatography. Using an instrument equipped with EPC, the chromatographer can perform separations under conditions that yield column flows resulting from programmed changes in head pressure similar to programs more conventionally used to vary column temperature during analysis. The advantages and potential difficulties associated with EPC will be dealt with in greater detail later.

5.7   Average Linear Velocity and Chromatographic Efficiency

Despite a number of subsequent refinements, the original van Deemter equation is still most commonly presented to introduce the basic relationship between chromatographic efficiency and the average linear velocity of carrier gas:

$$H = A + B/\bar{u} + C\bar{u} \tag{5.17}$$

where $\bar{u}$ is the average linear velocity of carrier gas; H is the height equivalent to a theoretical plate; and A, B, and C are all terms expressed so as to reflect the influence of phenomena that reduce chromatographic efficiency by contributing to the magnitude of H. The first of these terms, A, represents the contribution of a phenomenon called "eddy diffusion." Eddy diffusion occurs as solute molecules travel through interstitial regions of the stationary phase or stationary phase support found in a packed column. In such columns, the random arrangement and/or irregular shape of particles comprising this packing material reveal a variety of

Fig. 5.5. Schematic showing variations in routes, and the resulting lengths of paths, that individual molecules may take as they traverse a small cross section of column packing.

paths or flow streams along which any particular solute molecule might be carried by the mobile phase in which it is dissolved (Fig. 5.5).

Because some such paths or flow streams might ultimately prove longer or shorter than others, it may seem reasonable to assume that this alone is sufficient to account for the observation that molecules, otherwise expected to reach the end of the column at the same time, actually elute as a peak broadened to an extent reflected by the magnitude of A in Eq. (5.17). Although conceptually compelling, this theory fails in two significant regards as an explanation of eddy diffusion. The first is that it really has nothing to do with diffusion (i.e., the force compelling solute molecules to move from regions of higher to lower solute concentration). The second is that it overestimates the role of random chance in determining the course a particular solute molecule takes through a packing material in which channels are highly interconnected and at least as irregular in dimension as the particles composing their boundaries.

In any particular channel, the velocity of mobile phase flowing through that channel will vary as a function of distance from the channel wall. As a result, solute molecules finding themselves in faster flow streams near the center of a channel or slower flow streams near the channel wall will change velocity whenever a difference in solute concentration between two adjacent flow streams leads to diffusion from that flow stream in which solute concentration is higher to that in which solute concentration is lower. This, compounded by the fact that solute molecules traveling along any particular channel may at any moment experience a change in velocity as that channel empties out into a single wider channel or several narrower channels, suggests that the magnitude of A in Eq. (5.17) should depend more on variation in the time it takes particular solute molecules to travel the length of the column than on the actual distance traveled in doing so.

$$A = 2\lambda d_p \tag{5.18}$$

Although less rigorous treatments are frequently offered to justify the use of Eq. (5.18), in which A is given only as a function of particle diameter (d_p) and the "packing factor term" (λ), Giddings has shown that a more complete understanding of eddy diffusion can be obtained using Eq. (5.19), in which changes in the velocity of solute molecules due to flow effects are coupled to those due to diffusion [8].

$$A = \sum_i \frac{1}{1/2\lambda_i d_p + D_M/\omega_i v d_p^2} \tag{5.19}$$

In Eq. (5.19), where D_M is the coefficient of solute diffusion in the mobile phase, d_p is particle diameter, and v is mean mobile phase velocity, at least five specific contributions to A that account for changes in the velocity of solute molecules stemming from flow effects (i.e., $1/2\lambda_i d_p$) and diffusion (i.e., $D_m/\omega_i v d_p^2$) are summed to yield a result that collectively reflects the influence of eddy diffusion on the chromatographic efficiency of packed columns. In turn, λ_i and ω_i for each particular effect are related to the time it takes solute molecules to move from one velocity extreme to another across a single channel (transchannel effect), a single porous particle (transparticle effect), a few particle diameters (short-range interchannel effect), 5–10 particle diameters (long-range interchannel effect), or the entire diameter of the column (transcolumn effect). Finally, where $\lambda_i = \omega_\lambda \omega_\beta^2/2$ and $\omega_i = \omega_\alpha^2 \omega_\beta^2/2$, the values ω_α, ω_β, and ω_λ represent the following: ω_α is the number of particle diameters d_p equal to the distance over which a typical molecule must move to go from one velocity extreme to the other; ω_β is that fraction of the mean mobile phase velocity with which a solute molecule must move to travel the distance $\omega_\alpha d_p$ in the time it takes a typical solute molecule to move from one velocity extreme to the other; and ω_λ is the number of particle diameters over which a particular flow velocity persists. Since virtually all parameters required to calculate A in Eq. (5.19) must, themselves, typically be estimated rather than accurately measured or calculated, it remains fortunate that eddy diffusion can be disregarded for all but packed columns.

$$B = 2\gamma D_M \tag{5.20}$$

The second term, B, represents the contribution of longitudinal diffusion in the mobile phase. For packed columns, B depends on the coefficient of solute diffusion in the mobile phase (D_M) and a tortuosity factor (γ) [Eq. (5.21)]. For open tubular columns in which there are no obstacles to flow, this latter term is set equal to unity and so B is simply assumed proportional to D_M:

$$B = 2D_M \tag{5.21}$$

Although perhaps similar insofar as both A and B constitute contributions to H related to solute diffusion in the mobile phase, their distinction is important. It should be noted, for example, that longitudinal diffusion occurs along an axis that is parallel to flow through the column, while eddy diffusion occurs along an axis perpendicular to channel flow, but in no particular relationship to the direction of column flow.

The third term, C, actually represents the concerted effect of two phenomena: resistance to mass transfer in the stationary phase [Eq. (5.22)] and resistance to mass transfer in the mobile phase [Eq. (5.23)]:

$$C_M = \omega d_p^2 / D_M \tag{5.22}$$
$$C_M = [(1 + 6k + 11k^2)/(96(1 + k)^2)](d_c^2 / D_M) \tag{5.22b}$$

In Eq. (5.22), where C_M represents resistance to mass transfer in the mobile phase, ω is a factor to correct for radial diffusion, d_p is particle diameter, and D_M is the coefficient of solute diffusion in the mobile phase.

$$C_S = (q)[k/(1 + k)^2](d_f^2 / D_S) \tag{5.23}$$
$$C_S = \frac{2}{3}(k/(1 + k)^2)(d_f^2 / D_S) \tag{5.23b}$$

In Eq. (5.23), where C_S represents resistance to mass transfer in the stationary phase, q is a configuration factor depending on the uniformity of particle coating, k is the solute capacity or "separation" factor, d_f is stationary phase film thickness, and D_S is the coefficient of solute diffusion in the stationary phase.

After Golay, analogous equations uniquely well suited for use with open tubular columns can be derived. In Eq. (5.22b), k is solute retention factor, d_c is column ID, and D_M is again the coefficient of solute diffusion in the mobile phase. In Eq. (5.23b), a value of 2/3 for the configuration factor is assumed by virtue of the uniformity typical of stationary phase films found in most wall-coated open tubular columns.

In both the mobile phase and the stationary phase, resistance to mass transfer tends to reduce chromatographic efficiency (i.e., increase the magnitude of H) in direct proportion to $\bar{u}$ by virtue of the fact that solutes, once dissolved in the bulk of either phase, must depend on diffusion to reach the interface at which their equilibrium concentrations in both phases can be continuously reestablished. At low values of $\bar{u}$, the time it takes solute molecules to diffuse to the surface is relatively short in comparison to the rate at which mobile phase is being swept down the column. As a result, resistance to mass transport at low $\bar{u}$ values poses very little impediment to efficient reequilibration. Conversely, at higher values of $\bar{u}$, the time it take molecules to diffuse to the interface becomes longer relative to that rate at which the mobile phase is being swept from the column and so the efficiency of reequilibration is impeded.

When, for packed columns, A, B, and C in Eq. (5.17) are expanded to reflect the effects of eddy diffusion (A), longitudinal diffusion in the mobile phase (B), and resistance to mass transport ($C_S + C_M$), Eq. (5.24) describes the height equivalent to a theoretical plate (H) as a function of the average linear velocity of carrier gas ($\bar{u}$) illustrated in Fig. 5.6. When the effect of eddy diffusion can be disregarded, the tortuosity factor in Eq. (5.20) equals unity, and Eqs. (5.22b)–(5.23b) are used to account for resistance to mass transport, the Golay equation (Eq. 5.25) describes the

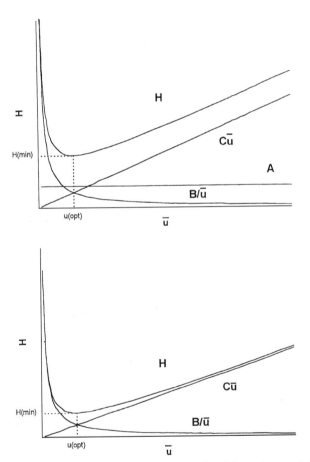

Fig. 5.6. Variation in H as a function of $\bar{u}$ for (top) packed, and (bottom) open tubular columns.

relationship between $\bar{u}$ and H that predicts the higher chromatographic efficiency (i.e., smaller γ as a function of $\bar{u}$) of open tubular columns:

$$H = 2\lambda d_p + \frac{2\lambda D_M}{\bar{u}} + \frac{\omega d_p^2 \bar{u}}{D_M} + q \frac{k}{(1+k)^2} \frac{d_f^2}{D_S} \bar{u} \tag{5.24}$$

$$H = \frac{2D_M}{\bar{u}} + \frac{1 + 6k + 11k^2}{96(1+k)^2} \frac{d_c^2}{D_M} \bar{u} + \frac{2}{3} \frac{k}{(1+k)^2} \frac{d_f^2}{D_S} \bar{u} \tag{5.25}$$

In Fig. 5.6, optimum chromatographic efficiency (i.e., H_{min}) is observed when chromatographic columns are operated at an average linear velocity of carrier gas equal to $\bar{u}_{opt}$. Although many practitioners of modern capillary column gas chromatography tend to resist the consideration and interpretation of van Deemter curves, they can with minimal effort generally prove extremely useful in both the qualitative and quantitative evaluation of chromatographic performance.

When Eq. (5.17) is differentiated with respect to $\bar{u}$, it can be shown that H_{min} and $\bar{u}_{opt}$ can be calculated directly from either known or estimated values of A, B, and C:

$$\bar{u}_{opt} = \sqrt{B/(C_M + C_S)} \tag{5.26}$$

$$H_{min} = A + 2\sqrt{B/(C_M + C_S)} \tag{5.27}$$

Values obtained in this way can be readily compared to values of $\bar{u}$ determined from measurements of column holdup time and values of H inferred from measurements of peak width:

$$\bar{u} = L/t_M \tag{5.28}$$

$$H = L/[5.545(t_R/W_h)^2] \tag{5.29}$$

In Eq. (5.28), L is column length (typically in centimeters) and t_M is column holdup time, $\bar{u}$ is the average linear velocity of carrier gas. In Eq. (5.29), where t_R is peak retention time, w_h is peak width at half height, and L is again column length (this time typically in mm), H is the height equivalent to a single theoretical plate.

In practice, when $\bar{u}$ from Eq. (5.28) is lower than $\bar{u}_{opt}$ from Eq. (5.26), the analyst will want to increase column head pressure to effect a corresponding reduction in H. This will yield sharper chromatographic peaks and will also reduce analysis time. When $\bar{u}$ from Eq. (5.28) is higher than $\bar{u}_{opt}$ from Eq. (5.26), the analyst *may* wish to reduce column head pressure, but only when the corresponding reduction in H seems to justify a longer analysis time. To achieve the best balance between chromatographic efficiency and analysis time, experienced chromatographers typically operate their columns at approximately 110% of $\bar{u}_{opt}$.

Similarly, analysts who observe a suspiciously large discrepancy between H from Eq. (5.29) and H_{min} from Eq. (5.27) when operating at $\bar{u}_{opt}$ will want to first verify their estimates of B and C by dividing $\bar{u}_{opt}$ from Eq. (5.26) by column hold-up time. If this yields a value for column length that differs significantly (i.e., >1 m) from what is known to be the case, parameters used to estimate A, B, and C should be reconsidered. If this fails to reduce or eliminate the difference between H and H_{min}, the analyst may want to investigate potential "extracolumn" effects that could be degrading peak shape (e.g., poor injection technique, possible coelution, column or inlet activity). When H from Eq. (5.29) is significantly lower than H_{min}, it is highly probable that an error has been made in estimating A, B, and/or C that has probably resulted in an erroneous value for $\bar{u}_{opt}$ in Eq. (5.28).

5.8 Calculating Reliable Estimates A, B, and C

More quantitative treatments of the relationship between the average linear velocity of carrier gas and chromatographic efficiency require reliable estimates of A, B, and C in the van Deemter equation and B and C in the Golay equation.

In the van Deemter equation for packed columns, estimating reasonable values for the effect of eddy diffusion is complicated by the need to know values for λ and d_p in Eq. (5.19) or λ_i, ω_i, d_p, and D_m in Eq. (5.20). For all practical purposes, estimates of the particle diameter (d_p) can be obtained directly from the name or description accompanying most commercially available packing materials. When the mean particle diameter of a packing material is cited in micrometers, this value can generally be used directly after conversion to the units in which the height equivalent to a theoretical plate will be expressed (usually millimeters). When d_p must be inferred from a "mesh size" (e.g., 80/100), this measurement must be interpreted as a range of particles sizes small enough to pass through the screen having fewer holes or meshes per linear inch (i.e., 80), but too large to pass through the screen having more holes per linear inch (i.e., 100). As a result, estimates of d_p derived from mesh sizes are typically based on an average particle size halfway between the size of the largest particle expected to pass through the first screen and the size of the smallest particle expected to be held up by the second screen. It is worth noting, however, that this number does not necessarily constitute a measurement of the actual mean particle size determined by more reliable means (microscopy, light scattering, etc.). In Table 5.2, various mesh sizes are compared to the size of the largest particles expected to pass through them.

Whereas values for the packing factor (λ) in Eq. (5.19) are most typically just estimated at somewhere between 1 and 2, estimates for $\lambda_i = \omega_\lambda \omega_\beta^2 / 2$ and $\omega_i = \omega_\alpha^2 \omega_\beta^2 / 2$ in Eq. (5.20) have been provided by Giddings (Table 5.3) [10]. In Table 5.3, where m is the number of particle diameters in one column diameter, r_c is column radius, and R_o is the coil radius, the transcolumn effect has been subclassified to separate what might otherwise be more typically considered the transcolumn effect (I) from an effect related to column coiling (II).

In providing these estimates of λ_i, ω_i, and the parameters on which their values depend, Giddings cautions against their interpretation as more than an aid

TABLE 5.2

**Typical Mesh Sizes and d_p for the Largest Particles
Expected to Pass**

Mesh size	Particle size (μm)	Mesh size	Particle size (μm)
4	4750	45	355
5	4000	50	300
6	3350	60	250
7	2800	70	212
8	2360	80	180
10	2000	100	150
12	1700	120	125
14	1400	140	106
16	1180	170	90
18	1000	200	75
20	850	230	63
25	710	270	53
30	600	325	45
35	500	400	38
40	425	625	20

to understanding. Nevertheless, when such estimates are used to calculate values for *A* from Eq. (5.19), several important observations can are made. Insofar as Eq. (5.19) yields estimates of *A* that compare reasonably well to those observed experimentally, substituting Eq. (5.19) for Eq. (5.18) in Eq. (5.24) reveals that at low mobile phase velocities, the degree to which eddy diffusion degrades the chromatographic efficiency of packed columns is smaller, but can be attributed to both flow and diffusion effects. At higher mobile phase velocities, flow effects dominate, while diffusion effects become less important (Fig. 5.7).

In Eq. (5.19), the coefficient of solute diffusion in the mobile phase can be calculated using the Fuller–Schettler–Giddings equation [Eq. (5.30)] [11, 12].

TABLE 5.3

Typical Values for Estimating the Degree to Which Eddy Diffusion Degrades Chromatographic Resolution of Packed Columns

Effect	λ_i	ω_i	ω_α	ω_β	ω_λ
Transchannel	0.5	0.01	0.1667	1	1
Transparticle	10,000	0.1	0.5	1	10,000
Short-range interparticle	0.5	0.5	1.25	0.8	1.5
Long-range interparticle	0.1	2	10	0.2	5
Transcolumn (I)	$0.02m^2$	$0.001m^2$	$m/2$	0.1	$5m^2$
Transcolumn (II)	$40m^2 r_c^2/R_0^2$	$2m^2 r_c^2/R_0^2$	m	$2r_c/R_0$	$20m^2$

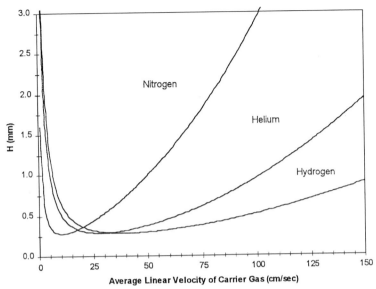

Fig. 5.7. Variation in H as a function of $\bar{u}$ for different carrier gases on a 30 m × 0.32 mm × 2.25 μm SE-30 column operated isothermally at 115°C. In all cases $k = 6.237$.

In Eq. (5.30), where D_M is the coefficient of solute diffusion in the mobile phase in square centimeters per second, T is absolute temperature, M_S is molecular mass of the solute, M_M is molecular mass of the mobile phase, P is pressure in atmospheres, $\sum v_S$ is the sum of atomic volume increments for the solute, and $\sum v_M$ is the sum of atomic volume increments for the mobile phase:

$$D_M = \frac{0.001 T^{1.75}(1/M_S + 1/M_M)^{1/2}}{P\left(\sum V_S^{1/3} + \sum V_M^{1/3}\right)^2} \tag{5.30}$$

As a first approximation, $\sum v_S$ can be calculated for from the atomic and structural diffusion volume increments presented in Table 5.4 [13]. For simple molecules,

TABLE 5.4

Atomic and Structural Volume Increments

Atom	V	Atom	V
C	16.5	Cl	19.5
H	1.98	S	17.0
O	5.48	Aromatic ring	−20.2
N	5.69	Heterocyclic ring	−20.2

TABLE 5.5

Diffusion Volumes for Simple Molecules

Molecule	Σv	Molecule	Σv
H_2	7.07	CO	18.9
He	2.88	CO_2	26.9
N_2	17.9	N_2O	35.9
O_2	16.6	NH_3	14.9
Ar	16.1	H_2O	12.7
Cl_2	37.7	SO_2	41.1
Br_2	67.2	SF_6	69.7

more reliable values of D_M can be obtained using the values presented in Table 5.5 [13].

In Eq. (5.19), the tortuosity factor (γ) needed to calculate the contribution of longitudinal diffusion in packed columns typically ranges from 0.6 to 0.8 and can, itself, be calculated as the product of two ratios. Unfortunately, neither is readily accessible. The first is the ratio of the effective or apparent diffusion coefficient to the coefficient of solute diffusion in the mobile phase (i.e., D/D_M). The second is a constriction factor equal to the one over the mean cross-sectional area of a channel times the mean reciprocal of that area. Where $D/D_M = F^2$ and F accounts for tortuosity, the product of these two ratios

$$\gamma = \{F^2[\bar{A}/(\overline{1/A})]\}^{-1} \tag{5.31}$$

yields an expression that actually accounts for both tortuosity and constriction [Eq. (5.31)]. This is the tortuosity factor from Eq. (5.20). Stated more simply, tortuosity can be thought of as the average ratio of the real to the actual path length of a molecule through a column packing. In packings dominated by channels that are not very tortuous, F^2 assumes a values close to unity because, on average, the path a molecule takes as it traverses the column is very close in length to the shortest path, that is, a straight line extending from one end of the column to the other. Similarly, the more uniform the diameter of channels through which molecules travel, the closer the tortuosity factor approaches unity, and the closer the constriction factor will be to unity. Thus, in the extreme, γ will never be greater than one and will never be less than zero. For open tubular columns γ always equals unity.

In Eq. (5.22), values for ω (i.e., the factor to correct for radial diffusion) can be obtained for each of the same five effects described earlier with respect longitudinal diffusion in the mobile phase. In this case, however, diffusion is understood to mean diffusion in a direction perpendicular to that of overall mobile phase flow.

In Eq. (5.23), the configuration factor, q, will typically assume a value ranging from slightly less than 0.1 to about 1.0 and will depend on the geometric configuration of the stationary phase (uniform film, spherical bead, etc.). For a uniform stationary phase film deposited on the interior wall of an open tubular column, $q = 2/3$ is usually assumed. For packed columns in which a liquid stationary phase has been deposited on a volume of spherical beads, $q = 1/12$ can be assumed. In other circumstances, it becomes necessary to distinguish between stationary phase deposited in narrow pores [Eq. (5.32)] from that deposited in wide pores [Eq. (5.33)]:

$$q = 2/(n + 1)(n + 3) \tag{5.32}$$

$$q = 4/(n + 2)(n + 3) \tag{5.33}$$

In both Eqs. (5.32) and (5.33), n is the taper factor used to describe variation in the cross-sectional area of the average pore as a function of its normalized depth [Eq. (5.34)]. In Eq. (5.23), values for the coefficient of solute diffusion in a typical liquid stationary phase can be calculated from any number of theoretical models (e.g., Stokes–Einstein, Eyring, Wilke–Chang).

$$a = a_0(x/d)^n \tag{5.34}$$

Unfortunately, none of these models appears to yield results sufficiently accurate to subsequently permit the calculation of reliable values for the height equivalent to a theoretical plate. Nevertheless, for many solutes, reasonable D_S values can be obtained using the empirical relationships of Kong and Hawkes [14]:

$$\ln D_S = K_0 + K_1 C + (K_2 + K_3 C)/RT \tag{5.35}$$

where R is the universal gas constant in calories per degree-mole, T is absolute temperature in kelvins, C is solute carbon number, and all three K terms are fitting constants (Table 5.6) [14].

TABLE 5.6

Fitting Parameters for Calculation of D_S Values According to Kong–Hawkes Relationships for Several Commercially Available Stationary Phases. Data from [14]

Phase	K_0	K_1	K_2	K_3	R
SE-30	−8.870	3.836×10^{-1}	4.461×10^2	-4.981×10^2	0.876
Viscasil	−4.609	-2.548×10^{-1}	-4.192×10^3	1.044×10^2	0.995
SF-96-2000	−5.966	7.466×10^{-2}	-3.034×10^3	-1.880×10^2	0.994
SF-96-200	−9.353	3.052×10^{-1}	-6.384×10^2	-3.784×10^2	0.926
DC-550	−3.997	-2.023×10^{-1}	-5.155×10^3	5.869×10^1	0.992
DC-710	−3.084	-2.419×10^{-1}	-6.260×10^3	1.017×10^2	0.998
OV-210	−6.343	5.570×10^{-1}	-1.507×10^3	-7.217×10^2	0.918
OV-225	−4.323	1.053×10^{-2}	-5.817×10^3	-1.027×10^2	0.998
SP-2401	−5.881	1.675×10^{-1}	-3.686×10^3	-2.588×10	0.997
OV-25	−4.282	-5.930×10^{-2}	-6.496×10^3	-7.631	0.998

5.9 Theory and Practice

Despite the utility of equations presented throughout the preceding sections of this chapter, the work of either calculating or estimating parameters required to formulate a reliable relationship between plate height and the average linear velocity of carrier gas remains considerable for any particular chromatographic system (i.e., solute, column, carrier gas, and temperature combination). So considerable, in fact, that some have compared it to the task of determining such relationships experimentally. This, compounded by the fact that such calculations still typically ignore the effects of injection efficiency, column inertness, and various mechanical properties related to the uniformity and stability of actual stationary phase coatings, is usually sufficient to convince most analysts that characterizing their chromatographic systems at all is probably a waste of time, particularly when the results they obtain seem "good enough." With time, however, three significant disadvantages are derived from this latter approach:

1. Analysts never develop an intuition for how seemingly subtle characteristics such as stationary phase diffusivity influence the efficiency of existing methods and the choices they make about systems selected for new method development.

2. There remains no consistent means by which the poor performance of a particular system can be expediently attributed to a particular cause (e.g., column dimensions, choice of carrier gas, or column quality).

3. Data that might otherwise facilitate the more accurate and expedient characterization of both new and existing systems remain hopelessly scarce.

Although at present, there seems to be no truly reliable means by which these disadvantages can be readily overcome, a number of computer models have been used to illustrate the influence of several more familiar chromatographic parameters [15, 16]. In the sections that follow, this same approach will be taken with a new program that correctly accounts for the effects neglected by Ingraham *et al.* [15], yet offers distinct advantages over the program of Leclercq and Cramers [16] by permitting the simultaneous comparison of multiple systems and accommodating the use of more familiar units of measure. In each case, the curves presented below are based on the basic Golay equation presented above as Eq. (5.25), but as modified to incorporate the pressure-drop correction factor of Giddings and a factor accounting for carrier gas compressibility:

$$H = \frac{2D_M j j''}{\bar{u}} + \frac{1 + 6k + 11k^2}{96(1+k)^2} \frac{d_c^2 j}{D_M j''} \bar{u} + \frac{2}{3} \frac{k}{(1+k)^2} \frac{d_f^2}{D_S} \bar{u} \qquad (5.36)$$

where, D_M is the coefficient of solute diffusion in the mobile phase, D_S is the coefficient of solute diffusion in the stationary phase, d_c is column ID, d_f is stationary phase film thickness, and k is solute retention factor, and f_1 and f_2 have the following definitions (in each case, p_i represent pressure at the column inlet and p_o represents pressure at the column outlet):

$$j = \frac{3}{2} \frac{\left(\dfrac{p_i}{p_o}\right)^2 - 1}{\left(\dfrac{p_i}{p_o}\right)^3 - 1} \tag{5.37}$$

$$j'' = \frac{9}{8} \frac{\left[\left(\dfrac{p_i}{p_o}\right)^4 - 1\right]\left[\left(\dfrac{p_i}{p_o}\right)^2 - 1\right]}{\left[\left(\dfrac{p_i}{p_o}\right)^3 - 1\right]^2} \tag{5.38}$$

5.10   Choice of Carrier Gas

In practice, the choice of carrier gas influences chromatographic efficiency as a measure of the height equivalent to a theoretical plate in two important regards. The first is by virtue of carrier gas viscosity. The second is by virtue of its role in determining the coefficient of solute diffusion in the mobile phase. In Fig. 5.7, the analysis of a solute for which $k = 6.237$ under isothermal conditions at 115°C on a 30 m × 0.320 mm × 0.25 μm SE-30 wall-coated open tubular column reveals that higher chromatographic efficiency (i.e., smaller H at $\bar{u}_{opt}$) can be achieved using nitrogen at a $\bar{u}_{opt}$ of 10.445 cm/sec than can be achieved using helium at a $\bar{u}_{opt}$ of 28.074 cm/sec or hydrogen at a $\bar{u}_{opt}$ of 38.906 cm/sec. The cost of such high efficiency, however, comes at the expense of analysis time which for nitrogen at 10.445 cm/sec would be about 2.7 times longer than for helium at 38.906 cm/sec and about 3.7 times longer than for hydrogen at 38.906 cm/sec. Although this analysis time might be shortened by operating at an average linear velocity higher than $\bar{u}_{opt}$, it is worth noting that the height equivalent to a single theoretical plate also increases more sharply with nitrogen than with helium or hydrogen.

5.11   The Effect of Solute Retention Factors

Figure 5.8 illustrates the influence of solute retention factors on the relationship between H and $\bar{u}$. Under isothermal conditions, H_{min} appears to shift toward higher values and $\bar{u}_{opt}$ appears to shift toward lower values as k increases. This occurs as a direct result of the fact that low-k solutes elute more rapidly from the column and typically exhibit larger D_S values than do high-k solutes.

From Fig. 5.9, it is apparent that this trend can be generalized to illustrate how both H_{min} and $\bar{u}_{opt}$ respond to variation in k under isothermal conditions. As k increases, H_{min} increases sharply and $\bar{u}_{opt}$ decreases sharply until becoming relatively stable beyond a k of about 5.5. Although no single $\bar{u}$ can ever deliver truly maximum chromatographic efficiency for all solutes in a particular sample, it

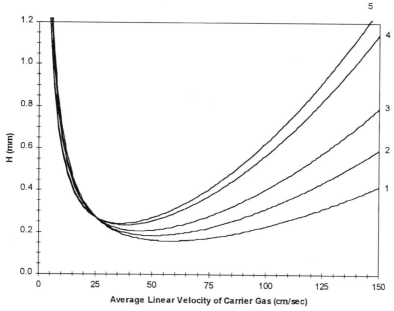

Fig. 5.8. Variation in H as a function of $\bar{u}$ for solutes of different retention factors on a 30 m × 0.25 mm × 0.50 μm column operated under isothermal conditions at 120°C with hydrogen carrier gas: 1) $k = 0.742$; 2) $k = 1.311$; 3) $k = 2.315$; 4) $k = 7.215$; and 5) $k = 12.739$.

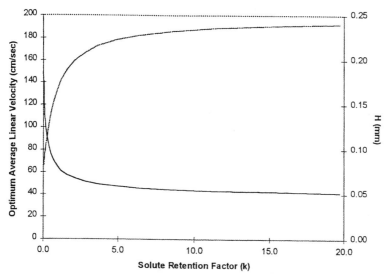

Fig. 5.9. Variation in H at $\bar{u}_{opt}$ as a function of the solute retention factor. Parameters for calculation included a 30 m × 0.25 mm × 0.50 SE-30 μm WCOT column operating isothermally at 250°C with hydrogen carrier gas. The upper curve illustrates the variation in H, while the lower curve shows the variation in u_{opt}.

is worth noting that the selection of a $\bar{u}$ near $\bar{u}_{opt}$ for a "mid-k" solute will typically yield the best overall results.

5.12   The Effect of Column Length and Inside Diameter

Although it never appears directly in either the van Deemter or Golay equations, column length does impact chromatographic efficiency by virtue of the relationship between column flow and the drop in pressure across a column required to induce that flow [Eqs. (5.1) and (5.2)]. While columns differing only in length rarely exhibit minimum plate heights differing by more than about 5% at $\bar{u}_{opt}$, shorter columns typically offer higher $\bar{u}_{opt}$ values and can actually exhibit smaller minimum plates heights than longer columns. For example, the 10-m column in Fig. 5.10 delivers 4577 theoretical plates per meter at a $\bar{u}_{opt}$ of 40.213 cm/sec, while the 100-m column operated at a $\bar{u}_{opt}$ of 19.846 cm/sec delivers only 4340 plates per meter. At average linear velocities just 20% higher than optimum, the 10-m column delivers 4463 plates per meter, while the 100-m column delivers only 4134

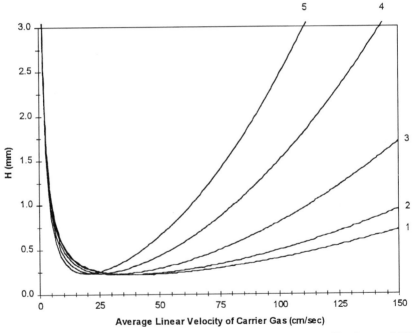

Fig. 5.10. Variation in H as a function of $\bar{u}$ for 0.25 mm $\times$ 0.30 μm SE-30 columns of different length operated under isothermal conditions at 115°C with helium carrier gas; 1) $L = 10$ m; 2) $L = 15$ m; 3) $L = 30$ m; 4) $L = 60$ m; 5) $L = 100$ m. In all cases, $k = 5.343$.

plates per meter (i.e., a difference of just more than 7%). As the average linear velocity of carrier gas is increased still further, the efficiency advantage of shorter columns becomes increasingly pronounced and is typically compounded by the fact that the higher $\bar{u}_{opt}$ values of shorter columns also help shorten analysis times.

In general, shorter columns are always preferable to longer columns of the same inside diameter provided they offer a sufficient number of total plates to yield the acceptable resolution of all critical pairs (i.e., closely eluting solutes, the separation of which differentiates a successful analysis from an unsuccessful analysis). In selecting the best column length for a particular separation, recall that resolution improves not at a function of column length, but as a function of the square root of column length [Eq. (1.22)].

Inside Diameter

The inside diameter of a column influences chromatographic efficiency in two important regards. First, it dramatically affects column flow by virtue of Eqs. (5.1) and (5.2), and this yields an effect similar to that observed for variation in column length. Second, it affects mobile phase volume. When the stationary phase film thickness (d_f) is held constant, this also affects the phase ratio (i.e., mobile phase volume/stationary phase volume) which, in turn, affects k.

In Fig. 5.11, variation in H as a function of $\bar{u}$ is plotted for a series of 45-m columns operated under isothermal conditions at 115°C with helium carrier gas. Although it may be true that smaller-ID columns having the same stationary phase film thickness will typically deliver higher chromatographic efficiency at $\bar{u}_{opt}$, several additional considerations warrant further attention. The first is that operating at $\bar{u}_{opt}$ for longer columns having extremely small inside diameters may prove impractical. In Fig. 5.10, for example, a head pressure of about 350 psig is required to operate the 0.05-mm-ID column at its $\bar{u}_{opt}$ of just 14.006 cm/sec (note that for a H_2 carrier operating at a $\bar{u}_{opt}$ of 23.047 cm/sec, a head pressure of about 259 psig would be required) and few commercial instruments can safely deliver such high head pressures. Those who have modified their instruments to operate at such high head pressure still have the problem of employing standard syringes to inject against that massive resistance.

It is also important to recognize the relationship between inside column diameter and the phase ratio, β = mobile phase volume/stationary phase volume. Since, under isothermal conditions, $K_c = \beta k$, smaller-ID columns having the same film thickness will have smaller β values and so exhibit proportionally larger k values. For example, in Fig. 5.9, the solute exhibiting a $k = 4.202$ on the 0.53-mm-ID column exhibits a $k = 6.955$ on the 0.32-mm-ID column, a $k = 8.898$ on the 0.25-mm-ID column, a $k = 22.178$ on the 0.10-mm-ID column, and a $k = 44.133$ on the 0.05-mm-ID column. Although this may seem to suggest a "hidden" advantage of

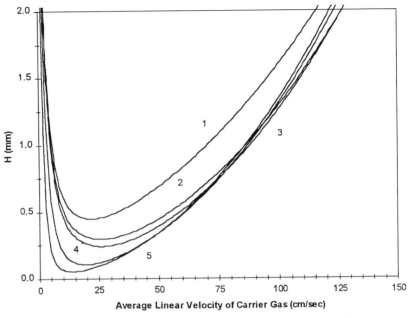

Fig. 5.11. Variation in H as a function of $\bar{u}$ for 45 m columns of different diameter operated isothermally at 115°C with helium carrier gas: 1) 0.53 mm I.D.; 2) 0.32 mm I.D.; 3) 0.25 mm I.D.; 4) 0.10 mm I.D.; and 5) 0.05 mm I.D. In all cases $d_f = 0.50 \, \mu$m.

smaller-ID columns when longer retention times are desired, it worth noting that, despite larger k values, the absolute volume of stationary phase found in smaller-ID columns will be considerably less. As a result, sample loading capacity (i.e., how much sample can be deposited on the column without incurring an overload) will typically suffer and this will reduce sensitivity unless more concentrated samples can be used.

If, as in Fig. 5.12, the stationary phase film thickness is varied with inside diameter to ensure a constant phase ratio, k values will remain the same under the same isothermal conditions, but sample loading capacities will drop off still more sharply by virtue of the concomitant reduction in absolute stationary phase volume. That this is occurring is most commonly observed as peak fronting and suggests that the column has simply been "overloaded." Nevertheless, Fig. 5.12 does indicate that, when the phase ratio is held constant, the efficiency advantage of smaller-ID columns is typically more pronounced at values of $\bar{u}$ well above the optimum. At the optimum, the difference is fairly negligible. In Fig. 5.11, minimum plate heights of 0.0445 mm, 0.0292 mm, 0.0238 mm, 0.0105 mm, 0.0054 mm are observed for the 0.53-mm-, 0.32-mm-, 0.25-mm-, 0.10-mm-, and 0.05-mm-ID columns, respectively. In Fig. 5.12, minimum plate heights of 0.0489 mm,

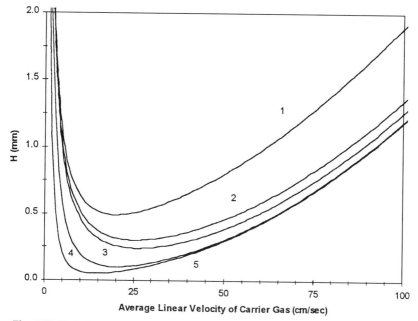

Fig. 5.12. Variation in H as a function of $\bar{u}$ for several 45 m columns operated under isothermal conditions at 115°C with helium carrier gas: 1) 0.53 mm I.D., $d_f = 1.060\ \mu$m; 2) 0.32 mm I.D., $d_f = 0.640\ \mu$m; 3) 0.25 mm I.D., $d_f = 0.500\ \mu$m; 4) 0.10 mm I.D., $d_f\ 0.200\ \mu$m; and 5) 0.05 mm I.D., $d_f = 0.100\ \mu$m. In all cases $\beta = 125$.

0.0299 mm, 0.0238 mm, 0.0099 mm, and 0.0050 mm are observed for the 0.53-mm-, 0.32-mm-, 0.25-mm-, 0.10-mm-, and 0.05-mm-ID columns, respectively.

Small-diameter columns are less "forgiving" with respect to column quality, injection efficiency, and sample loading capacity. Nevertheless, such columns are finding increasingly broader application in "high-speed GC." As analysts become more familiar with the limitations of these high-plate generators, there are certain analyses for which their advantages make them an extremely valuable commodity.

5.13   The Effect of Stationary Phase Film Thickness

To a large extent, the effect of stationary phase film thickness is apparent from the preceding discussion of the effect insider diameter has on the phase ratio both when film thickness is held constant and when it is allowed to vary so as to insure constant β. Because the stationary phase film thickness influences both the phase ratio and resistance to mass transport in the stationary phase, thicker stationary phase films in otherwise identical columns will typically yield somewhat lower chromatographic efficiencies at $\bar{u}_{opt}$, and considerably lower chromatographic efficiencies at higher average linear velocities of carrier. Furthermore, optimum velocities for thicker-film columns will typically occur at lower average linear velocities suggesting that thicker-film columns are best operated at slower column flow rates. In practice, the effects of stationary phase film thickness are confounded by practical limitations implicit to both the stability of extremely thick stationary phase films and the comparably poorer inertness of extremely thin stationary phase films. When applicable, such complications tend to reduce chromatographic efficiencies below that predicted strictly from theory.

In selecting the best stationary phase film thickness for a particular analysis, several additional factors require attention. Although thicker stationary phase films do offer higher solute retention factors under equivalent isothermal conditions, they do so by virtue of the fact that they constitute considerably larger absolute stationary phase volumes. Although in contrast to columns containing extremely thin stationary phase films, this will tend to increase sample loading capacities, it will also typically lead to higher column bleed levels. Conversely, thinner stationary phase films may offer shorter analysis times by virtue of smaller solute retention factors and higher values of $\bar{u}_{opt}$, but they also constitute smaller absolute stationary phase volumes which typically cause columns containing such films to exhibit higher levels of activity observed as peak tailing. So, too, will columns containing thinner stationary phase films be more easily overloaded.

5.14   The Effect of Stationary Phase Diffusivity

The effects of stationary phase diffusivity are all too frequently ignored when comparing the performance of different chromatographic systems. One plausible explanation for this is the relative scarcity of reliable data for many of the more typical stationary phases found in commercially available wall-coated open tubular columns. Nevertheless, the effect of stationary phase diffusivity can be significant, particularly at higher than optimum average linear velocities.

In general, columns containing stationary phases in which the solutes of interest have high diffusion coefficients are also columns that will exhibit higher chromatographic efficiency when all other parameters are held constant. So, too, can it be said that, in any particular stationary phase, larger solute molecules will typically exhibit smaller diffusion coefficients under the same isothermal conditions. To some this may seem counterintuitive since larger coefficients of solute diffusion in the stationary phase would seem to imply a greater propensity for band broadening, but it is important to recall the role such coefficients play in both the van Deemter and Golay equations. In each case, these coefficients affect the term describing the contribution of resistance to mass transport in the stationary phase. Consequently, when such coefficients are large, resistance to mass transport actually decreases (i.e., molecules move more freely throughout the stationary phase layer). Although this "freedom" occurs both parallel and perpendicular to the direction of column flow, the latter effect ultimately proves far more critical within the scope of time required for solutes to reestablish the equilibrium concentrations on which partitioning between the stationary and mobile phases is based. In other words, large coefficients of solute diffusion in the stationary phase which facilitate reequilibration of solute concentrations in both the mobile and stationary phases do so in a way that far outweigh their contribution to band broadening brought about by longitudinal rather than radial diffusion.

Unfortunately, this is a difficult trend to illustrate in light of the fact that most stationary phases in which a particular solute might exhibit significantly different coefficients of diffusion are also stationary phases exhibiting significantly different selectivities for the same solute. As a result, it may not always be possible to separate the effects of stationary phase diffusivity from k without more extensive study. In Fig. 5.13, the height equivalent to a theoretical plate as a function of average linear velocity is plotted for several columns in which the effect of stationary phase selectivity has essentially been ignored (i.e., solutes have been assumed to exhibit the same retention factors under the same isothermal conditions on all

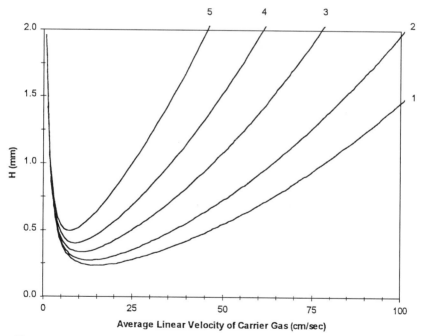

Fig. 5.13. Variation in H as a function of $\bar{u}$ for several 60 m SE-30 columns operated isothermally at 150°C with nitrogen carrier gas: 1) $d_f = 0.15\ \mu m$, $k = 0.323$; 2) $d_f = 0.25\ \mu m$, $k = 0.538$; 3) $d_f = 0.50\ \mu m$, $k = 1.075$; 4) $d_f = 1.00\ \mu m$, $k = 2.147$; and 5) $d_f = 3.00\ \mu m$, $k = 6.417$.

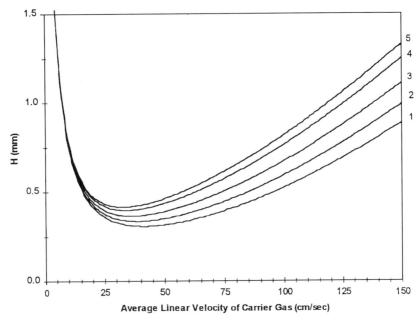

Fig. 5.14. Variation in H as a function of $\bar{u}$ for several 30 m $\times$ 0.32 mm $\times$ 1.00 μm columns operating isothermally at 150°C with hydrogen carrier gas: 1) $D_S = 1.292 \times 10^{-5}$ cm^2/sec; 2) $D_S = 6.252 \times 10^{-6}$ cm^2/sec; 3) $D_S = 3.941 \times 10^{-6}$ cm^2/sec; 4) $D_S = 2.752 \times 10^{-6}$ cm^2/sec; and 5) $D_S = 2.342 \times 10^{-6}$ cm^2/sec.

five phases). What can be gleaned from Fig. 5.13 (see also Fig. 5.14) is both the importance of D_S at values of $\bar{u}$ higher than optimum and that lower coefficients of solute diffusion in the stationary phase tend to drive the values of $\bar{u}_{\text{opt}}$ toward slower column flow rates.

5.15   The Effects of Temperature

Much like column length, temperature never appears directly in either the van Deemter or Golay equations. Nevertheless, temperature plays several important roles in determining the chromatographic efficiency of a particular system by virtue of its effect on mobile phase viscosity, D_M, D_S, and k. In Fig. 5.15, the net effect of temperature on the chromatographic efficiency of a particular 30 m $\times$ 0.25 mm $\times$ 0.5 μm column is illustrated.

As temperature increases, $\bar{u}_{\text{opt}}$ appears to shift to higher values, while H_{min} appears to shift to lower values. This reflects the fact that, at higher temperatures, k becomes smaller, while carrier gas viscosity and coefficients of solute diffusion in both the mobile phase and the stationary phase will increase. This is compounded

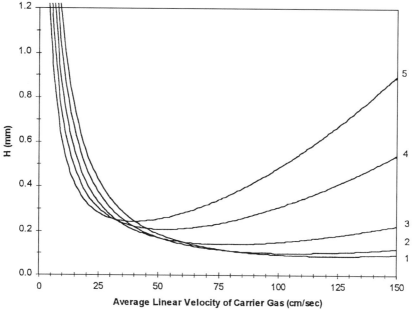

Fig. 5.15. Variation in H as a function of $\bar{u}$ for a particular solute on several 30 m × 0.250 mm × 0.50 μm columns operated under isothermal conditions with hydrogen carrier gas. 1) $T = 300°C$, $k = 0.044$, $D_S = 1.039 \times 10^{-4}$ cm^2/sec, $D_M = 0.677$ cm^2/sec; 2) $T = 250°C$, $k = 0.126$, $D_S = 6.267 \times 10^{-5}$ cm^2/sec, $D_M = 0.577$ cm^2/sec; 3) $T = 200°C$, $k = 0.460$, $D_S = 3.395 \times 10^{-5}$ cm^2/sec, $D_M = 0.484$ cm^2/sec; 4) $T = 150°C$, $k = 0.2.276$, $D_S = 1.592 \times 10^{-5}$ cm^2/sec, $D_M = 0.398$ cm^2/sec; 5) $T = 100°C$, $k = 0.17.263$, $D_S = 6.089 \times 10^{-6}$ cm^2/sec, $D_M = 0.319$ cm^2/sec.

by the fact that, as carrier gas viscosity increases, higher head pressures will be required to achieve optimum column flows and this will tend to cause efficiency to drop off more sharply at higher than optimum average linear velocities.

In general, decisions regarding column temperature should be made with the understanding that the effect temperature has on solute retention factors far outweighs the impact it has on chromatographic efficiency. When separations can only be achieved under isothermal conditions at extremely low temperatures on a particular column, the use of different columns dimensions, a thinner stationary phase film, or the feasibility of temperature programming should be considered.

5.16 Optimum Practical Gas Velocity

In each of the preceding sections, great care has been taken to illustrate the effect on chromatographic efficiency when columns are or are not operated at $\bar{u}_{opt}$. Unfortunately, conditions that provide for maximum chromatographic efficiency (i.e.,

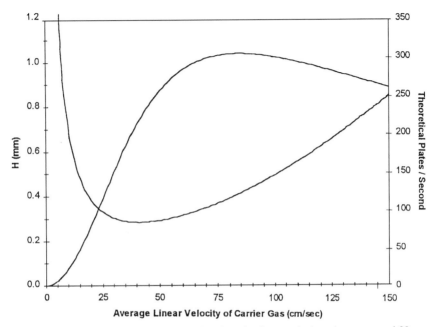

Fig. 5.16. Variation in H and N/sec as a function of $\bar{u}$ for a particular solute on several 30 m × 0.32 mm × 0.50 μm columns under isothermal conditions at 120°C with hydrogen carrier gas.

highest number of theoretical plates or H_{min}) almost always result in analysis times that are typically longer than necessary. For this reason, it has become the practice of more experienced chromatographers to evaluate the efficiency of their chromatographic systems in a slightly different way. In Fig. 5.16 the height equivalent to a single theoretical plate is plotted as a function of average linear velocity for a particular solute on a standard 30 m × 0.32 mm × 0.5 μm column under isothermal conditions at 120°C with hydrogen carrier gas. Also plotted in Fig. 5.16 is the number of theoretical plates per second generated by this column for the same solute under the same conditions.

At a $\bar{u}_{opt}$ of 40.537 cm/sec, an H_{min} of 0.285 mm is realized. For a solute exhibiting a $k = 5.6391$ under these conditions, elution can be expected in 8.189 min, and so it can be said that this particular column is generating 214.2 theoretical plates per second [i.e., $(30,000/0.285)/(8.189 \times 60)$]. Similarly, at an average linear velocity of 84.560 cm/sec or about twice $\bar{u}_{opt}$, this same column exhibits an H of 0.418 mm and the solute elutes in 3.926 mins. Under these latter conditions, this column delivers 304.6 theoretical plates per second. In other words, the faster than optimum average linear velocity, although it delivers an overall fewer total number of theoretical plates, actually delivers a higher number of theoretical plates per second.

When the efficiency of chromatographic systems is evaluated in terms of theoretical plates per second rather than theoretical plates per meter or total theoretical plates, the optimum average linear velocity of carrier gas no longer yields the best result. Instead, the best result is achieved by operating at a $\bar{u}$ referred to as the "optimum practical gas velocity" (OPGV). Since with this latter approach, it is actually the number of plates per second that determines efficiency, the OPGV will always be found at that $\bar{u}$ for which the ratio N/t_R has been maximized. Interpreted another way, the OPGV is the average linear velocity beyond which the loss in chromatographic efficiency can be expected to outweigh the advantage of any corresponding decrease in analysis time.

In general, the OPGV of any particular chromatographic system will shift to smaller $\bar{u}$ as the plot of H versus $\bar{u}$ grows steeper and will shift to larger $\bar{u}$ as the contribution of resistance to mass transport terms becomes smaller.

5.17 Temperature-Programmed Conditions

Until now the subject of temperature-programmed conditions has been carefully avoided. The reason for this, however, has less to do with the utility of temperature programs than with the additional dimension of complexity such programs add to consideration of the effects addressed thus far. Throughout subsequent sections of this chapter, the issue of temperature programmed conditions will be dealt with by demonstrating how equations for each of the temperature dependent parameters dealt with thus far can be rewritten to reflect the influence of variation in column temperature as a function of time during temperature-programmed analyses.

5.18 Column Flow under Temperature-Programmed Conditions

Under isothermal conditions at constant p_i and p_o, the volumetric flow of carrier gas as measured at the column outlet can be calculated using Eq. (5.1) [repeated below for convenience as Eq. (5.39)]:

$$F_0 = \left\{ \left[(p_i^2 - p_o^2)\pi r_c^4 \right] / [16\eta(T_c)p_o L] \right\} (T_a/T_c) \qquad (5.39)$$

where T_c is replaced by $T_c(t)$, Eq. (5.40) gives F_0 as a function of time during a temperature-programmed analysis (Fig. 5.17).

$$F_0(t) = \left(\left[(p_i^2 - p_o^2)\pi r_c^4 \right] / \{16\eta[T_c(t)]p_o L\} \right) [T_a/T_c(t)] \qquad (5.40)$$

When, as is frequently the case, more complex temperature programs are used (i.e., programs consisting of multiple ramps and/or ramps punctuated by isothermal segments), an analytical form for $T_c(t)$ may be difficult to formulate. This problem can be solved by treating each segment of such programs separately and

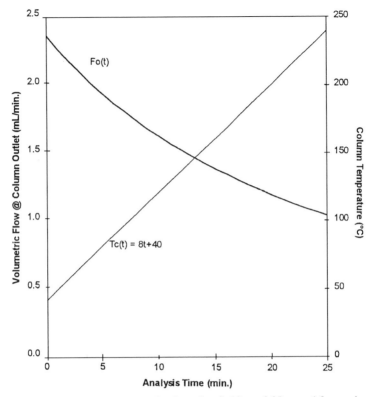

Fig. 5.17. Volumetric flow as measured at the outlet of a 30 m × 0.25 mm × 1.0 μm column under programmed conditions (40°C to 250°C @ 8°C/min) with helium carrier gas. Inlet pressure constant @ 22 psig, outlet pressure constant @ 14.69 psig. Outlet flows were measured at 20°C.

subsequently combining their individual $F_0(t)$ values to give a more complete picture of $F_0(t)$ for the entire analysis.

5.19   Average Linear Velocity under Temperature-Programmed Conditions

Just as the volumetric flow of carrier gas measured at the outlet varies with time as a function of column temperature during a temperature programmed analysis at constant p_i and p_o (see Section 5.17), so will the average linear velocity of carrier gas vary with $T_c(t)$. Obviated by the relationship between $\bar{u}$ and F_0 in Eq. (5.15), $\bar{u}(t)$ can now be expressed as in Eq. (5.41) for temperature-programmed analyses.

$$\bar{u}(t) = F_0(t)\,j/\pi r_c^2 \tag{5.41}$$

Among the implications of Eq. (5.41) is the fact that, if $\bar{u}$ now varies with time as a consequence of the fact that T_c varies with time, it stands to reason that chromatographic efficiency will also vary with time by virtue of the relationship between $\bar{u}$ and H (see Section 5.21).

5.20 D_S and D_M under Temperature-Programmed Conditions

Under temperature-programmed conditions, equations for both the coefficient of solute diffusion in the mobile phase (D_M) and the coefficient of solute diffusion in the stationary phase D_S assume time-dependent forms analogous to those presented in the preceding two sections for volumetric flow and the average linear velocity of carrier gas. Originally presented in Section 5.7 as Eq. (5.30), the time-dependent form of the equation for D_M is presented as

$$D_M = \frac{0.001 T_c(t)^{1.75}(1/M_S + 1/M_M)^{1/2}}{P\left(\sum V_S^{1/3} + \sum V_M^{1/3}\right)^2} \qquad (5.42)$$

Notice that, as before, this equation does not reflect the fact that pressure (P) varies as a function of distance along the column [Eq. (5.11)] because such variation is later accounted for in Eq. (5.36) through the use of f_1 and f_2. Consequently, under temperature-programmed conditions at constant p_i and p_o, there is no need to modify this equation further.

Originally presented in Section 5.7 as Eq. (5.35), the time-dependent analog of the equation for calculating D_S is given as

$$\ln D_S = K_0 + K_1 C + (K_2 + K_3 C)/R T_c(t) \qquad (5.43)$$

In using Eq. (5.43), particular attention must be paid to the fact that Eqs. (5.35) and (5.43) both constitute empirical relationships that may ultimately prove inappropriate for some more exotic stationary phase materials.

5.21 Solute Retention under Temperature-Programmed Conditions

In Chapter 1, the concept of equilibrium partitioning was introduced through reference to the equilibrium distribution constant $K_c = \beta k$. At equilibrium K_c defines the ratio of solute concentration in the stationary phase to solute concentration in the mobile phase. With the phase ratio β defined as the ratio of mobile phase volume to stationary phase volume, the definition for k becomes a ratio of the amount of solute dissolved in the stationary phase divided by the amount of solute dissolved in the mobile phase. Since, for a particular column, β remains constant, k is ultimately determined by K_c, which, in turn, is ultimately determined by temperature and the nature of the stationary phase. Like all such equilibrium distribution coefficients, the relationship between K_c and absolute temperature can be expressed as

$$\Delta G = RT \ln(K_c) \tag{5.44}$$

where R is the universal gas constant, T is absolute temperature, K_c is the equilibrium distribution constant at temperature T, and ΔG is free energy. Since, as in Eq. (5.45), it is also true that ΔG equals enthalpy (ΔH) less absolute temperature times entropy (ΔS), substitution of Eq. (5.44) into Eq. (5.45)

$$\Delta G = \Delta H - T\Delta S \tag{5.45}$$

$$RT \ln(K_c) = \Delta H - T\Delta S \tag{5.46}$$

yields the result presented below as Eq. (5.46): In Eq. (5.46) K_c is replaced by the product βk, rearrangement of the resulting equation yields Eq. (5.47), which expresses the dependence of temperature on k for any particular solute. For presentation purposes (i.e., charts and figures), Eq. (5.47) is often rearranged one more time to yield Eq. (5.48):

$$\ln(k) = \Delta H/RT - \Delta S/R - \ln(\beta) \tag{5.47}$$

$$-T \ln(k) = -\Delta H/R + T(\Delta S/R + \ln(\beta)) \tag{5.48}$$

For any range of T over which ΔH and ΔS remain constant, Eq. (5.48) yields a straight line of slope $[(\Delta S/R - \ln(\beta))]$ and intercept $\Delta H/R$. Thus, the effect of temperature on the k of any particular solute under isothermal conditions can be predicted with reasonable accuracy whenever accurate values of ΔH and ΔS are available. Although the calculation of such values is well beyond the scope of this work, it is possible to extract this information from a series of experimental measurements for each solute of interest under isothermal conditions at several different temperatures.

Under temperature-programmed conditions, T_c [and so T in Eq. (5.48)] varies as some function of time and so the concomitant variation in k must be properly taken into account. What makes this difficult is the fact that at any particular time and temperature [i.e., $T_c(t_i)$], Eq. (5.48) does not, in its present form, account for the degree to which a given solute has already been retained by the stationary phase at $T_c(t_{i-1})$, $T_c(t_{i-2})$, etc. Correcting for this requires derivation of an equation capable of integrating the effect of solute retention over the entire span of time it takes a particular solute to elute from the column.

$$k = (1/\beta)e^{(\Delta H/RT - \Delta S/R)} \tag{5.49}$$

where solving Eq. (5.48) for k gives Eq. (5.49), $(t_r - t_0)/t_0$ can be substituted for k to get Eq. (5.50):

$$(t_R - t_M)/t_M = (1/\beta)e^{(\Delta H/RT - \Delta S/R)} \tag{5.50}$$

$$t_R = t_M + (t_M/\beta)e^{(\Delta H/RT - \Delta S/R)} \tag{5.51}$$

$$dz/dt = L/t_R \tag{5.52}$$

In turn, solving Eq. (5.50) for t_R gives Eq. (5.51) which permits substitution of Eq. (5.51) into Eq. (5.52) relating the distance dt a solute travels along a particular column of length L in time dt. Rearrangement

$$dz/L = dt/[t_M + (t_M/\beta)e^{(\Delta H/RT - \Delta S/R)}] \tag{5.53}$$

after substitution gives Eq. (5.53), suggesting the subsequent integration in Eq. (5.54). Although Eq. (5.54) serves well when the right-hand side is integrated over time, the goal is to find an equation that can be integrated over temperature. This can be accomplished by changing the limits of integration, but not before doing something about $(1/t_M)$, which, as we know, also depends on temperature.

$$\int_0^z \frac{dz}{L} = \frac{1}{t_M} \int_0^{t_R} \frac{dt}{1 + \left(\dfrac{1}{\beta}\right)e^{(\Delta H/RT - \Delta S/R)}} \tag{5.54}$$

Because $t_M = L/u$ and $\bar{u} = F_0 j/\pi r^2$, $1/t_M = F_0 j/L\pi r^2$. A comparison of this result to Eq. (5.2) [repeated below as Eq. (5.55)] reveals that variation in t_0 with temperature depends solely on the temperature dependence of carrier gas viscosity provided F_0 is always measured at some standard temperature T_a [Eq. (5.56)]. This is important because it suggests that if t_M is accurately known at one temperature, it can be readily calculated at any other temperature [Eq. (5.57)].

$$F_0 = \left\{ \left[(p_i^2 - p_o^2) pr_c^4 \right] / [16\eta(T_c)p_oL] \right\}(T_a/T_c) \tag{5.55}$$

$$1/t_M = \left\{ \left[(p_i^2 - p_o^2) r_c^2 j \right] / [16\eta(T_c)p_oL^2] \right\}(T_a/T_c) \tag{5.56}$$

In Eq. (5.57), T_{co} is the column temperature at which $t_M(T_{co})$ is observed, T_{ci} is the column temperature at which $t_M(T_{ci})$ will be observed for the same column operated at constant p_i and p_o, and x assumes a value that depends on the choice of carrier gas. More specifically, x is equal to the values listed for B in Table 5.1.

From the temperature dependence of t_M in Eqs. (5.55)–(5.56), Eq. (5.54) can now be recast to permit the use of more appropriate limits of integration [Eq. (5.57)]. Integration of the left-hand side of Eq. (5.57) gives z/L, and integration of the right-hand side of Eq. (5.57) yields solute retention time under temperature-programmed conditions beginning at T_0 and ending at T_r (or the temperature at which the solute actually elutes).

$$t_M(T_{ci}) = t_M(T_{co})(T_{ci}/T_{co})^{(1+x)} \tag{5.57}$$

For multistep programs, use Eq. (5.58) for each segment during which column temperature is varied and use Eq. (5.54) for segments during which column temperature is held constant (Fig. 5.17). As neither Eq. (5.54) nor Eq. (5.57) have analytical solutions, their evaluation must typically be accomplished by less direct numerical means.

$$\int_0^z \frac{dz}{L} = \frac{1}{t_M(T_0)} \int_{T_0}^{T_R} \left(\frac{T_0}{T}\right)^{(1+x)} \frac{dT}{1 + \left(\frac{1}{\beta}\right) e^{(\Delta H/RT - \Delta S/R)}} \qquad (5.58)$$

5.22   Chromatographic Efficiency under Temperature-Programmed Conditions

Despite the temperature dependence of various factors affecting the height equivalent to a single theoretical plate in both the Golay and van Deemter equations, the width of any particular chromatographic peak in the absence of significant "extracolumn effects" will ultimately depend on how fast the solute corresponding to that peak exits the column and enters the detector. Consequently, accurate estimates of chromatographic efficiency for a particular solute can typically be obtained from Golay or van Deemter equations in which various temperature-dependent parameters assume values suggested by the temperature at which the same solute elutes under temperature programmed conditions (i.e., T_R). This is particularly true whenever the pressure drop across the column is small.

Although perhaps not intuitively obvious, this can be best understood by recalling the temperature dependence of k [Eq. (5.49)] and the nature of K_c as an equilibrium distribution constant [Eq. (5.44)]. As the temperature increases under temperature-programmed conditions, solute retention factors decrease and corresponding solute equilibrium distribution constants favor higher solute concentrations in the mobile phase. As a result, the rate at which solutes elute off the stationary phase and are swept out of the column is enhanced and the chromatographic efficiency of the column appears to improve.

Conversely, if the temperature is caused to decrease as a function of time during a temperature-programmed run, solute retention factors become larger and equilibrium distribution constants favor lower solute concentrations in the mobile phase. This impedes the rate at which solutes can be swept out of the column and so the chromatographic efficiency of the column appears to degrade.

When the pressure drop across the column is large, variation in the linear velocity of carrier gas as a function of distance from the column inlet [Eq. (5.12)] can become extreme. This complicates matters somewhat by requiring the need to consider the degree to the height equivalent to a single theoretical plate at T_r may gradually become a function of the linear velocity of carrier gas near the outlet rather than the average linear velocity of carrier gas through the column as a whole.

5.23   Changes in Solute Elution Order

A surprising number of inquiries from practicing chromatographers are concerned with the fact that the solute elution order of a given sample changed with the substitution of another column coated with the same stationary phase. Solute–stationary phase interactions that influence solute volatility are considered in Chapter 4; the nature of the solute (and its functional groups) would be expected to affect the degree to which a temperature increase altered those interactions and hence the degree to which the volatility of the solute was affected by the temperature increase.

In a column coated with a stationary phase capable of a high degree of hydrogen bonding, a short-chain alcohol will be retained and elute after a hydrocarbon of much greater chain length when the mixture is chromatographed on that column at some given temperature T_1. This is because the alcohol is more susceptible to hydrogen bonding. If the same mixture is chromatographed on the same column at a significantly higher temperature T_2, the degree of hydrogen bonding is reduced and the alcohol may now elute before the hydrocarbon. Hence, the elution order at T_1 is alcohol–hydrocarbon, and at T_2 it becomes hydrocarbon–alcohol.

In a programmed temperature separation that begins at a temperature at or below T_1 and proceeds at a steady rate to a final temperature at or above T_2, the elution order will be determined by the length of time the solutes are subjected to the chromatographic process (i.e., the temperature profile that they experience). Anything that reduced the analysis time (higher gas velocities, shorter columns, larger-diameter columns at constant d_f, thinner stationary phase films at constant column diameter) would discharge the solutes at lower temperatures and tend to preserve the elution order hydrocarbon–alcohol. Anything that increased the analysis time (lower gas velocities, longer columns, smaller-diameter columns at constant d_f, or thicker stationary phase films at constant column diameter) would retain the solutes until the program had advanced further; the solutes would be exposed to higher temperatures and would tend to elute in the order alcohol–hydrocarbon. Obviously, a different program rate or different initial or final temperatures could also affect elution order.

Analysis of a mixture containing the pesticides parathion and aldrin can serve as an example. In a series of isothermal analyses conducted on an apolar (DB-5) column, the elution order on this column under the operational conditions employed was aldrin–parathion at 185°C. The two solutes coeluted as a single peak at 190°C, and at 195°C the elution order was parathion–aldrin. In a programmed run, their elution order would depend on the factors discussed above.

With programmed temperature separations, mixtures containing dissimilar solutes not infrequently exhibit changes in the elution order of those solutes when changes are made in the length, diameter, or stationary phase film thickness of the column; in the carrier gas velocity (as happens when optimizing with hydrogen versus helium); in the pressure drop through the column; or in any of the program parameters.

References

1. L. S. Ettre, *Chromatographia* **18**:243 (1984).
2. S. J. Hawkes, *Chromatographia* **37**:399 (1993).
3. Y. S. Touloukian, S. C. Saxena, and P. Hestermans, *Thermo-physical Properties of Matter*, Vol. 11, Plenum Press, New York, 1970.
4. J. Kestin, K. Knierim, E. A. Mason, B. Najafi, S. T. Ro, and M. Waldman, *J. Phys. Chem. Ref. Data* **13**:229 (1984).
5. E. Vogel, T. Strehlow, J. Millat, and W. A. Wakeham, *Z. Phys. Chemie, Leipzig* **270**:1145 (1989).
6. A. T. James and A. J. P. Martin, *Biochem. J.* **50**:679 (1952).
7. G. Takeoka, H. M. Richard, M. F. Mehran, and W. Jennings, *J. High Res. Chromatogr.* **6**:145 (1983).
8. C. J. Giddings, *Dynamics of Chromatography*, Vol. 1, pp. 13–118, Dekker, Inc., New York, 1965.
9. R. C. Weast, *Handbook of Chemistry and Physics*, 64th ed., p. F-114, CRC Press, Boca Raton, Florida, 1984.
10. C. J. Giddings, *Dynamics of Chromatography*, Vol. 1, pp. 40–63, Dekker, New York, 1965.
11. E. N. Fuller and C. J. Giddings, *J. Gas Chromatogr.* **3**:222 (1965).
12. E. N. Fuller, P. F. Schettler, and J. C. Giddings, *Ind. Eng. Chem.* **58**:19 (1966).
13. R. H. Perry, D. W. Green, and J. O. Maloney (eds.), *Perry's Chemical Engineer's Handbook*, 6th ed., pp. 3–285, McGraw-Hill, New York, 1984.
14. J. M. Kong and S. J. Hawkes, *J. Chromatogr. Sci.* **14**:279 (1976).
15. D. F. Ingraham, C. F. Shoemaker, and W. Jennings, *J. High Res. Chromatogr.* **5**:227 (1982).
16. P. A. Leclercq and C. A. Cramers, *J. High Res. Chromatogr.* **8**:764 (1985).

CHAPTER 6

COLUMN SELECTION, INSTALLATION, AND USE

6.1 General Considerations

Column selection is all too often historically based: Someone once performed a certain separation on a given stationary phase, ergo that stationary phase should be used for that separation. Although the knowledge of separations that have been achieved in the past can provide some useful information, the conditions specified for those separations should usually be considered as points of departure. Progress in gas chromatography has been rapid, and the analyst cognizant of more recent developments can often produce results superior to those obtained in earlier works. Modern columns are more efficient and more inert. A wider range of lower-bleed stationary phases, available in different film thicknesses and column diameters, extends our capabilities to manipulate analytical temperatures and solute retention factors. Newly developed stationary phases permit greater selectivity through adjustments of the separation factors of key solutes. New techniques of injection may also be appealing, and some of these are practical only with bonded phase columns.

It can be useful to perform a preliminary separation on a short apolar column, swept to its upper temperature limit at a high program rate under conditions of high carrier gas velocities. Routes to improving the analysis can then be deduced from those preliminary results. If early-eluting components are bunched together, lowering the initial temperature and selecting a slower initial program rate will prove beneficial. Establishing the temperature necessary to produce well formed peaks for the final compounds indicates the necessary upper temperature limit. Lower program rates will improve overall separation, and gas velocities can be more nearly optimized at $\bar{u}_{opt}$ or OPGV for the most critical region. In short, some preliminary information in the form of an initial chromatographic separation is required as a point of departure from which one can postulate the effect(s) that increasing column length, decreasing column diameter, changing d_f, or changing the stationary phase will have on the various areas of that original chromatogram. Other changes, such as a multistep temperature program, changing the carrier gas velocity, or the use of electronic pressure control, may be indicated. The rationale in proceeding should be based on theoretical considerations that can be gleaned from Chapters 1, 4, and (especially) 5. As used in this section, those considerations have been complemented by long exposure to a vast number of chromatographic problems, questions, and observations from a multitude of users over a wide range of endeavor. It should also be pointed out that the installation of the column, including the type of seals and hanger design, can affect both the performance and the longevity of the column.

6.2   Selecting the Stationary Phase

Solute–Stationary Phase Incompatibility

Some columns have been specifically prepared for the separation of certain classes of solute and may be unsuitable for some other classes of solutes. This mismatch is sometimes attributable to incompatibility of the solutes and the stationary phase and at other times to the use of surface pretreatments or deactivation procedures that render that column unsatisfactory for certain solutes or types of analysis. The "Carbowax–amine-deactivated" (CAM) column can serve as an example. A strongly alkaline column is required for amine analysis. The tubing, which may or may not be deactivated [polyethylene glycol (PEG) is sometimes allowed to serve as its own deactivating agent], is usually flushed with aqueous potassium hydroxide (KOH), dried, and coated with Carbowax 20M. It is obvious that acidic solutes will be unable to negotiate these strongly alkaline columns. Columns of this type should always be flame sealed when not in use: KOH is hygroscopic, and Carbowax is water-soluble. Even under the best of conditions, the lifetime of a standard CAM column will be relatively short. While some crosslinked and surface-bonded forms of PEG do not dissolve in water, the strongly alkaline substrate inhibits the bonding reactions. Bonded CAM columns are not feasible within our present state of the art.

An analogous incompatibility can be demonstrated between columns coated with the "free-fatty-acid phase" (FFAP) or OV-351 stationary phases and even slightly alkaline solutes. These columns are usually prepared by refluxing Carbowax 20M with terephthalic acid under conditions where the terminal —OH groups of the polyethylene glycol condense with one of the carboxyl moieties of the difunctional acid, leaving the other carboxyl free. The resulting polymer is acidic, and alkaline solutes are unable to negotiate the column. Again, the acidic reaction inhibits bonding. In order to utilize nonextractable columns for acidic solutes, however, some users pretreat columns coated with one of the bonded polyethylene glycols to temporarily enhance their performance with slightly acidic solutes. Typically, the column is rinsed with a few milliliters of dichloromethane containing ~10 ppm FFAP or OV-351 (dilute phosphoric acid and acetic acid have also been employed) and dried under carrier flow to impart a slightly acidic reaction to the column. The eventual departure of the acidic residues dissolved in the bonded PEG phase is accelerated by exposure to higher temperatures or by prolonged use, and the procedure must be repeated.

In some cases, a typical "overloaded" peak actually testifies to incompatibility between the solute and the stationary phase. Overloaded peaks, which are discussed in Chapter 10, are characterized by a sloping peak front and a more abrupt drop at the rear of the peak. Such peaks are generated when the equilibrium K_c cannot be established because the solute vapor pressure is too low to permit vaporization of all of that solute not dissolved by the stationary phase [1]. The problem can be corrected by increasing the solute vapor pressure (higher temperature, which also lowers K_c) or by increasing the solute-dissolving capacity of the stationary phase (thicker films of stationary phase lower the amount that must be accommodated by the gas phase at the equilibrium K_c). When equal amounts of an extended homologous series are subjected to an isothermal separation, the later peaks are often overloaded. Vapor pressures of the higher-boiling (later-eluting) solutes limit the amount of those solutes that can be in the vapor phase. Solutes that are essentially insoluble in a given stationary phase are normally forced by that incompatibility to generate overloaded peaks on that phase. Even small amounts of free acids typically yield overload peaks on methyl siloxane phases (Fig. 4.1), because of the very limited solubility of acids in that phase.

Nonbonded Stationary Phases

If precise retention indices [2] are of major concern, the stationary phase has been partially defined by that requirement. Most precision indices utilize one of the polydimethylsiloxane stationary phases (OV-1, SE-30) as one standard and a polyethylene glycol such as Carbowax 20M for the other [3]. Indices determined on the former are, of course, far more consistent and usually expressed with greater precision. The PEG phases are based on *average* molecular weights [(Carbowax 20M, average molecular weight (MW_{av}) 20,000; range not specified), and batch-to-batch variations in the concentration and availability of functional groups are the rule rather than the exception. In addition, PEG reacts readily with oxygen (and with some solutes), and the retention characteristics of these columns often change with use and with their operational history. Because these changes can affect both the mode and degree of interaction with solutes, the retentions for apolar solutes relative to the retentions for polar solutes (*the basis for retention indices*) can vary with use and over time. Hence, retention indices for solutes on the PEG phases are usually more loosely specified.

Because of these considerations, retention index windows can be quite narrow ($+/-0.1\%$) on the polysiloxane phase and must be appreciably wider (e.g., $+/-0.5$ to 1%) on a PEG phase. Some bonding reactions require the incorporation of vinyl into the stationary phase oligimers prior to polymerization. Crosslinking consumes some of the vinyl moieties, and others participate in bonding with pretreated silanol groups on the silica surface. Slight batch-to-batch variations may occur in the extent of these two reactions, and there is also a slight possibility of residual vinyl functionality. These variations would affect the retention characteristics. Retention indices on bonded phase columns are consequently somewhat less precise and usually vary at least slightly from those obtained on the nonbonded counterpart. But even where retention characteristics are of primary importance, the advantages conferred by a bonded-phase column (see below) often outweigh the fact that retention windows must be broadened to employ the bonded phase.

In most practical situations and certainly whenever splitless or on-column injections will be employed, there are definite advantages to bonded phase columns. Not only will the bonded phase column resist the localized phase stripping that inevitably accompanies splitless and on-column injections, but nonvolatile and high-boiling residues can be cleansed from a bonded phase column by washing with an appropriate solvent. This "cleanability" helps in prolonging column lifetime and is an important preliminary step to column rejuvenation treatments that involve end-capping reactions (Chapter 10).

It is also good general practice to use the least polar stationary phase that is capable of producing the required separation. Apolar columns usually exhibit higher efficiencies (lower H), have lower minimum and higher maximum operating temperature limits, have longer lifetimes, and are more amenable to rejuvenation treatments (Chapter 10). Columns coated with bonded forms of polydimethylsiloxane or SE-54 have traditionally been the first choice of most experienced chromatographers, but the newer silarylenesiloxane-based polymers are also becoming favorites. In cases where separation on that column is inadequate, all factors that influence component separation (lower temperatures or lower program rates, gas flow set at $\bar{u}_{opt}$ for the average retention factors of the most critical solutes; Chapter 5) should be carefully optimized for the solutes in question. Only after the selectivity of the column under these optimized conditions has been proved inadequate is it advisable to consider using a more selective stationary phase. Increased selectivity often requires a stationary phase of greater polarity, and column efficiencies, high- and low-temperature tolerances, and general longevities usually vary inversely with the polarity of the stationary phase.

6.3   Stationary Phase Selectivity

In the early days of gas chromatography, judgments concerning the most appropriate stationary phase were entirely empirical and based largely on trial and error. Columns were prepared with almost anything from detergent chemicals to ill-defined lipid materials derived from various plant and animal sources. The occasional successes from these hit-or-miss efforts resulted in a bewildering array of poorly defined materials used as stationary phases, some of which endure to this day (e.g., the Apiezons and Carbowaxes, Ethofat, Fluorolube, the Igepals, THEED, Tergitol).

Selectivity decisions today are usually better formulated and based on qualitative judgments that try to relate the dispersive, acid–base, and dipole potentials of the solutes with those of specific stationary phases (Chapter 4). The hydrogen-bonding potential of alcohols, for example, would be exploited by the stationary phases in Table 4.1 in the order

$$PEG > \text{cyanopropyl} > \text{trifluoropropyl} \cong \text{methyl siloxane}$$

whereas the trifluoropropyl siloxane would exhibit improved selectivity for solutes that might be better differentiated on the basis of their dipole moments. The key word here is "differentiation"; although the hydrogen-bonding potential of the stationary phase would be beneficial in differentiating an alcohol from a hydrocarbon of similar boiling point, it would not help in differentiating 2-methyl-butanol from 3-methylbutanol, whose hydrogen-bonding characteristics would be quite similar.

The most elegant approach to stationary phase selectivity is based on predictions of the precise functionality that would be required to maximize the separation factors of all solutes in a given mixture, followed by the synthesis of that stationary phase if it does not already exist (e.g., [4]). Alternatively, dissimilar columns can be serially coupled, with or without "selectivity tuning," to achieve this goal (e.g., [5–7]). Some of these methods are discussed in Chapter 8.

6.4   Selecting the Column Diameter

It is generally advisable to use the smallest-diameter column with which the analyst is comfortable, commensurate with other facets of the analysis. Ease of operation and user friendliness vary inversely with column diameter. As the column diameter decreases, the system-posed challenges increase for both the analyst and the equipment. The 530 μm column is valuable to those first converting from packed to open tubular columns because it permits the transition to be accomplished by one simple step at a time rather than one precipitous leap of total commitment. These large-diameter open tubular columns still surpass packed columns in yielding improved component separation at higher sensitivities in shorter analysis times. Where maximum theoretical plate numbers are not required, the high-capacity large-diameter column also yields dividends in improved quantitation, both be-cause of its superior inertness and because it can accept the total injected sample without fractionation. The ruggedness of these columns is such that they can be used by unskilled technicians for process line analysis on the production floor. They are also useful in those applications where the lower gas flows of smaller-diameter columns could cause problems. These include some valve-switched analyses; the internal volumes of some valves are sufficiently large that they deteriorate the chromatographic performance of small-diameter columns, but work well with the higher flows of 530-μm columns. Similarly, most "purge and trap" systems that are widely used in some environmental analyses generate better results with these larger higher flow columns.

More recently, 450-μm columns became available. These are also high flow columns, and offer all the advantages of the 530-μm columns while generating higher efficiencies. These are described in greater detail in Chapter 7.

Where true capillary chromatography is desired, a smaller-diameter column will be required. In general, theoretical plate numbers vary inversely with column diameter, and the smaller the diameter of the column, the shorter the length of col-umn that will be required for a given degree of separation (see Table 6.1). The most widely used diameters in open tubular gas chromatography are 320 and 250 μm, because they are compatible with most modern chromatographs, most analysts can readily develop the skills to use these more demanding columns, and they combine reasonably high efficiencies with practical sample loading capacities. Columns of 200-, 180-, and 150-μm diameters are also available, and although they require a little more care, they can be very useful, particularly where the experienced chro-matographer requires improved resolution, and/or shorter analysis times. As the column diameter decreases into the 100- and 50-μm regions, the propensity for sample overloading and the difficulties of connecting the column with the injec-tor and the detector increase. It is difficult to interface 50- and 100-μm diameter columns to most commercial chromatographs, and their intelligent use usually involves a "learning curve." Nevertheless, highly skilled chromatographers can use both of these small-diameter columns to good advantage in selected cases.

6.5 Selecting the Column Length

If the column is destined for a specific analysis, the length can be selected to achieve the desired degree of resolution of that pair of solutes that is most challenging in the mixture. To utilize this route, some basic information will be required, such as the separation factor of those solutes (which is a function of the stationary phase and the temperature), the retention factor of the more retained solute of the pair (which can be regulated to some degree by manipulating the column temperature and the thickness of the stationary phase film; Table 1.1), and the degree to which those solutes must be resolved (Fig. 1.5). Equation (1.22) can then be employed to estimate N_{req}.

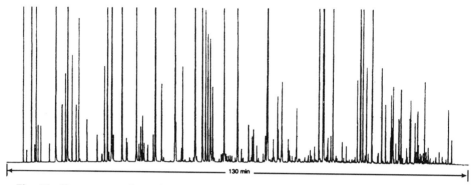

Fig. 6.1. Chromatogram of an unleaded gasoline on a $100\,m \times 0.25\,mm$ column, d_f $1.0\,\mu m$. One μL injection, split $1:300$. Oven, $0°C$ (15 min) to $50°C$ @ $1°C/min$, to $130°C$ @ $2°C/min$, to $180°C$ @ $4°C/min$, 20 min hold. The low starting temperature and $1.0\,\mu m$ d_f are required to hold and separate C_1, C_2s, and C_3s. See text for discussion.

For an all-round general-purpose column, 30 m is usually the most useful length. Shorter columns are frequently adequate for a given analysis, and because $\bar{u}_{opt}$, OPGV, and the "flatness" of the van Deemter curve all vary inversely with column length (Chapter 4), shorter columns should be used where possible.

One exception to this occurs in the analysis of gasolines. Gasolines are highly complex, and contain about 450 compounds, ranging from C_1 to C_{14}, plus aromatics. There are many groups of isomers and closely related compounds, which precludes improving their chromatographic differentiation by manipulating separation factors (α) or retention factors (k). We are left with the number of theoretical plates, N. Figure 6.1 shows an unleaded gasoline on the "DB-Petro 100 column," $100\,m \times 0.25\,mm$, d_f $0.5\,\mu m$, which generates some 500,000 theoretical plates. Obviously, these analyses require extended analysis times. With a few exceptions of this type, very long columns rarely serve a useful purpose.

A case in point is the 105-m column used in environmental purge-and-trap applications (Chapter 3), ostensibly to permit separation of gaseous components without resorting to cryogenic trapping. It has been suggested that this is accomplished because of the greater number of theoretical plates offered by the longer column. In fact, the advantage can be traced to the higher pressure at the head of the column necessitated by the high pressure drop across this column. As trapped components are desorbed into this area, they tend to expand, which increases the length of the starting bands. The longer column requires a higher pressure in this area, limits the expansion of the desorbing bands, and yields shorter starting bands. A similar separation can achieved by a 30-m column, if a short restrictor (10–20 cm of 50-μm fused silica tubing) is attached to its outlet end, thus forcing its operation at higher column head pressures [8].

Later decisions concerning which stationary phase should be used and the stationary phase film thickness will also play roles here:

1. *Stationary phase.* Apolar columns usually exhibit lower values of H_{min} and are therefore capable of generating more theoretical plates per meter than columns containing more polar stationary phases (DB-WAX is an exception to this generalization).

2. *Column diameter.* If all other factors are equal, H_{min} varies directly with the diameter of the column. The disadvantages of smaller-diameter columns must be weighed against their greater separation potential per unit length and are discussed in Section 6.5.

3. *Stationary phase film thickness.* Aspects of this were considered in connection with Table 1.1. The negative aspects are normally of little concern until d_f exceeds 0.4 μm and are also discussed in Chapter 5 and in Section 6.6.

6.6   Selecting the Stationary Phase Film Thickness

The advent of securely bonded phases has made available a wide range of stationary phase film thicknesses. For general-purpose use, a standard film thickness (0.2–0.35 μm) is usually preferred because both column efficiency and sample carrying capacity are reasonable, and retention factors of a wide range of solutes can be adjusted (via temperature) to yield good separation and high sensitivity. Thin-film columns ($d_f = 0.1 \mu$m) offer utility for very high-boiling solutes, but such columns are more easily overloaded and are in many cases less inert. Film thicknesses $\geq 1 \mu$m result in a significant increase in capacity and offer advantages in cold trapping of headspace volatiles [9, 10]. Retention factors become larger, which mandates either higher operating temperatures or a loss in sensitivity (Chapters 1 and 4). Much thicker-film columns (up to 8 μm) have recently become available. Although these show great utility for the analysis of very low-boiling solutes, they should never be regarded as general-purpose columns, because they sacrifice one of the major advantages of open tubular columns [11]. Theoretical plate numbers of thick-film columns are usually significantly lower, and for intermediate- or higher-k solutes the negative effects on separation, analysis times, and sensitivities can be substantial. Stationary phase diffusivity becomes increasingly important with thicker-film columns. In general, high-phenyl, cyanopropyl, and PEG phases are characterized by lower diffusivities and their use as thick films should be avoided [11, 12].

6.7   Column Installation

This section envisions installation of a column in an open tubular–compatible instrument that has previously been used for that purpose. Conversion of the packed column instrument to capillary capability, inlet and detector adaptation, and the desirability and selection of a makeup gas are all considered in Chapter 7.

The fused silica column should be installed in a configuration that does not subject the column to a high degree of strain. Strained areas are more subject to surface corrosion, static fatigue, and breakage. The importance of avoiding strained mounting configurations increases with column diameter, because the degree of strain is directly proportional to the diameter of the tubing for any given coiling diameter. It is also important that the column support hardware contact the column support cage and not the column itself. Most commercial support cages are electropolished to minimize abrasion of the polyimide coating. In cases where the column comes in contact with other support hardware, movement encouraged by air currents within the oven may lead to abrasion of the polyimide coating and breakage will occur at those points. The very low thermal mass of the fused silica column means that it will respond "instantaneously" to radiant heat. There should not be a "line of sight" between the oven heater and the column. On instruments where the column can "see" the heating element, an aluminum foil barrier, which can be suspended or draped over the column support cage, will prevent the peak splitting and "Christmas tree" effects (see Chapter 7) that can be caused by exposure to sporadic radiant heat.

After freeing approximately 30 cm of column from each end of the support cage, the cage should be hung on the supporting hardware with the freed ends oriented toward the inlet and detector. Ascertain the length required for one end to reach the inlet in an unstrained configuration, with the column terminating at the proper point in the inlet (see instructions of the inlet manufacturer or Chapter 3). In some cases, the distance the column should extend into the inlet beyond the base of the securing nut is specified. The proposed point of attachment (base of the ferrule) can be marked with typewriter correction fluid. When the column is held in that position, there should be no possibility of the column coming in contact with the oven wall. The column should now be cut at a point where it will be 2–4 cm too long.

A mounting nut and ferrule should now be slipped onto the column. For normal use, graphite ferrules are preferred, but for some special applications, graphitized Vespel is required (see below). Vespel ferrules are denser and less permeable, but because they do not compress as readily as the graphite ferrule, their tolerance for discrepancies between the diameter of the column and the size of the hole in the ferrule is quite small. The excess length (2–4 cm) can now be trimmed from the end of the column. This procedure eliminates the possibility of depositing graphite scrapings inside the column.

To cut the column, use a ceramic or silicon carbide wafer, a diamond file or a carbide or diamond pencil (a very sharp razor blade can be used less satisfactorily), and scribe lightly through the polyimide. It is not necessary to scratch the silica itself. Grasping the column on both sides of and as close to the scribe mark as possible, increase tension at that point by pulling and bending. The column should snap cleanly. Especially with the larger-diameter open tubular columns, harsher cutting methods (e.g., snipping with scissors) can result in lines of fracture that extend longitudinally through an appreciable length of the tubing and lead to repetitive breakage.

After examining the end of the column (at 10–20 × magnification) to verify a clean, square cut and the absence of chips of fused silica or polyimide, it is wise to activate the flow of carrier gas to flush from the inlet air that would otherwise be forced through the column. The inlet end can be secured while the carrier is still flowing. With graphite ferrules the nut should normally be tightened only one-half turn beyond the fingertight position.

At this point it is advisable to insert the still unsecured detector end of the column into a small beaker of solvent and verify that the carrier is flowing through the column. A flow of approximately 2 cm^3/min is desirable. This flow can be continued during the rest of the installation procedure and should be continued for at least 30–40 min before raising the oven temperature.

The length of column necessary to reach to the detector in a smooth and un-strained configuration with the column end terminating at the proper point within the sensing zone (see below) should now be determined. Again, the column should be cut so that it is 2–4 cm too long, the securing nut and ferrule installed before the column is cut to the proper length, and the end examined carefully under magnification.

Several factors determine the ideal location of the end of the column within the detector. It is theoretically desirable to detect components at the instant they emerge from the column and to remove them immediately after detection. This would preserve the separation achieved by the column on closely eluting solutes. In practice, some degree of compromise is usually required. Columns that are positioned too close to the flame in a flame ionization detector (FID) sometimes suffer pyrolysis or carbonization of both the stationary phase and the outer polyimide coating on the final few millimeters of column. The pyrolyzed material can be extremely active, leading to reversible adsorption of some solutes (tailing peaks) and irreversible adsorption of others (total or partial component abstraction). Earlier recommendations sometimes suggested that where possible, the column should be inserted completely through the jet, then pushed back so that it is flush with the jet orifice, and finally retracted an additional 2 mm. Retracting it farther than this can expose solutes to active sites, whose presence in both metal and quartz flame jets is readily demonstrable. It should be pointed out that when inserting the column into or through narrow portions of the flame jet, there is a danger of leaving deposits (e.g., scrapings of polyimide) in on the jet, which could have serious consequences.

Some older instruments utilize transfer lines to conduct the eluted solutes from the end of the column to the detector. Where possible, the potential activity of that configuration should be circumvented by extending the column completely through the transfer line. Other instruments incorporate sharp bends at the detector connection or restrictions at or within the flame jet that prohibit ideal positioning of the column end. Positioning is also critical with other types of detector. The column should extend as far as possible into the detector to eliminate as many active sites as possible, but a loss in sensitivity results if it is positioned too deeply in the electron capture (EC), nitrogen/phosphorus (NP), or flame photometric (FP) detector or in the ion source of a mass spectrometer.

An electronic leak detector is the most satisfactory means of verifying the integrity of column connections. Where leak test solutions are employed, they are invariably aspirated into the system, where they contaminate the column, a problem that is exacerbated by surface activity in those solutions. Test solution residues that contaminate the column and/or the fittings can generate anomalous results, including baseline drift and high-sensitivity "noise." A column that has been leak-tested with one of the proprietary formulations may give satisfactory service with FID, but its use with NPD may be forever precluded. If a solution must be employed, it should be limited to volatile solvents that are compatible with the stationary phase and that produce a minimum detector response.

Before the oven temperature is increased, the detector should be activated and a few microliters of an extremely volatile solute injected to verify that carrier gas is flowing through the column at a reasonable rate. With FID, methane (or propane/butane from a disposable cigarette lighter) is suitable; air, difluorodichloromethane (Freon 12), or sulfur hexafluoride can be used with ECD; air can also be employed with GC/MS. Heating the column in the absence of carrier flow can seriously damage and may destroy the column. At this low temperature, the carrier gas velocity should be at least twice the value desired for $\bar{u}_{opt}$ at some higher temperature. This high flow also helps ensure that all air is flushed from both the inlet and the column before the temperature is increased. Unless the methane peak is needle-sharp, that problem should be corrected before proceeding further. The column securing nut at the detector should be loosened slightly, the column retracted another 1–2 mm, and the methane injection repeated. If the methane peak still remains broad or tails, the flow rate through the inlet (e.g., the split ratio) should be rechecked. The position, seal, and unbroken state of the inlet liner should be verified. The position of the column in the inlet should also be rechecked.

Most manufacturers supply a test chromatogram with the column, and it is wise to inject the same test sample under the manufacturer's test conditions, specifically to duplicate the test chromatogram. It is not uncommon to find that an active solute exhibits severe tailing that was not demonstrated on the test chromatogram. This alerts the user to the fact that there are extracolumn sources of activity in the system, most probably in the inlet but possibly in the detector (see Chapter 10).

Other installation systems have been suggested. One commercially available system utilizes spring pressure, which is released by means of a lever, to compress into a fitting a standard graphite ferrule that is first properly positioned on the column end (Fig. 6.2).

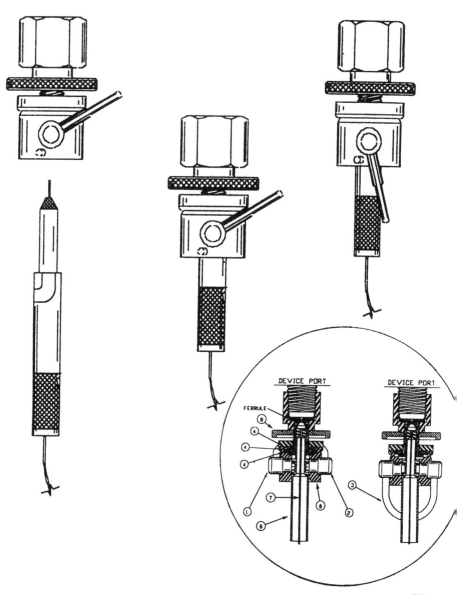

Fig. 6.2. A "Quick Connector" for column installation, using standard graphitized Vespel™ ferrules. *Courtesy of Quadrex, Inc.*

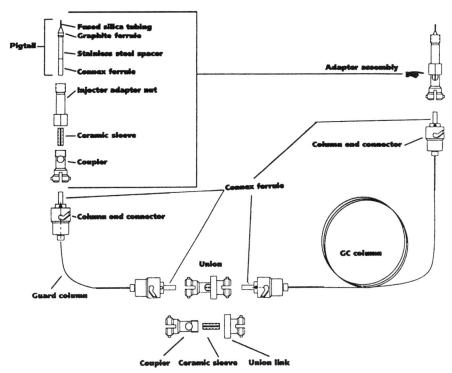

Fig. 6.3. The "Connex" connector. Based on spring-loaded abutment of two zirconia ceramic ferrules, various versions are available: column to injector and/or detector; column to retention gap or guard column; column to MS transfer line; and a closure for capping the MS transfer line during column changing.

Another device utilizes zirconia ferrules, which, butted together under slight spring pressure, deform to produce an inert leaktight seal (Fig. 6.3). The latter offers a quick connect–disconnect that is useful for the installation and replacement of retention gaps, and that permits the GC/MS operator to leave the fused silica transfer line in position while replacing the column. Future developments may well lead to a "column cassette" in which the column ends terminate in slightly protruding and prepositioned zirconia ferrules. The instrument could be designed to automatically load the cassette (much like a VCR), butting the column ferrules to matching ferrules in the instrument.

6.8   Column Conditioning

Packed columns normally require extensive high-temperature conditioning. Open tubular columns rarely need the same degree of conditioning and may suffer damage when subjected to well-intentioned but ill-advised treatments.

Following installation and after all air is purged from the system, the column should be conditioned briefly at a temperature slightly above the maximum temperature at which it will be used. Exposure to higher temperatures places the column under stress and shortens the lifetime. Such treatments should be avoided unless they serve some useful purpose. Samples that are run at moderate temperatures may also contain higher-boiling solutes that gradually accumulate on the column, and a periodic higher-temperature hold may be the easiest method to remove those residues from the column. All stationary phases are susceptible to stress from both oxygen and exposure to high temperatures. The two stresses may have a synergistic effect on the stationary phase—levels of oxygen that could be tolerated for several weeks at a lower temperature may destroy a column in a few minutes at higher temperatures. The most oxygen tolerant stationary phases are the new polysilphenylenesiloxanes and related phases described in Chapter 4. With this exception, the oxygen tolerance of stationary phases in general decreases in the order

$$\text{Polymethylsiloxane} \gg \text{phenylsiloxane} > \text{cyanopropyl}$$
$$= \text{trifluoropropylsiloxane} \gg \text{PEG}$$

With oxygen-free carrier, the high-methyl polysiloxanes are usable at temperatures of 340–350°C, but even traces of oxygen can cause problems at temperatures above 200°C, in that C—Si bonds projecting from the polysiloxane backbone cleave to yield —OH (silanol) groups [13]. Free-silanol groups lead to a temperature-dependent degradation of the polysiloxane polymer [14–17] that will persist even in the absence of oxygen contamination. Methods that have been used to rectify this problem are described later in Chapter 10.

6.9 Optimizing Operational Parameters for Specific Columns

The graphs in Figs. 6.4–6.81 were generated to permit optimization of specific systems. In most cases, the analyst has a column on hand and wishes to optimize the operational conditions. Although that column has a certain length, there may be an advantage to employing a shorter segment of column. The column diameter, on the other hand, is usually fixed. Hence, for a given column, the stationary phase film thickness and the type of stationary phase are also fixed. The most appropriate curves are probably best selected by proceeding in the following order: (1) the diameter of the column to be used, (2) the length of that column available, (3) the stationary phase film thickness, (4) the type of stationary phase, (5) the carrier gas employed, and (6) the retention factors of the solutes for which the best separation is desired. The influence of the solute k on $\bar{u}_{\text{opt}}$ is most pronounced for low values of k. For $k < 0.5$, optimum velocities are significantly higher than for $k = 2$. The difference between $k = 2$ and $k = 5$ covers a much smaller range, and at values of $k > 5$, the variation can be ignored (Fig. 1.10).

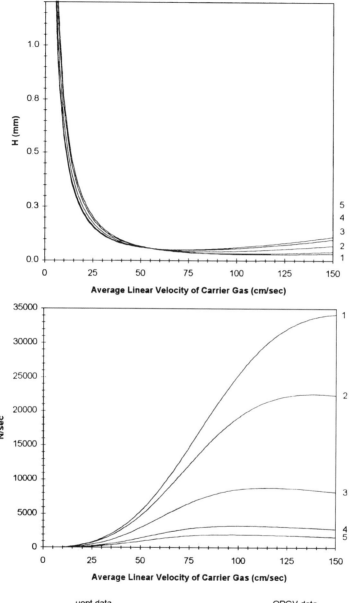

Plot	k	Hmin (mm)	uopt (cm/s)	N	time (s)	OPGV (cm/s)	N/s (max)	N	time (s)

uopt data   OPGV data

Plot	k	Hmin (mm)	uopt (cm/s)	N	time (s)	OPGV (cm/s)	N/s (max)	N	time (s)
1	0.399	0.028	113.9	181,574	6.14	153.5	34,087	155,341	4.56
2	0.649	0.032	102.5	156,504	8.04	138.6	22,467	133,626	5.95
3	1.717	0.041	84.9	121,108	16.00	115.4	8,764	103,208	11.78
4	4.542	0.048	74.6	104,115	37.13	101.4	3,247	88,747	27.33
5	7.387	0.050	71.4	99,955	58.74	96.9	1,971	85,234	43.25

Fig. 6.4. Column: d_c 50 μm, L 5 m, d_f 0.2 μm; hydrogen carrier. Values assume ideality and are calculated for isothermal conditions. Values of H in mm, u_{opt} and OPGV in cm/sec. Values for "time" are values of t_R for that solute calculated at u_{opt} or OPGV.

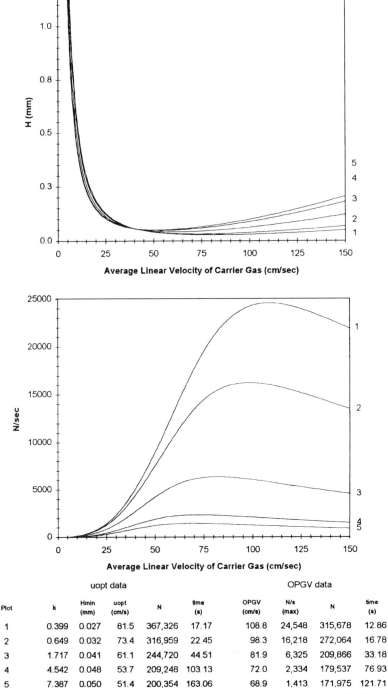

Plot	k	Hmin (mm)	uopt (cm/s)	N	time (s)	OPGV (cm/s)	N/s (max)	N	time (s)
				uopt data			OPGV data		
1	0.399	0.027	81.5	367,326	17.17	108.8	24,548	315,678	12.86
2	0.649	0.032	73.4	316,959	22.45	98.3	16,218	272,064	16.78
3	1.717	0.041	61.1	244,720	44.51	81.9	6,325	209,866	33.18
4	4.542	0.048	53.7	209,248	103.13	72.0	2,334	179,537	76.93
5	7.387	0.050	51.4	200,354	163.06	68.9	1,413	171,975	121.71

Fig. 6.5. Column: d_c 50 μm, L 10 m, d_f 0.2 μm; hydrogen carrier. Values assume ideality and are calculated for isothermal conditions. Values of H in mm, u_{opt} and OPGV in cm/sec. Values for "time" are values of t_R for that solute calculated at u_{opt} or OPGV.

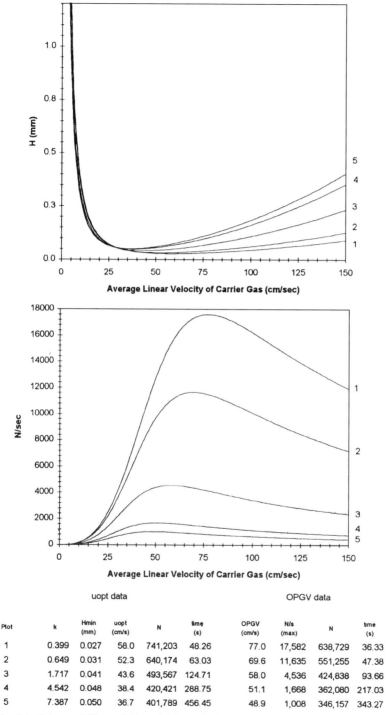

Plot	k	Hmin (mm)	uopt (cm/s)	N	time (s)	OPGV (cm/s)	N/s (max)	N	time (s)
				uopt data			OPGV data		
1	0.399	0.027	58.0	741,203	48.26	77.0	17,582	638,729	36.33
2	0.649	0.031	52.3	640,174	63.03	69.6	11,635	551,255	47.38
3	1.717	0.041	43.6	493,567	124.71	58.0	4,536	424,838	93.66
4	4.542	0.048	38.4	420,421	288.75	51.1	1,668	362,080	217.03
5	7.387	0.050	36.7	401,789	456.45	48.9	1,008	346,157	343.27

Fig. 6.6. Column: d_c 50 μm, L 20 m, d_f 0.2 μm; hydrogen carrier. Values assume ideality and are calculated for isothermal conditions. Values of H in mm, u_{opt} and OPGV in cm/sec. Values for "time" are values of t_R for that solute calculated at u_{opt} or OPGV.

171

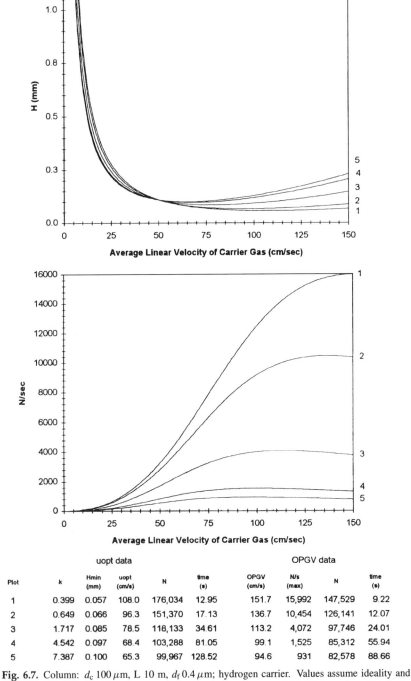

Plot	k	H_{min} (mm)	u_{opt} (cm/s)	N	time (s)	OPGV (cm/s)	N/s (max)	N	time (s)
		uopt data				OPGV data			
1	0.399	0.057	108.0	176,034	12.95	151.7	15,992	147,529	9.22
2	0.649	0.066	96.3	151,370	17.13	136.7	10,454	126,141	12.07
3	1.717	0.085	78.5	118,133	34.61	113.2	4,072	97,746	24.01
4	4.542	0.097	68.4	103,288	81.05	99.1	1,525	85,312	55.94
5	7.387	0.100	65.3	99,967	128.52	94.6	931	82,578	88.66

Fig. 6.7. Column: d_c 100 μm, L 10 m, d_f 0.4 μm; hydrogen carrier. Values assume ideality and are calculated for isothermal conditions. Values of H in mm, u_{opt} and OPGV in cm/sec. Values for "time" are values of t_R for that solute calculated at u_{opt} or OPGV.

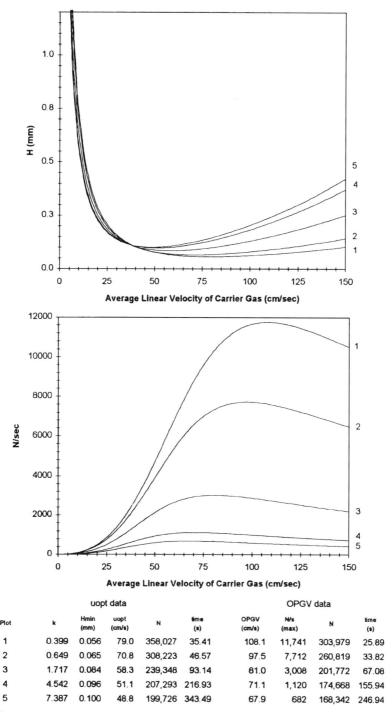

		uopt data					OPGV data		
Plot	k	Hmin (mm)	uopt (cm/s)	N	time (s)	OPGV (cm/s)	N/s (max)	N	time (s)
1	0.399	0.056	79.0	358,027	35.41	108.1	11,741	303,979	25.89
2	0.649	0.065	70.8	308,223	46.57	97.5	7,712	260,819	33.82
3	1.717	0.084	58.3	239,348	93.14	61.0	3,008	201,772	67.08
4	4.542	0.096	51.1	207,293	216.93	71.1	1,120	174,668	155.94
5	7.387	0.100	48.8	199,726	343.49	67.9	682	168,342	246.94

Fig. 6.8. Column: d_c 100 μm, L 20 m, d_f 0.4 μm; hydrogen carrier. Values assume ideality and are calculated for isothermal conditions. Values of H in mm, u_{opt} and OPGV in cm/sec. Values for "time" are values of t_R for that solute calculated at u_{opt} or OPGV.

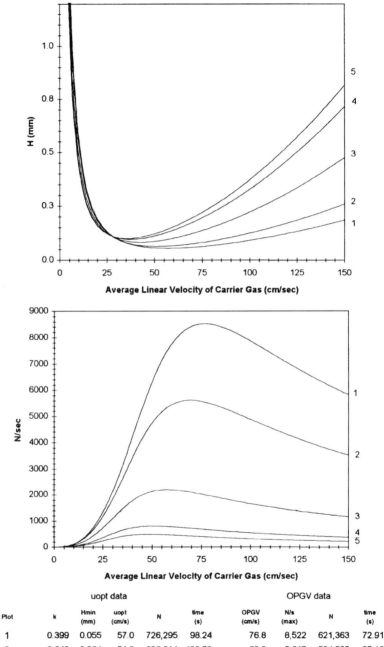

		uopt data					OPGV data		
Plot	k	Hmin (mm)	uopt (cm/s)	N	time (s)	OPGV (cm/s)	N/s (max)	N	time (s)
1	0.399	0.055	57.0	726,295	98.24	76.8	8,522	621,363	72.91
2	0.649	0.064	51.2	626,014	128.72	69.3	5,617	534,505	95.16
3	1.717	0.083	42.5	484,432	255.94	57.7	2,191	412,832	188.42
4	4.542	0.096	37.3	416,458	594.14	50.7	812	354,986	437.29
5	7.387	0.100	35.7	399,819	939.87	48.5	493	340,936	692.06

Fig. 6.9. Column: d_c 100 μm, L 40 m, d_f 0.4 μm; hydrogen carrier. Values assume ideality and are calculated for isothermal conditions. Values of H in mm, u_{opt} and OPGV in cm/sec. Values for "time" are values of t_R for that solute calculated at u_{opt} or OPGV.

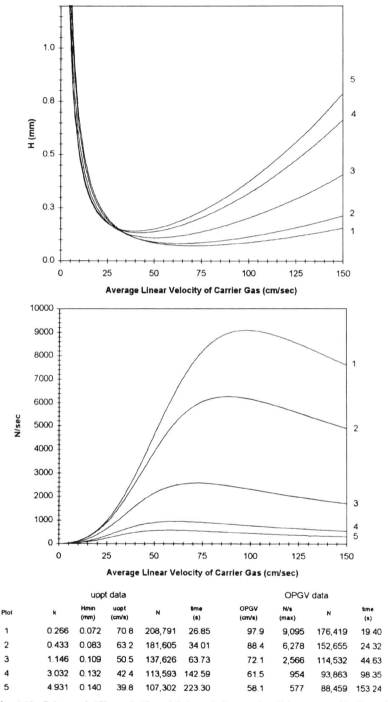

		uopt data				OPGV data			
Plot	k	Hmin (mm)	uopt (cm/s)	N	time (s)	OPGV (cm/s)	N/s (max)	N	time (s)
1	0.266	0.072	70.8	208,791	26.85	97.9	9,095	176,419	19.40
2	0.433	0.083	63.2	181,605	34.01	88.4	6,278	152,655	24.32
3	1.146	0.109	50.5	137,626	63.73	72.1	2,566	114,532	44.63
4	3.032	0.132	42.4	113,593	142.59	61.5	954	93,863	98.35
5	4.931	0.140	39.8	107,302	223.30	58.1	577	88,459	153.24

Fig. 6.10. Column: d_c 150 μm, L 15 m, d_f 0.4 μm; helium carrier. Values assume ideality and are calculated for isothermal conditions. Values of H in mm, u_{opt} and OPGV in cm/sec. Values for "time" are values of t_R for that solute calculated at u_{opt} or OPGV.

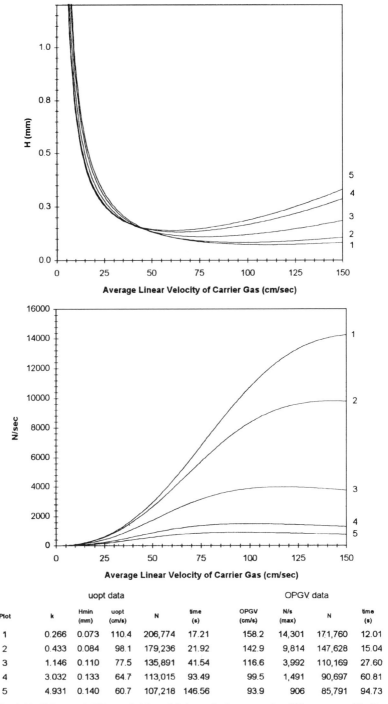

			uopt data				OPGV data		
Plot	k	Hmin (mm)	uopt (cm/s)	N	time (s)	OPGV (cm/s)	N/s (max)	N	time (s)
1	0.266	0.073	110.4	206,774	17.21	158.2	14,301	171,760	12.01
2	0.433	0.084	98.1	179,236	21.92	142.9	9,814	147,628	15.04
3	1.146	0.110	77.5	135,891	41.54	116.6	3,992	110,169	27.60
4	3.032	0.133	64.7	113,015	93.49	99.5	1,491	90,697	60.81
5	4.931	0.140	60.7	107,218	146.56	93.9	906	85,791	94.73

Fig. 6.11. Column: d_c 150 μm, L 15 m, d_f 0.4 μm; hydrogen carrier. Values assume ideality and are calculated for isothermal conditions. Values of H in mm, u_{opt} and OPGV in cm/sec. Values for "time" are values of t_R for that solute calculated at u_{opt} or OPGV.

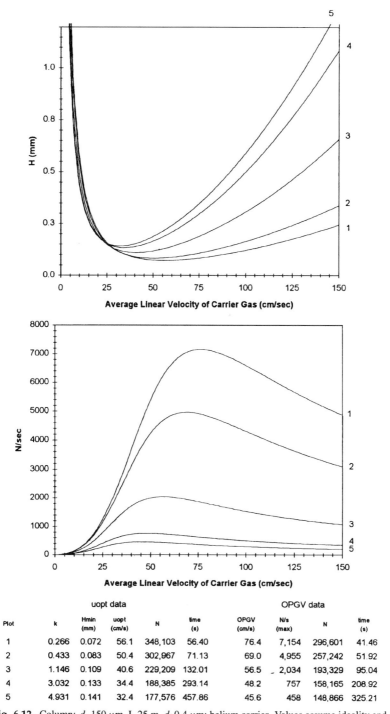

Plot	k	Hmin (mm)	uopt (cm/s)	N	time (s)	OPGV (cm/s)	N/s (max)	N	time (s)
			uopt data					OPGV data	
1	0.266	0.072	56.1	348,103	56.40	76.4	7,154	296,601	41.46
2	0.433	0.083	50.4	302,967	71.13	69.0	4,955	257,242	51.92
3	1.146	0.109	40.6	229,209	132.01	56.5	2,034	193,329	95.04
4	3.032	0.133	34.4	188,385	293.14	48.2	757	158,165	208.92
5	4.931	0.141	32.4	177,576	457.86	45.6	458	148,866	325.21

Fig. 6.12. Column: d_c 150 μm, L 25 m, d_f 0.4 μm; helium carrier. Values assume ideality and are calculated for isothermal conditions. Values of H in mm, u_{opt} and OPGV in cm/sec. Values for "time" are values of t_R for that solute calculated at u_{opt} or OPGV.

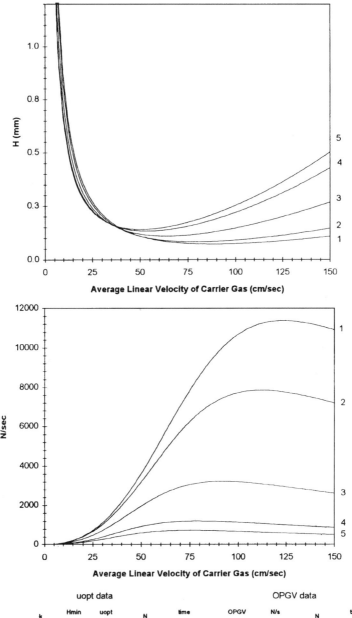

		uopt data				OPGV data			
Plot	k	Hmin (mm)	uopt (cm/s)	N	time (s)	OPGV (cm/s)	N/s (max)	N	time (s)
1	0.266	0.072	88.9	345,009	35.61	123.8	11,363	290,608	25.57
2	0.433	0.083	79.5	299,425	45.06	112.0	7,840	250,751	31.98
3	1.146	0.110	63.7	226,662	84.24	91.7	3,211	187,786	58.48
4	3.032	0.133	53.7	187,596	187.83	78.5	1,201	154,351	128.48
5	4.931	0.141	50.5	177,535	293.61	74.2	729	145,779	199.93

Fig. 6.13. Column: d_c 150 μm, L 25 m, d_f 0.4 μm; hydrogen carrier. Values assume ideality and are calculated for isothermal conditions. Values of H in mm, u_{opt} and OPGV in cm/sec. Values for "time" are values of t_R for that solute calculated at u_{opt} or OPGV.

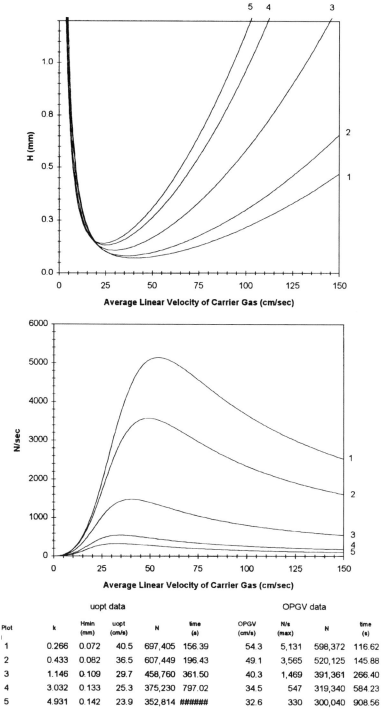

Fig. 6.14. Column: d_c 150 μm, L 50 m, d_f 0.4 μm; helium carrier. Values assume ideality and are calculated for isothermal conditions. Values of H in mm, u_{opt} and OPGV in cm/sec. Values for "time" are values of t_R for that solute calculated at u_{opt} or OPGV.

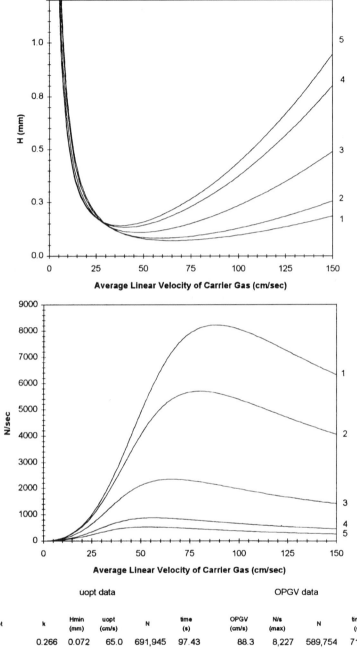

			uopt data				OPGV data		
Plot	k	Hmin (mm)	uopt (cm/s)	N	time (s)	OPGV (cm/s)	N/s (max)	N	time (s)
1	0.266	0.072	65.0	691,945	97.43	88.3	8,227	589,754	71.69
2	0.433	0.083	58.5	601,347	122.51	80.1	5,706	510,777	89.52
3	1.146	0.110	47.5	454,308	226.11	65.8	2,351	383,517	163.13
4	3.032	0.134	40.4	373,658	499.35	56.4	880	314,249	357.25
5	4.931	0.142	38.1	352,469	778.04	53.4	533	296,112	555.21

Fig. 6.15. Column: d_c 150 μm, L 50 m, d_f 0.4 μm; hydrogen carrier. Values assume ideality and are calculated for isothermal conditions. Values of H in mm, u_{opt} and OPGV in cm/sec. Values for "time" are values of t_R for that solute calculated at u_{opt} or OPGV.

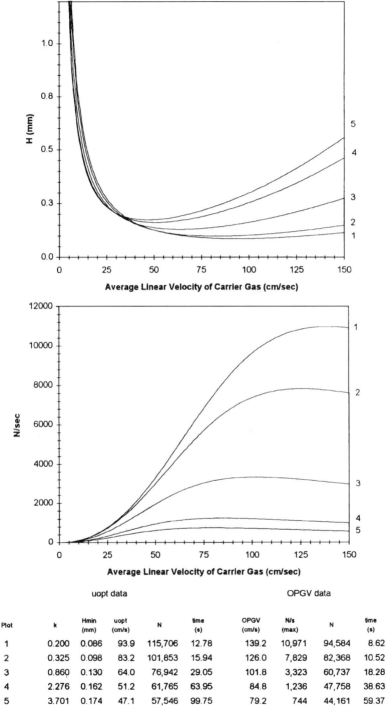

Plot	k	Hmin (mm)	uopt (cm/s)	N	time (s)	OPGV (cm/s)	N/s (max)	N	time (s)
1	0.200	0.086	93.9	115,706	12.78	139.2	10,971	94,584	8.62
2	0.325	0.098	83.2	101,853	15.94	126.0	7,829	82,368	10.52
3	0.860	0.130	64.0	76,942	29.05	101.8	3,323	60,737	18.28
4	2.276	0.162	51.2	61,765	63.95	84.8	1,236	47,758	38.63
5	3.701	0.174	47.1	57,546	99.75	79.2	744	44,161	59.37

Fig. 6.16. Column: d_c 200 μm, L 10 m, d_f 0.4 μm; helium carrier. Values assume ideality and are calculated for isothermal conditions. Values of H in mm, u_{opt} and OPGV in cm/sec. Values for "time" are values of t_R for that solute calculated at u_{opt} or OPGV.

181

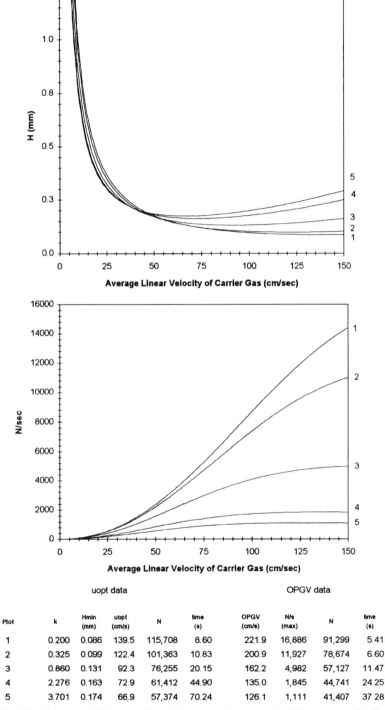

Fig. 6.17. Column: d_c 200 μm, L 10 m, d_f 0.4 μm; hydrogen carrier. Values assume ideality and are calculated for isothermal conditions. Values of H in mm, u_{opt} and OPGV in cm/sec. Values for "time" are values of t_R for that solute calculated at u_{opt} or OPGV.

Plot	k	Hmin (mm)	uopt (cm/s)	N	time (s)	OPGV (cm/s)	N/s (max)	N	time (s)
1	0.200	0.086	139.5	115,708	8.60	221.9	16,886	91,299	5.41
2	0.325	0.099	122.4	101,363	10.83	200.9	11,927	78,674	6.60
3	0.860	0.131	92.3	76,255	20.15	162.2	4,982	57,127	11.47
4	2.276	0.163	72.9	61,412	44.90	135.0	1,845	44,741	24.25
5	3.701	0.174	66.9	57,374	70.24	126.1	1,111	41,407	37.28

182

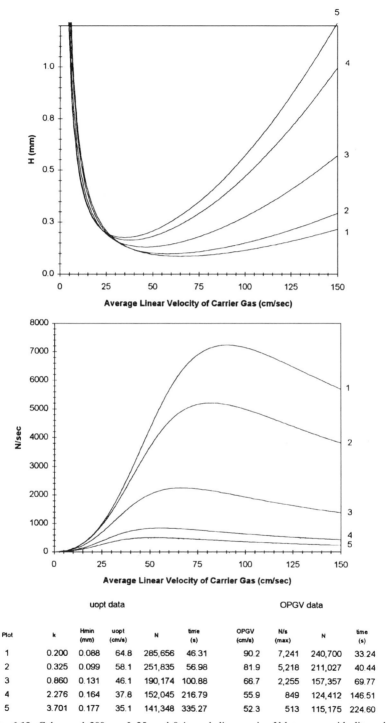

Plot	k	Hmin (mm)	uopt (cm/s)	N	time (s)	OPGV (cm/s)	N/s (max)	N	time (s)
1	0.200	0.088	64.8	285,656	46.31	90.2	7,241	240,700	33.24
2	0.325	0.099	58.1	251,835	56.98	81.9	5,218	211,027	40.44
3	0.860	0.131	46.1	190,174	100.88	66.7	2,255	157,357	69.77
4	2.276	0.164	37.8	152,045	216.79	55.9	849	124,412	146.51
5	3.701	0.177	35.1	141,348	335.27	52.3	513	115,175	224.60

Fig. 6.18. Column: d_c 200 μm, L 25 m, d_f 0.4 μm; helium carrier. Values assume ideality and are calculated for isothermal conditions. Values of H in mm, u_{opt} and OPGV in cm/sec. Values for "time" are values of t_R for that solute calculated at u_{opt} or OPGV.

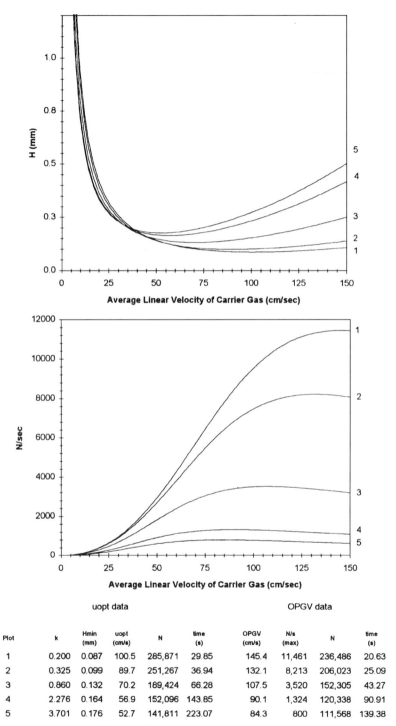

Plot	k	Hmin (mm)	uopt (cm/s)	N	time (s)	OPGV (cm/s)	N/s (max)	N	time (s)
			uopt data					OPGV data	
1	0.200	0.087	100.5	285,871	29.85	145.4	11,461	236,486	20.63
2	0.325	0.099	89.7	251,267	36.94	132.1	8,213	206,023	25.09
3	0.860	0.132	70.2	189,424	66.28	107.5	3,520	152,305	43.27
4	2.276	0.164	56.9	152,096	143.85	90.1	1,324	120,338	90.91
5	3.701	0.176	52.7	141,811	223.07	84.3	800	111,568	139.38

Fig. 6.19. Column: d_c 200 μm, L 25 m, d_f 0.4 μm; hydrogen carrier. Values assume ideality and are calculated for isothermal conditions. Values of H in mm, u_{opt} and OPGV in cm/sec. Values for "time" are values of t_R for that solute calculated at u_{opt} or OPGV.

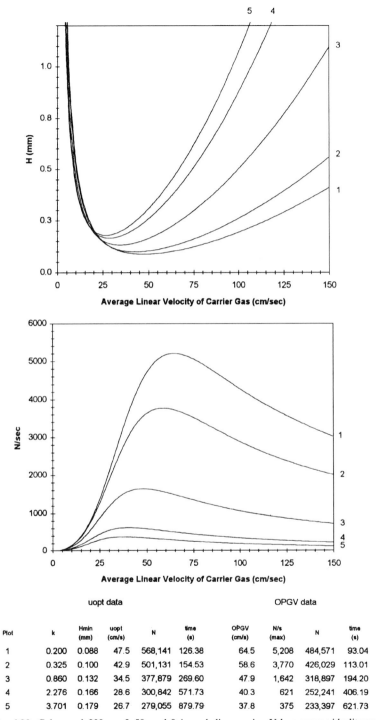

| | | uopt data | | | | | OPGV data | | |
Plot	k	Hmin (mm)	uopt (cm/s)	N	time (s)	OPGV (cm/s)	N/s (max)	N	time (s)
1	0.200	0.088	47.5	568,141	126.38	64.5	5,208	484,571	93.04
2	0.325	0.100	42.9	501,131	154.53	58.6	3,770	426,029	113.01
3	0.860	0.132	34.5	377,879	269.60	47.9	1,642	318,897	194.20
4	2.276	0.166	28.6	300,842	571.73	40.3	621	252,241	406.19
5	3.701	0.179	26.7	279,055	879.79	37.8	375	233,397	621.73

Fig. 6.20. Column: d_c 200 μm, L 50 m, d_f 0.4 μm; helium carrier. Values assume ideality and are calculated for isothermal conditions. Values of H in mm, u_{opt} and OPGV in cm/sec. Values for "time" are values of t_R for that solute calculated at u_{opt} or OPGV.

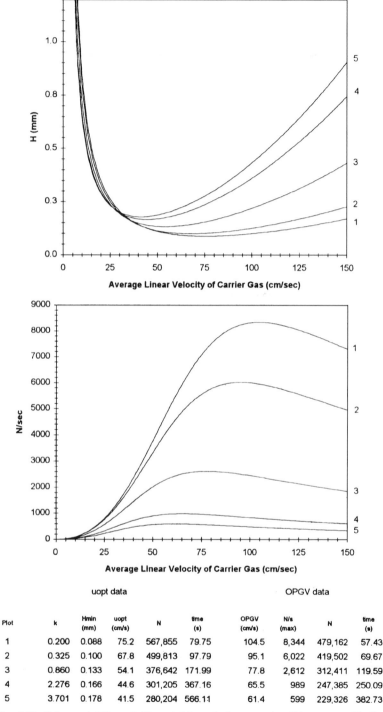

Plot	k	H_min (mm)	uopt (cm/s)	N	time (s)	OPGV (cm/s)	N/s (max)	N	time (s)
			uopt data				OPGV data		
1	0.200	0.088	75.2	567,855	79.75	104.5	8,344	479,162	57.43
2	0.325	0.100	67.8	499,813	97.79	95.1	6,022	419,502	69.67
3	0.860	0.133	54.1	376,642	171.99	77.8	2,612	312,411	119.59
4	2.276	0.166	44.6	301,205	367.16	65.5	989	247,385	250.09
5	3.701	0.178	41.5	280,204	566.11	61.4	599	229,326	382.73

Fig. 6.21. Column: d_c 200 μm, L 50 m, d_f 0.4 μm; hydrogen carrier. Values assume ideality and are calculated for isothermal conditions. Values of H in mm, u_{opt} and OPGV in cm/sec. Values for "time" are values of t_R for that solute calculated at u_{opt} or OPGV.

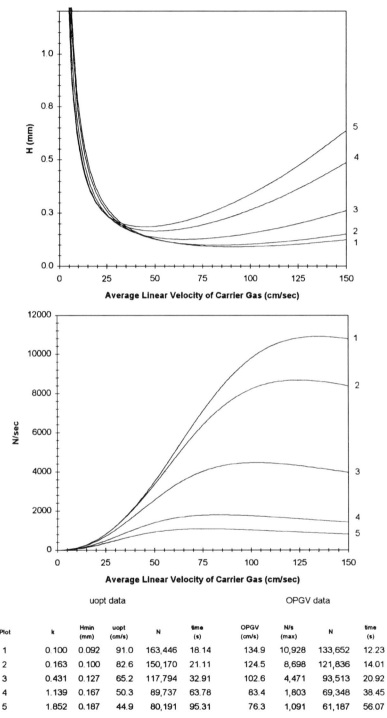

Plot	k	Hmin (mm)	uopt (cm/s)	N	time (s)	OPGV (cm/s)	N/s (max)	N	time (s)
			uopt data					OPGV data	
1	0.100	0.092	91.0	163,446	18.14	134.9	10,928	133,652	12.23
2	0.163	0.100	82.6	150,170	21.11	124.5	8,698	121,836	14.01
3	0.431	0.127	65.2	117,794	32.91	102.6	4,471	93,513	20.92
4	1.139	0.167	50.3	89,737	63.78	83.4	1,803	69,348	38.45
5	1.852	0.187	44.9	80,191	95.31	76.3	1,091	61,187	56.07

Fig. 6.22. Column: d_c 250 μm, L 15 m, d_f 0.25 μm; helium carrier. Values assume ideality and are calculated for isothermal conditions. Values of H in mm, u_{opt} and OPGV in cm/sec. Values for "time" are values of t_R for that solute calculated at u_{opt} or OPGV.

187

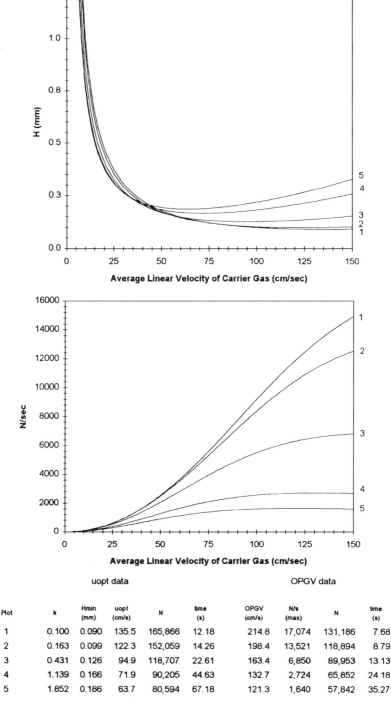

Plot	k	Hmin (mm)	uopt (cm/s)	N	time (s)	OPGV (cm/s)	N/s (max)	N	time (s)
			uopt data					OPGV data	
1	0.100	0.090	135.5	165,866	12.18	214.8	17,074	131,186	7.68
2	0.163	0.099	122.3	152,059	14.26	198.4	13,521	118,894	8.79
3	0.431	0.126	94.9	118,707	22.61	163.4	6,850	89,953	13.13
4	1.139	0.166	71.9	90,205	44.63	132.7	2,724	65,852	24.18
5	1.852	0.186	63.7	80,594	67.18	121.3	1,640	57,842	35.27

Fig. 6.23. Column: d_c 250 μm, L 15 m, d_f 0.25 μm; hydrogen carrier. Values assume ideality and are calculated for isothermal conditions. Values of H in mm, u_{opt} and OPGV in cm/sec. Values for "time" are values of t_R for that solute calculated at u_{opt} or OPGV.

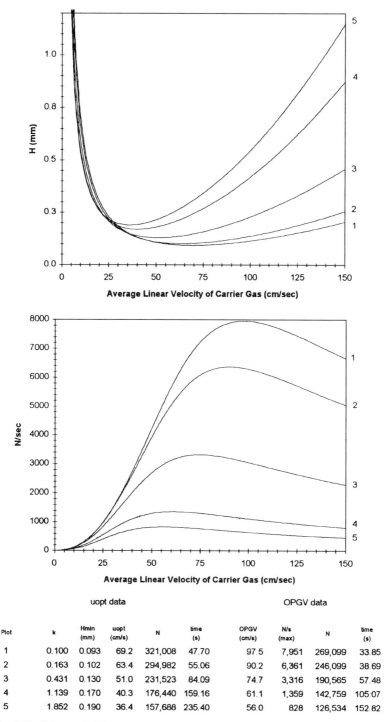

Plot	k	Hmin (mm)	uopt (cm/s)	N	time (s)	OPGV (cm/s)	N/s (max)	N	time (s)
			uopt data					OPGV data	
1	0.100	0.093	69.2	321,008	47.70	97.5	7,951	269,099	33.85
2	0.163	0.102	63.4	294,982	55.06	90.2	6,361	246,099	38.69
3	0.431	0.130	51.0	231,523	84.09	74.7	3,316	190,565	57.48
4	1.139	0.170	40.3	176,440	159.16	61.1	1,359	142,759	105.07
5	1.852	0.190	36.4	157,688	235.40	56.0	828	126,534	152.82

Fig. 6.24. Column: d_c 250 μm, L 30 m, d_f 0.25 μm; helium carrier. Values assume ideality and are calculated for isothermal conditions. Values of H in mm, u_{opt} and OPGV in cm/sec. Values for "time" are values of t_R for that solute calculated at u_{opt} or OPGV.

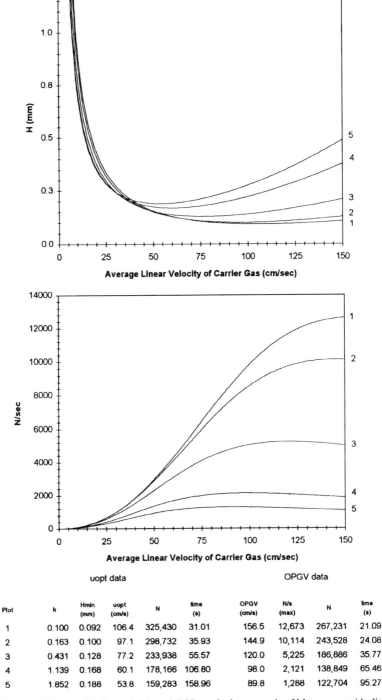

Plot	k	Hmin (mm)	uopt (cm/s)	N	time (s)	OPGV (cm/s)	N/s (max)	N	time (s)
1	0.100	0.092	106.4	325,430	31.01	156.5	12,673	267,231	21.09
2	0.163	0.100	97.1	298,732	35.93	144.9	10,114	243,528	24.08
3	0.431	0.128	77.2	233,938	55.57	120.0	5,225	186,886	35.77
4	1.139	0.168	60.1	178,166	106.80	98.0	2,121	138,849	65.46
5	1.852	0.188	53.8	159,283	158.96	89.8	1,288	122,704	95.27

uopt data   OPGV data

Fig. 6.25. Column: d_c 250 μm, L 30 m, d_f 0.25 μm; hydrogen carrier. Values assume ideality and are calculated for isothermal conditions. Values of H in mm, u_{opt} and OPGV in cm/sec. Values for "time" are values of t_R for that solute calculated at u_{opt} or OPGV.

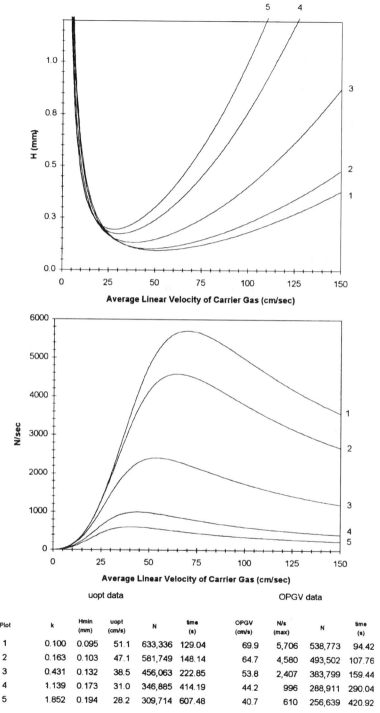

Plot	k	Hmin (mm)	uopt (cm/s)	N	time (s)	OPGV (cm/s)	N/s (max)	N	time (s)
1	0.100	0.095	51.1	633,336	129.04	69.9	5,706	538,773	94.42
2	0.163	0.103	47.1	581,749	148.14	64.7	4,580	493,502	107.76
3	0.431	0.132	38.5	456,063	222.85	53.8	2,407	383,799	159.44
4	1.139	0.173	31.0	346,885	414.19	44.2	996	288,911	290.04
5	1.852	0.194	28.2	309,714	607.48	40.7	610	256,639	420.92

Fig. 6.26. Column: d_c 250 μm, L 60 m, d_f 0.25 μm; helium carrier. Values assume ideality and are calculated for isothermal conditions. Values of H in mm, u_{opt} and OPGV in cm/sec. Values for "time" are values of t_R for that solute calculated at u_{opt} or OPGV.

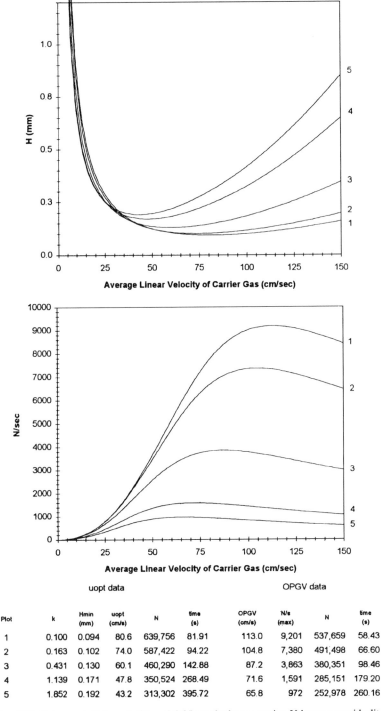

			uopt data			OPGV data			
Plot	k	Hmin (mm)	uopt (cm/s)	N	time (s)	OPGV (cm/s)	N/s (max)	N	time (s)
1	0.100	0.094	80.6	639,756	81.91	113.0	9,201	537,659	58.43
2	0.163	0.102	74.0	587,422	94.22	104.8	7,380	491,498	66.60
3	0.431	0.130	60.1	460,290	142.88	87.2	3,863	380,351	98.46
4	1.139	0.171	47.8	350,524	268.49	71.6	1,591	285,151	179.20
5	1.852	0.192	43.2	313,302	395.72	65.8	972	252,978	260.16

Fig. 6.27. Column: d_c 250 μm, L 60 m, d_f 0.25 μm; hydrogen carrier. Values assume ideality and are calculated for isothermal conditions. Values of H in mm, u_{opt} and OPGV in cm/sec. Values for "time" are values of t_R for that solute calculated at u_{opt} or OPGV.

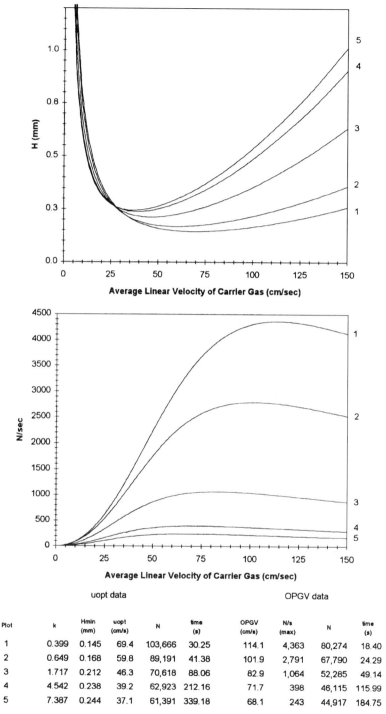

Plot	k	Hmin (mm)	uopt (cm/s)	N	time (s)	OPGV (cm/s)	N/s (max)	N	time (s)
1	0.399	0.145	69.4	103,666	30.25	114.1	4,363	80,274	18.40
2	0.649	0.168	59.8	89,191	41.38	101.9	2,791	67,790	24.29
3	1.717	0.212	46.3	70,618	88.06	82.9	1,064	52,285	49.14
4	4.542	0.238	39.2	62,923	212.16	71.7	398	46,115	115.99
5	7.387	0.244	37.1	61,391	339.18	68.1	243	44,917	184.75

Fig. 6.28. Column: d_c 250 μm, L 15 m, d_f 1.0 μm; helium carrier. Values assume ideality and are calculated for isothermal conditions. Values of H in mm, u_{opt} and OPGV in cm/sec. Values for "time" are values of t_R for that solute calculated at u_{opt} or OPGV.

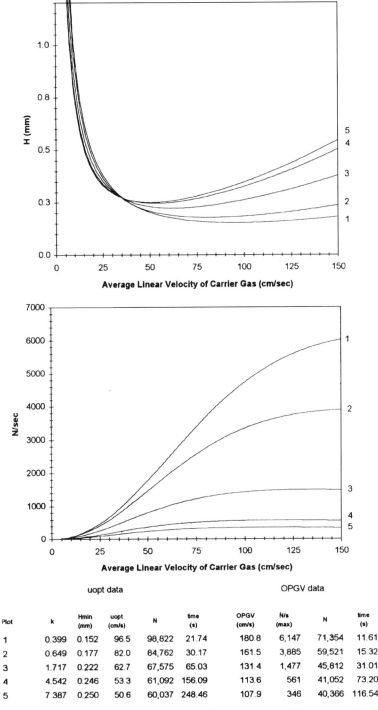

			uopt data			OPGV data			
Plot	k	Hmin (mm)	uopt (cm/s)	N	time (s)	OPGV (cm/s)	Ṅ/s (max)	N	time (s)
1	0.399	0.152	96.5	98,822	21.74	180.8	6,147	71,354	11.61
2	0.649	0.177	82.0	84,762	30.17	161.5	3,885	59,521	15.32
3	1.717	0.222	62.7	67,575	65.03	131.4	1,477	45,812	31.01
4	4.542	0.246	53.3	61,092	156.09	113.6	561	41,052	73.20
5	7.387	0.250	50.6	60,037	248.46	107.9	346	40,366	116.54

Fig. 6.29. Column: d_c 250 μm, L 50 m, d_f 1.0 μm; hydrogen carrier. Values assume ideality and are calculated for isothermal conditions. Values of H in mm, u_{opt} and OPGV in cm/sec. Values for "time" are values of t_R for that solute calculated at u_{opt} or OPGV.

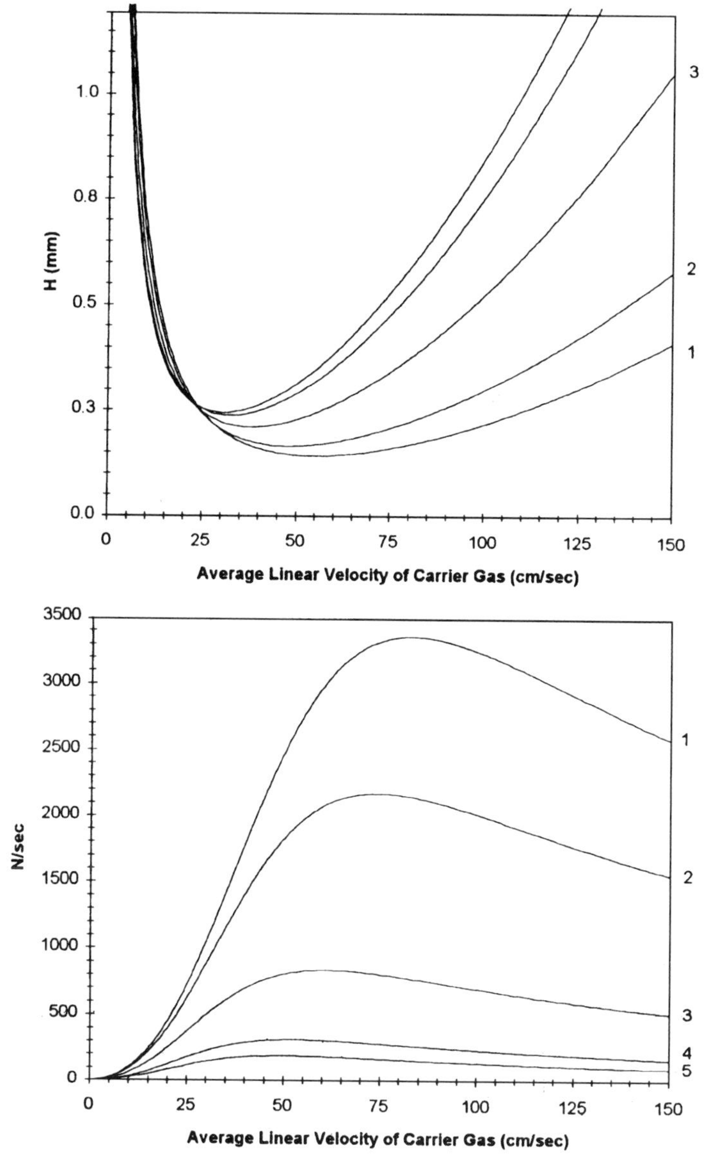

	uopt data					OPGV data			
Plot	k	Hmin (mm)	uopt (cm/s)	N	time (s)	OPGV (cm/s)	N/s (max)	N	time (s)
1	0.399	0.143	55.0	210,318	76.30	82.8	3,360	170,353	50.71
2	0.649	0.166	48.1	181,066	102.75	74.1	2,172	144,954	66.72
3	1.717	0.210	38.1	142,543	213.92	60.7	836	112,217	134.24
4	4.542	0.238	32.6	125,923	510.47	52.7	312	98,455	315.58
5	7.387	0.245	30.9	122,389	814.43	50.1	190	95,571	501.92

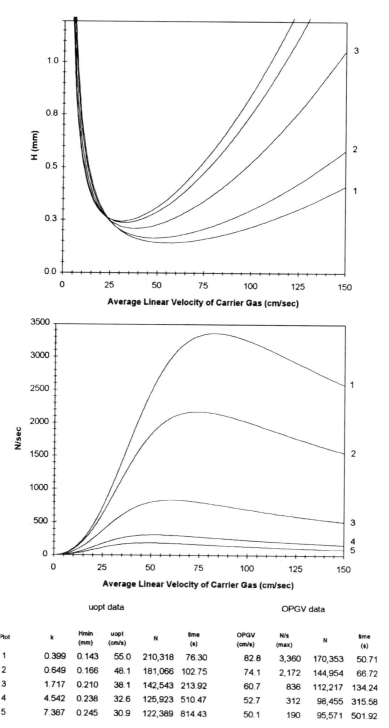

Fig. 6.30. Column: d_c 250 μm, L 30 m, d_f 1.0 μm; helium carrier. Values assume ideality and are calculated for isothermal conditions. Values of H in mm, u_{opt} and OPGV in cm/sec. Values for "time" are values of t_R for that solute calculated at u_{opt} or OPGV.

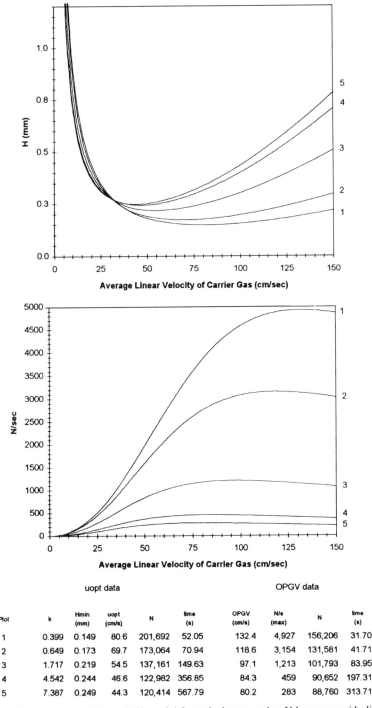

Plot	k	Hmin (mm)	uopt (cm/s)	N	time (s)	OPGV (cm/s)	N/s (max)	N	time (s)
			uopt data					OPGV data	
1	0.399	0.149	80.6	201,692	52.05	132.4	4,927	156,206	31.70
2	0.649	0.173	69.7	173,064	70.94	118.6	3,154	131,581	41.71
3	1.717	0.219	54.5	137,161	149.63	97.1	1,213	101,793	83.95
4	4.542	0.244	46.6	122,982	356.85	84.3	459	90,652	197.31
5	7.387	0.249	44.3	120,414	567.79	80.2	283	88,760	313.71

Fig. 6.31. Column: d_c 250 μm, L 30 m, d_f 1.0 μm; hydrogen carrier. Values assume ideality and are calculated for isothermal conditions. Values of H in mm, u_{opt} and OPGV in cm/sec. Values for "time" are values of t_R for that solute calculated at u_{opt} or OPGV.

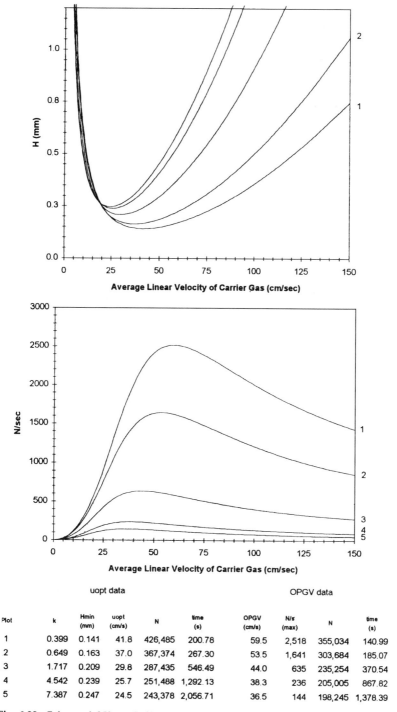

Plot	k	Hmin (mm)	uopt (cm/s)	N	time (s)	OPGV (cm/s)	N/s (max)	N	time (s)
1	0.399	0.141	41.8	426,485	200.78	59.5	2,518	355,034	140.99
2	0.649	0.163	37.0	367,374	267.30	53.5	1,641	303,684	185.07
3	1.717	0.209	29.8	287,435	546.49	44.0	635	235,254	370.54
4	4.542	0.239	25.7	251,488	1,292.13	38.3	236	205,005	867.82
5	7.387	0.247	24.5	243,378	2,056.71	36.5	144	198,245	1,378.39

Fig. 6.32. Column: d_c 250 μm, L 60 m, d_f 1.0 μm; helium carrier. Values assume ideality and are calculated for isothermal conditions. Values of H in mm, u_{opt} and OPGV in cm/sec. Values for "time" are values of t_R for that solute calculated at u_{opt} or OPGV.

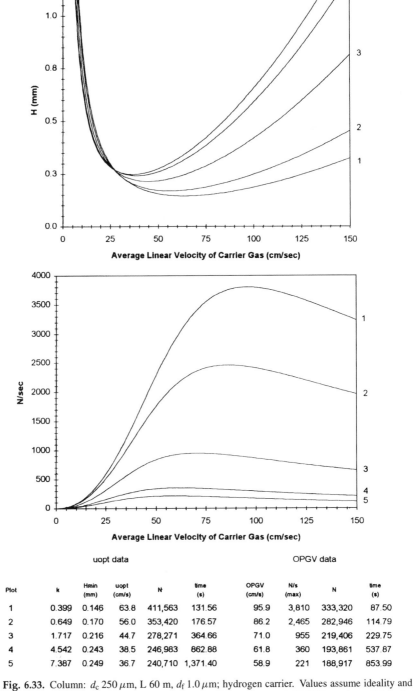

Plot	k	Hmin (mm)	uopt (cm/s)	N	time (s)	OPGV (cm/s)	N/s (max)	N	time (s)
1	0.399	0.146	63.8	411,563	131.56	95.9	3,810	333,320	87.50
2	0.649	0.170	56.0	353,420	176.57	86.2	2,465	282,946	114.79
3	1.717	0.216	44.7	278,271	364.66	71.0	955	219,406	229.75
4	4.542	0.243	38.5	246,983	862.88	61.8	360	193,861	537.87
5	7.387	0.249	36.7	240,710	1,371.40	58.9	221	188,917	853.99

Fig. 6.33. Column: d_c 250 μm, L 60 m, d_f 1.0 μm; hydrogen carrier. Values assume ideality and are calculated for isothermal conditions. Values of H in mm, u_{opt} and OPGV in cm/sec. Values for "time" are values of t_R for that solute calculated at u_{opt} or OPGV.

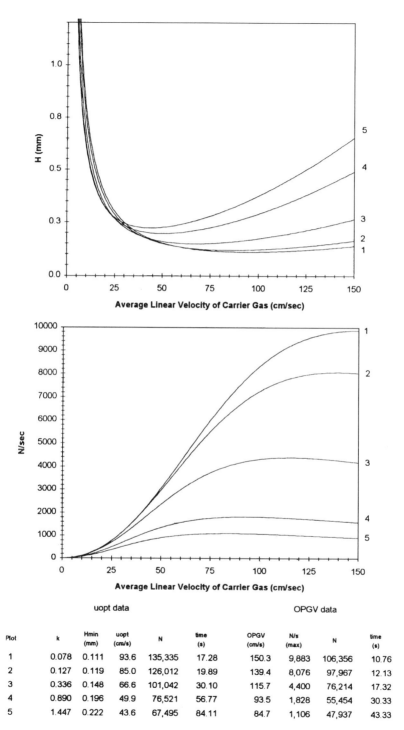

Plot	k	Hmin (mm)	uopt (cm/s)	N	time (s)	OPGV (cm/s)	N/s (max)	N	time (s)
			uopt data				OPGV data		
1	0.078	0.111	93.6	135,335	17.28	150.3	9,883	106,356	10.76
2	0.127	0.119	85.0	126,012	19.89	139.4	8,076	97,967	12.13
3	0.336	0.148	66.6	101,042	30.10	115.7	4,400	76,214	17.32
4	0.890	0.196	49.9	76,521	56.77	93.5	1,828	55,454	30.33
5	1.447	0.222	43.6	67,495	84.11	84.7	1,106	47,937	43.33

Fig. 6.34. Column: d_c 320 μm, L 15 m, d_f 0.25 μm; helium carrier. Values assume ideality and are calculated for isothermal conditions. Values of H in mm, u_{opt} and OPGV in cm/sec. Values for "time" are values of t_R for that solute calculated at u_{opt} or OPGV.

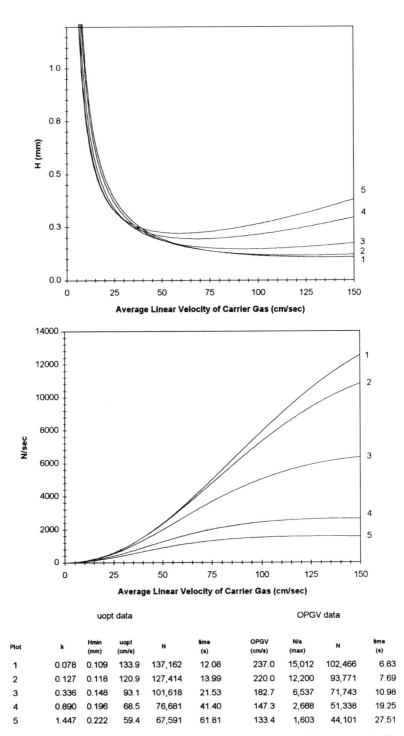

		uopt data				OPGV data			
Plot	k	Hmin (mm)	uopt (cm/s)	N	time (s)	OPGV (cm/s)	N/s (max)	N	time (s)
1	0.078	0.109	133.9	137,162	12.08	237.0	15,012	102,466	6.83
2	0.127	0.118	120.9	127,414	13.99	220.0	12,200	93,771	7.69
3	0.336	0.148	93.1	101,618	21.53	182.7	6,537	71,743	10.98
4	0.890	0.196	68.5	76,681	41.40	147.3	2,688	51,338	19.25
5	1.447	0.222	59.4	67,591	61.81	133.4	1,603	44,101	27.51

Fig. 6.35. Column: d_c 320 μm, L 15 m, d_f 0.25 μm; hydrogen carrier. Values assume ideality and are calculated for isothermal conditions. Values of H in mm, u_{opt} and OPGV in cm/sec. Values for "time" are values of t_R for that solute calculated at u_{opt} or OPGV.

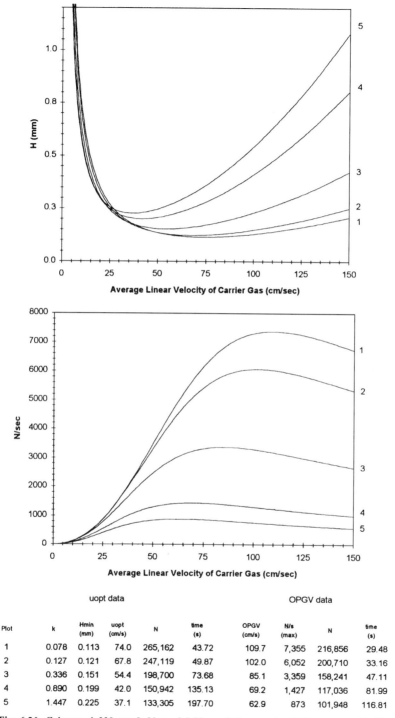

Plot	k	Hmin (mm)	uopt (cm/s)	N	time (s)	OPGV (cm/s)	N/s (max)	N	time (s)
			uopt data			OPGV data			
1	0.078	0.113	74.0	265,162	43.72	109.7	7,355	216,856	29.48
2	0.127	0.121	67.8	247,119	49.87	102.0	6,052	200,710	33.16
3	0.336	0.151	54.4	198,700	73.68	85.1	3,359	158,241	47.11
4	0.890	0.199	42.0	150,942	135.13	69.2	1,427	117,036	81.99
5	1.447	0.225	37.1	133,305	197.70	62.9	873	101,948	116.81

Fig. 6.36. Column: d_c 320 μm, L 30 m, d_f 0.25 μm; helium carrier. Values assume ideality and are calculated for isothermal conditions. Values of H in mm, u_{opt} and OPGV in cm/sec. Values for "time" are values of t_R for that solute calculated at u_{opt} or OPGV.

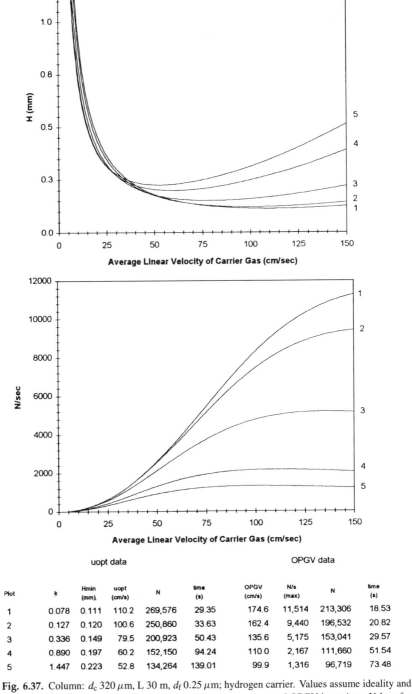

Plot	k	uopt data				OPGV data			
		Hmin (mm)	uopt (cm/s)	N	time (s)	OPGV (cm/s)	N/s (max)	N	time (s)
1	0.078	0.111	110.2	269,576	29.35	174.6	11,514	213,306	18.53
2	0.127	0.120	100.6	250,860	33.63	162.4	9,440	196,532	20.82
3	0.336	0.149	79.5	200,923	50.43	135.6	5,175	153,041	29.57
4	0.890	0.197	60.2	152,150	94.24	110.0	2,167	111,660	51.54
5	1.447	0.223	52.8	134,264	139.01	99.9	1,316	96,719	73.48

Fig. 6.37. Column: d_c 320 μm, L 30 m, d_f 0.25 μm; hydrogen carrier. Values assume ideality and are calculated for isothermal conditions. Values of H in mm, u_{opt} and OPGV in cm/sec. Values for "time" are values of t_R for that solute calculated at u_{opt} or OPGV.

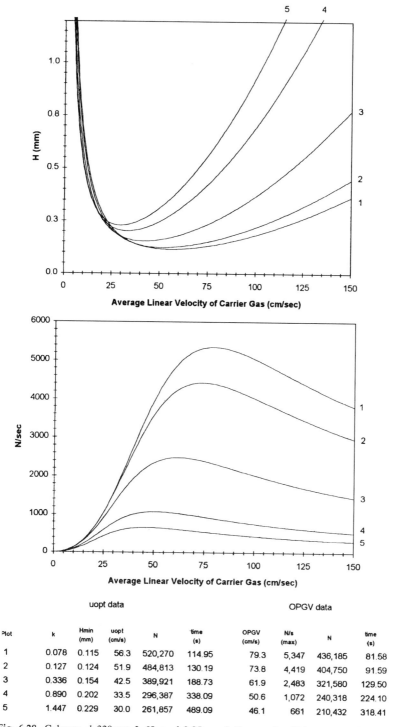

Fig. 6.38. Column: d_c 320 μm, L 60 m, d_f 0.25 μm; helium carrier. Values assume ideality and are calculated for isothermal conditions. Values of H in mm, u_{opt} and OPGV in cm/sec. Values for "time" are values of t_R for that solute calculated at u_{opt} or OPGV.

Plot	k	Hmin (mm)	uopt (cm/s)	N	time (s)	OPGV (cm/s)	N/s (max)	N	time (s)
1	0.078	0.115	56.3	520,270	114.95	79.3	5,347	436,185	81.58
2	0.127	0.124	51.9	484,813	130.19	73.8	4,419	404,750	91.59
3	0.336	0.154	42.5	389,921	188.73	61.9	2,483	321,580	129.50
4	0.890	0.202	33.5	296,387	338.09	50.6	1,072	240,318	224.10
5	1.447	0.229	30.0	261,857	489.09	46.1	661	210,432	318.41

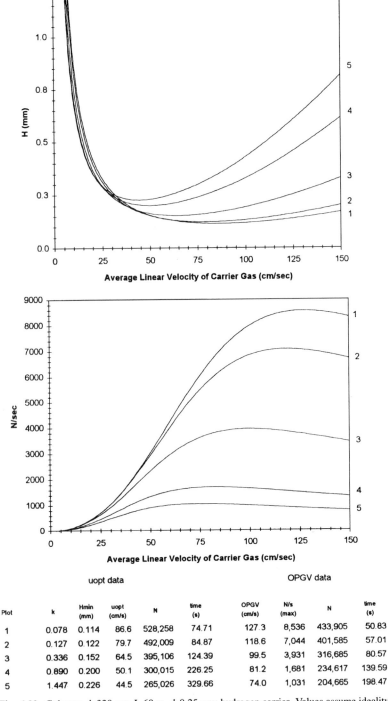

Plot	k	Hmin (mm)	uopt (cm/s)	N	time (s)	OPGV (cm/s)	N/s (max)	N	time (s)
1	0.078	0.114	86.6	528,258	74.71	127.3	8,536	433,905	50.83
2	0.127	0.122	79.7	492,009	84.87	118.6	7,044	401,585	57.01
3	0.336	0.152	64.5	395,106	124.39	99.5	3,931	316,685	80.57
4	0.890	0.200	50.1	300,015	226.25	81.2	1,681	234,617	139.59
5	1.447	0.226	44.5	265,026	329.66	74.0	1,031	204,665	198.47

uopt data   OPGV data

Fig. 6.39. Column: d_c 320 μm, L 60 m, d_f 0.25 μm; hydrogen carrier. Values assume ideality and are calculated for isothermal conditions. Values of H in mm, u_{opt} and OPGV in cm/sec. Values for "time" are values of t_R for that solute calculated at u_{opt} or OPGV.

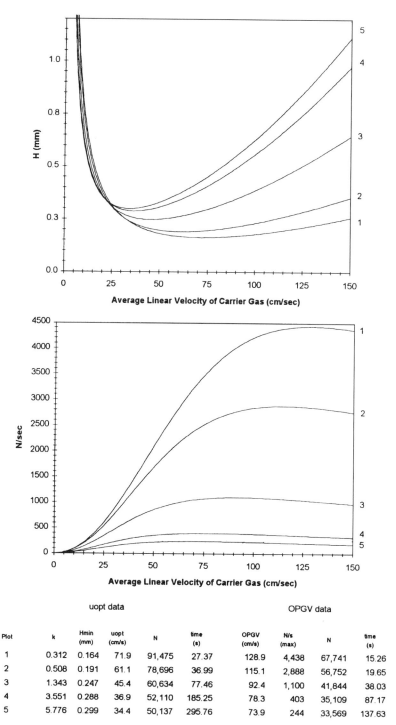

Plot	k	Hmin (mm)	uopt (cm/s)	N	time (s)	OPGV (cm/s)	N/s (max)	N	time (s)
			uopt data					OPGV data	
1	0.312	0.164	71.9	91,475	27.37	128.9	4,438	67,741	15.26
2	0.508	0.191	61.1	78,696	36.99	115.1	2,888	56,752	19.65
3	1.343	0.247	45.4	60,634	77.46	92.4	1,100	41,844	38.03
4	3.551	0.288	36.9	52,110	185.25	78.3	403	35,109	87.17
5	5.776	0.299	34.4	50,137	295.76	73.9	244	33,569	137.63

Fig. 6.40. Column: d_c 320 μm, L 15 m, d_f 1.0 μm; helium carrier. Values assume ideality and are calculated for isothermal conditions. Values of H in mm, u_{opt} and OPGV in cm/sec. Values for "time" are values of t_R for that solute calculated at u_{opt} or OPGV.

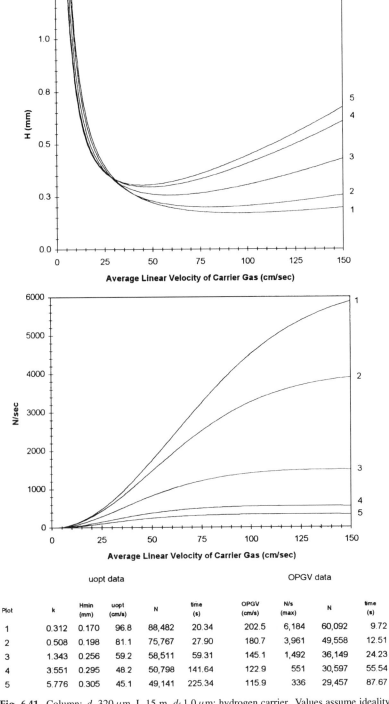

Plot	k	Hmin (mm)	uopt (cm/s)	N	time (s)	OPGV (cm/s)	N/s (max)	N	time (s)
1	0.312	0.170	96.8	88,482	20.34	202.5	6,184	60,092	9.72
2	0.508	0.198	81.1	75,767	27.90	180.7	3,961	49,558	12.51
3	1.343	0.256	59.2	58,511	59.31	145.1	1,492	36,149	24.23
4	3.551	0.295	48.2	50,798	141.64	122.9	551	30,597	55.54
5	5.776	0.305	45.1	49,141	225.34	115.9	336	29,457	87.67

Fig. 6.41. Column: d_c 320 μm, L 15 m, d_f 1.0 μm; hydrogen carrier. Values assume ideality and are calculated for isothermal conditions. Values of H in mm, u_{opt} and OPGV in cm/sec. Values for "time" are values of t_R for that solute calculated at u_{opt} or OPGV.

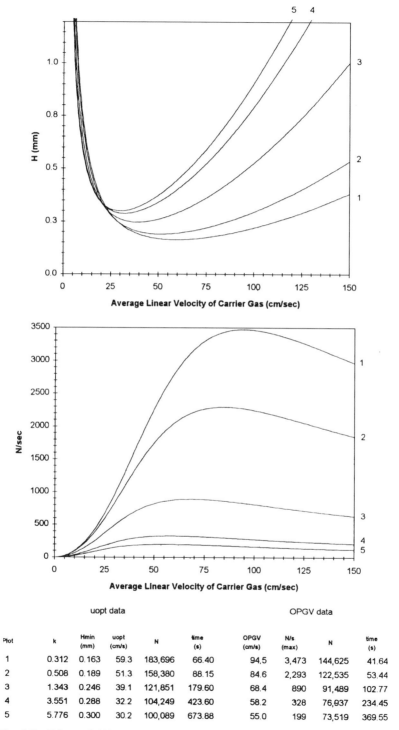

Plot	k	Hmin (mm)	uopt (cm/s)	N	time (s)	OPGV (cm/s)	N/s (max)	N	time (s)
			uopt data					OPGV data	
1	0.312	0.163	59.3	183,696	66.40	94.5	3,473	144,625	41.64
2	0.508	0.189	51.3	158,380	88.15	84.6	2,293	122,535	53.44
3	1.343	0.246	39.1	121,851	179.60	68.4	890	91,489	102.77
4	3.551	0.288	32.2	104,249	423.60	58.2	328	76,937	234.45
5	5.776	0.300	30.2	100,089	673.88	55.0	199	73,519	369.55

Fig. 6.42. Column: d_c 320 μm, L 30 m, d_f 1.0 μm; helium carrier. Values assume ideality and are calculated for isothermal conditions. Values of H in mm, u_{opt} and OPGV in cm/sec. Values for "time" are values of t_R for that solute calculated at u_{opt} or OPGV.

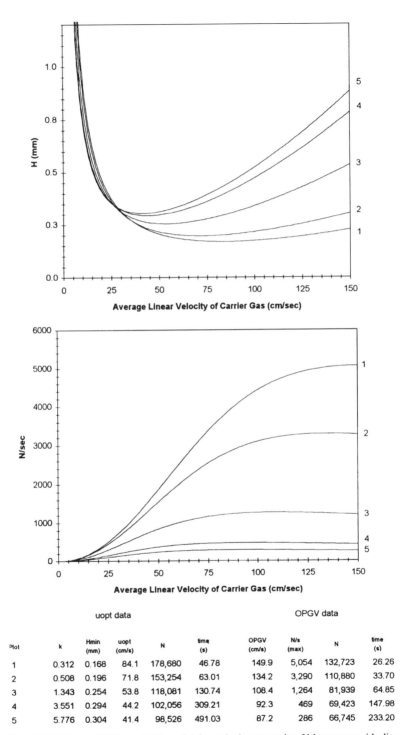

Plot	k	Hmin (mm)	uopt (cm/s)	N	time (s)	OPGV (cm/s)	N/s (max)	N	time (s)
				uopt data				OPGV data	
1	0.312	0.168	84.1	178,680	46.78	149.9	5,054	132,723	26.26
2	0.508	0.196	71.8	153,254	63.01	134.2	3,290	110,880	33.70
3	1.343	0.254	53.8	118,081	130.74	108.4	1,264	81,939	64.85
4	3.551	0.294	44.2	102,056	309.21	92.3	469	69,423	147.98
5	5.776	0.304	41.4	98,526	491.03	87.2	286	66,745	233.20

Fig. 6.43. Column: d_c 320 μm, L 30 m, d_f 1.0 μm; hydrogen carrier. Values assume ideality and are calculated for isothermal conditions. Values of H in mm, u_{opt} and OPGV in cm/sec. Values for "time" are values of t_R for that solute calculated at u_{opt} or OPGV.

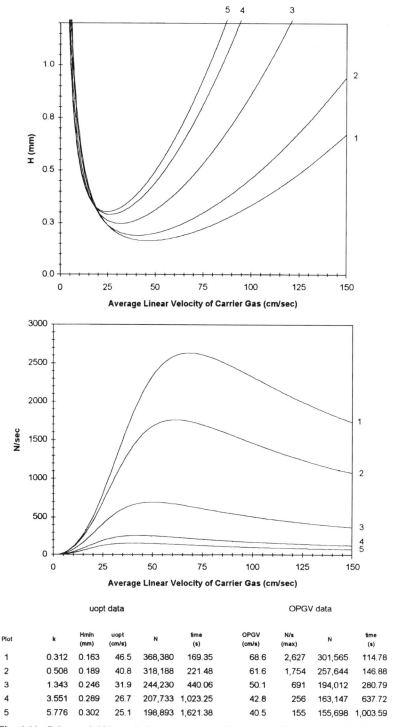

Plot	k	Hmin (mm)	uopt (cm/s)	N	time (s)	OPGV (cm/s)	N/s (max)	N	time (s)
1	0.312	0.163	46.5	368,380	169.35	68.6	2,627	301,565	114.78
2	0.508	0.189	40.8	318,188	221.48	61.6	1,754	257,644	146.88
3	1.343	0.246	31.9	244,230	440.06	50.1	691	194,012	280.79
4	3.551	0.289	26.7	207,733	1,023.25	42.8	256	163,147	637.72
5	5.776	0.302	25.1	198,893	1,621.38	40.5	155	155,698	1,003.59

Fig. 6.44. Column: d_c 320 μm, L 60 m, d_f 1.0 μm; helium carrier. Values assume ideality and are calculated for isothermal conditions. Values of H in mm, u_{opt} and OPGV in cm/sec. Values for "time" are values of t_R for that solute calculated at u_{opt} or OPGV.

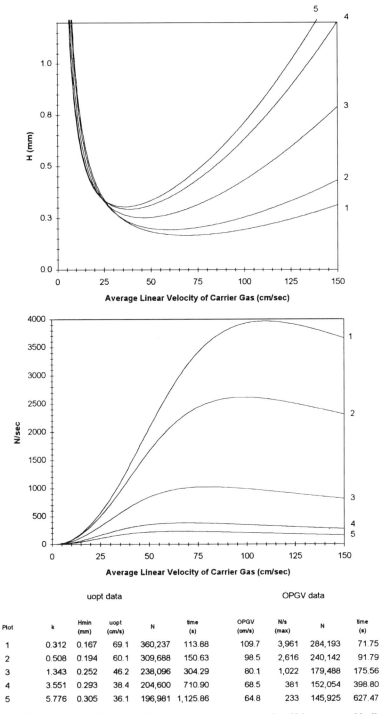

Plot	k	Hmin (mm)	uopt (cm/s)	N	time (s)	OPGV (cm/s)	N/s (max)	N	time (s)
1	0.312	0.167	69.1	360,237	113.88	109.7	3,961	284,193	71.75
2	0.508	0.194	60.1	309,688	150.63	98.5	2,616	240,142	91.79
3	1.343	0.252	46.2	238,096	304.29	80.1	1,022	179,488	175.56
4	3.551	0.293	38.4	204,600	710.90	68.5	381	152,054	398.80
5	5.776	0.305	36.1	196,981	1,125.86	64.8	233	145,925	627.47

Fig. 6.45. Column: d_c 320 μm, L 60 m, d_f 1.0 μm; hydrogen carrier. Values assume ideality and are calculated for isothermal conditions. Values of H in mm, u_{opt} and OPGV in cm/sec. Values for "time" are values of t_R for that solute calculated at u_{opt} or OPGV.

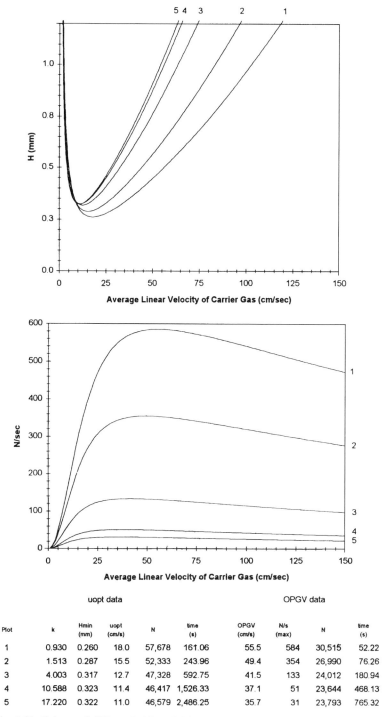

Fig. 6.46. Column: d_c 320 μm, L 15 m, d_f 3.0 μm; nitrogen carrier. Values assume ideality and are calculated for isothermal conditions. Values of H in mm, u_{opt} and OPGV in cm/sec. Values for "time" are values of t_R for that solute calculated at u_{opt} or OPGV.

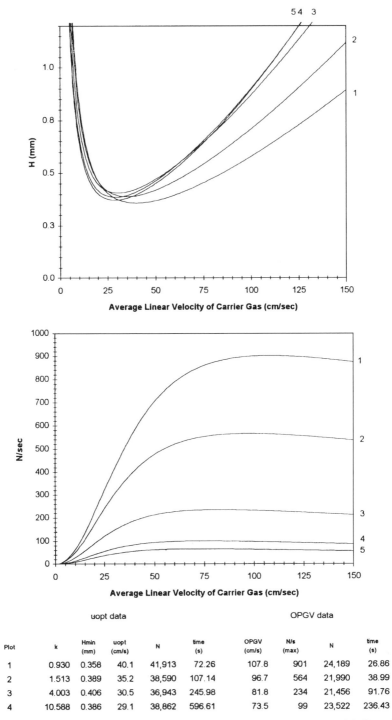

Plot	k	Hmin (mm)	uopt (cm/s)	N	time (s)	OPGV (cm/s)	N/s (max)	N	time (s)
1	0.930	0.358	40.1	41,913	72.26	107.8	901	24,189	26.86
2	1.513	0.389	35.2	38,590	107.14	96.7	564	21,990	38.99
3	4.003	0.406	30.5	36,943	245.98	81.8	234	21,456	91.76
4	10.588	0.386	29.1	38,862	596.61	73.5	99	23,522	236.43

Fig. 6.47. Column: d_c 320 μm, L 15 m, d_f 3.0 μm; helium carrier. Values assume ideality and are calculated for isothermal conditions. Values of H in mm, u_{opt} and OPGV in cm/sec. Values for "time" are values of t_R for that solute calculated at u_{opt} or OPGV.

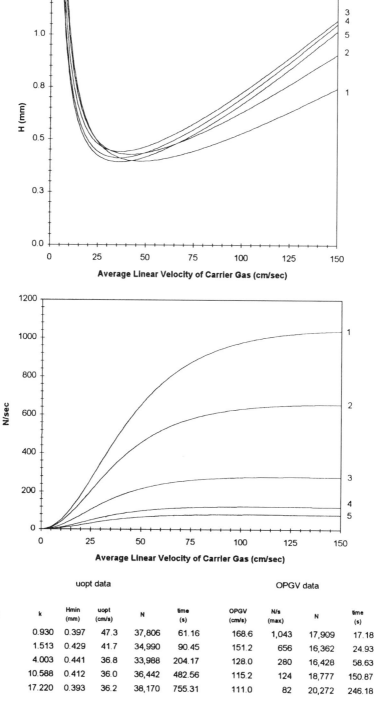

Plot	k	Hmin (mm)	uopt (cm/s)	N	time (s)	OPGV (cm/s)	N/s (max)	N	time (s)
					uopt data			OPGV data	
1	0.930	0.397	47.3	37,806	61.16	168.6	1,043	17,909	17.18
2	1.513	0.429	41.7	34,990	90.45	151.2	656	16,362	24.93
3	4.003	0.441	36.8	33,988	204.17	128.0	280	16,428	58.63
4	10.588	0.412	36.0	36,442	482.56	115.2	124	18,777	150.87
5	17.220	0.393	36.2	38,170	755.31	111.0	82	20,272	246.18

Fig. 6.48. Column: d_c 320 μm, L 15 m, d_f 3.0 μm; hydrogen carrier. Values assume ideality and are calculated for isothermal conditions. Values of H in mm, u_{opt} and OPGV in cm/sec. Values for "time" are values of t_R for that solute calculated at u_{opt} or OPGV.

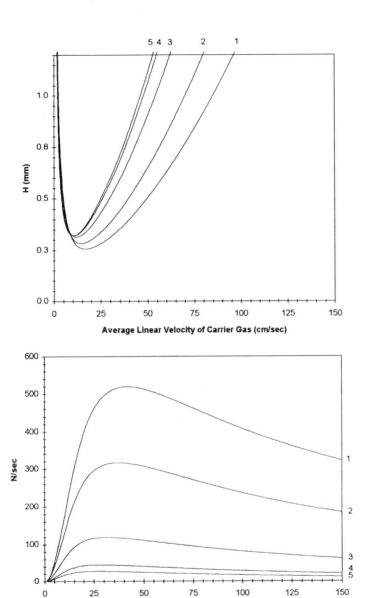

Plot	k	Hmin (mm)	uopt (cm/s)	N	time (s)	OPGV (cm/s)	N/s (max)	N	time (s)
			uopt data				OPGV data		
1	0.930	0.257	17.0	116,946	340.22	41.7	519	72,096	138.78
2	1.513	0.283	14.7	105,836	512.62	37.3	316	63,867	202.04
3	4.003	0.315	12.1	95,278	1,241.39	31.5	119	56,621	477.09
4	10.588	0.322	10.9	93,133	3,198.11	28.2	45	55,434	1,230.76
5	17.220	0.321	10.5	93,344	5,211.70	27.2	28	55,652	2,010.09

Fig. 6.49. Column: d_c 320 μm, L 30 m, d_f 3.0 μm; nitrogen carrier. Values assume ideality and are calculated for isothermal conditions. Values of H in mm, u_{opt} and OPGV in cm/sec. Values for "time" are values of t_R for that solute calculated at u_{opt} or OPGV.

214

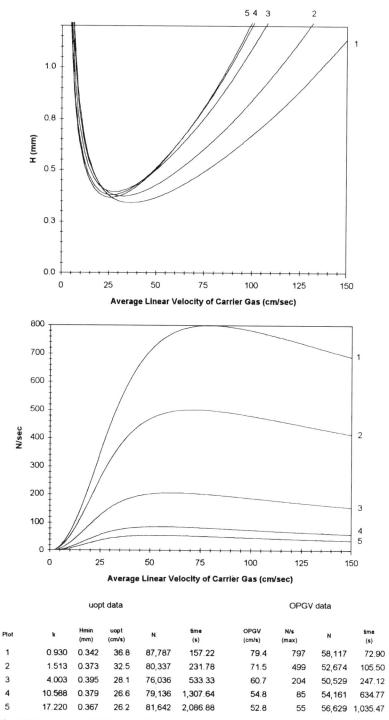

Plot	k	Hmin (mm)	uopt (cm/s)	N.	time (s)	OPGV (cm/s)	N/s (max)	N	time (s)
1	0.930	0.342	36.8	87,787	157.22	79.4	797	58,117	72.90
2	1.513	0.373	32.5	80,337	231.78	71.5	499	52,674	105.50
3	4.003	0.395	28.1	76,036	533.33	60.7	204	50,529	247.12
4	10.588	0.379	26.6	79,136	1,307.64	54.8	85	54,161	634.77
5	17.220	0.367	26.2	81,642	2,086.88	52.8	55	56,629	1,035.47

(above columns grouped: "uopt data" spanning k, Hmin, uopt, N., time; "OPGV data" spanning OPGV, N/s, N, time)

Fig. 6.50. Column: d_c 320 μm, L 30 m, d_f 3.0 μm; helium carrier. Values assume ideality and are calculated for isothermal conditions. Values of H in mm, u_{opt} and OPGV in cm/sec. Values for "time" are values of t_R for that solute calculated at u_{opt} or OPGV.

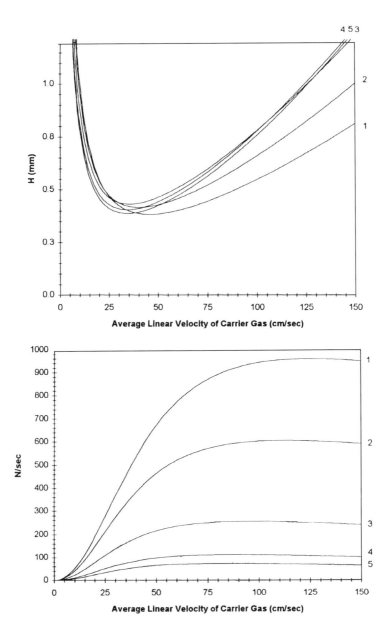

Fig. 6.51. Column: d_c 320 μm, L 30 m, d_f 3.0 μm; hydrogen carrier. Values assume ideality and are calculated for isothermal conditions. Values of H in mm, u_{opt} and OPGV in cm/sec. Values for "time" are values of t_R for that solute calculated at u_{opt} or OPGV.

		uopt data					OPGV data		
Plot	k	Hmin (mm)	uopt (cm/s)	N	time (s)	OPGV (cm/s)	N/s (max)	N	time (s)
1	0.930	0.383	45.5	78,227	127.36	125.4	960	44,334	46.17
2	1.513	0.416	40.1	72,080	187.84	112.9	605	40,388	66.81
3	4.003	0.432	35.3	69,521	425.24	96.0	255	39,954	156.38
4	10.588	0.405	34.3	74,070	1,014.49	86.6	112	44,743	401.23
5	17.220	0.388	34.2	77,337	1,596.56	83.6	73	47,814	654.05

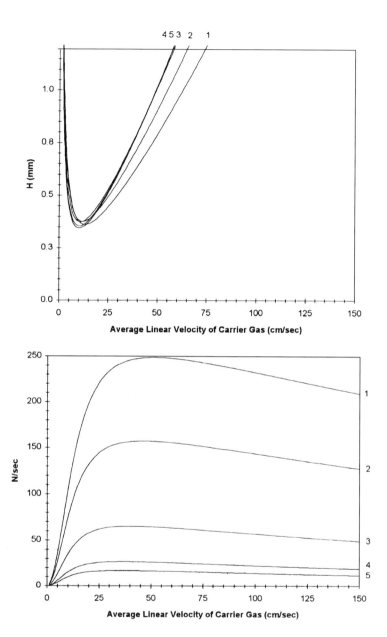

		uopt data				OPGV data			
Plot	k	Hmin (mm)	uopt (cm/s)	N	time (s)	OPGV (cm/s)	N/s (max)	N	time (s)
1	1.541	0.362	13.2	41,400	287.77	51.2	249	18,496	74.42
2	2.506	0.377	12.0	39,838	438.29	46.3	157	17,835	113.54
3	6.629	0.374	10.8	40,063	1,055.09	40.1	65	18,576	285.63
4	17.535	0.355	10.4	42,228	2,663.98	36.6	27	20,392	760.24
5	28.518	0.345	10.3	43,416	4,297.15	35.4	17	21,297	1,252.01

Fig. 6.52. Column: d_c 320 μm, L 15 m, d_f 5.0 μm; nitrogen carrier. Values assume ideality and are calculated for isothermal conditions. Values of H in mm, u_{opt} and OPGV in cm/sec. Values for "time" are values of t_R for that solute calculated at u_{opt} or OPGV.

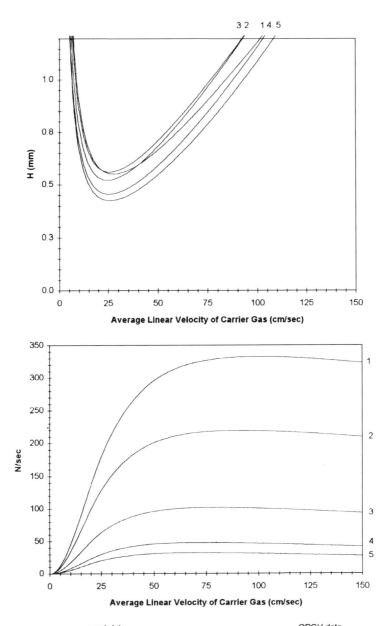

Plot	k	Hmin (mm)	uopt (cm/s)	N	time (s)	OPGV (cm/s)	N/s (max)	N	time (s)
		uopt data					OPGV data		
1	1.541	0.551	27.6	27,228	138.16	100.0	332	12,668	38.12
2	2.506	0.560	25.6	26,774	205.56	90.9	219	12,672	57.86
3	6.629	0.521	24.6	28,776	465.44	79.1	102	14,697	144.61
4	17.535	0.455	25.4	32,931	1,096.43	72.5	48	18,330	383.68
5	28.518	0.425	25.9	35,264	1,711.55	70.1	32	20,357	631.32

Fig. 6.53. Column: d_c 320 μm, L15 m, d_f 5.0 μm; helium carrier. Values assume ideality and are calculated for isothermal conditions. Values of H in mm, u_{opt} and OPGV in cm/sec. Values for "time" are values of t_R for that solute calculated at u_{opt} or OPGV.

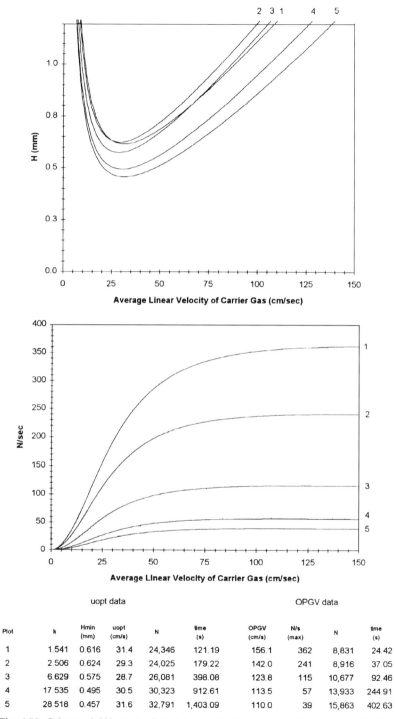

Plot	k	Hmin (mm)	uopt (cm/s)	N	time (s)	OPGV (cm/s)	N/s (max)	N	time (s)
1	1.541	0.616	31.4	24,346	121.19	156.1	362	8,831	24.42
2	2.506	0.624	29.3	24,025	179.22	142.0	241	8,916	37.05
3	6.629	0.575	28.7	26,081	398.08	123.8	115	10,677	92.46
4	17.535	0.495	30.5	30,323	912.61	113.5	57	13,933	244.91
5	28.518	0.457	31.6	32,791	1,403.09	110.0	39	15,863	402.63

Fig. 6.54. Column: d_c 320 μm, L 15 m, d_f 5.0 μm; hydrogen carrier. Values assume ideality and are calculated for isothermal conditions. Values of H in mm, u_{opt} and OPGV in cm/sec. Values for "time" are values of t_R for that solute calculated at u_{opt} or OPGV.

219

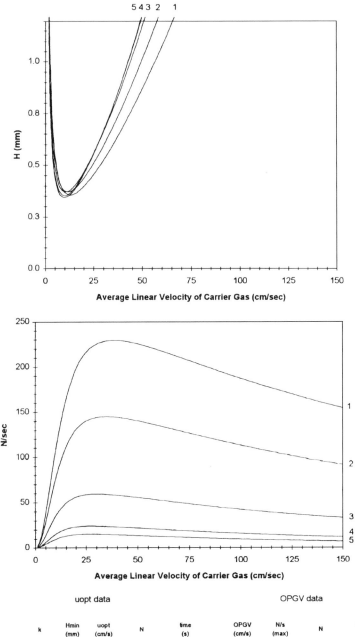

		uopt data				OPGV data			
Plot	k	Hmin (mm)	uopt (cm/s)	N	time (s)	OPGV (cm/s)	N/s (max)	N	time (s)
1	1.541	0.356	12.9	84,239	593.06	38.6	229	45,265	197.34
2	2.506	0.371	11.6	80,830	903.14	35.0	145	43,451	300.26
3	6.629	0.371	10.5	80,901	2,181.79	30.4	59	44,680	752.44
4	17.535	0.353	10.0	84,944	5,536.50	27.8	24	48,430	1,997.97
5	28.518	0.344	9.9	87,193	8,950.21	26.9	15	50,323	3,287.54

Fig. 6.55. Column: d_c 320 μm, L 30 m, d_f 5.0 μm; nitrogen carrier. Values assume ideality and are calculated for isothermal conditions. Values of H in mm, u_{opt} and OPGV in cm/sec. Values for "time" are values of t_R for that solute calculated at u_{opt} or OPGV.

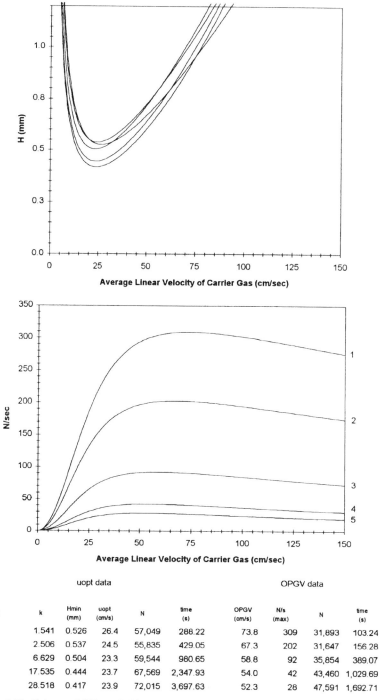

Plot	k	Hmin (mm)	uopt (cm/s)	N	time (s)	OPGV (cm/s)	N/s (max)	N	time (s)
1	1.541	0.526	26.4	57,049	288.22	73.8	309	31,893	103.24
2	2.506	0.537	24.5	55,835	429.05	67.3	202	31,647	156.28
3	6.629	0.504	23.3	59,544	980.65	58.8	92	35,854	389.07
4	17.535	0.444	23.7	67,569	2,347.93	54.0	42	43,460	1,029.69
5	28.518	0.417	23.9	72,015	3,697.63	52.3	28	47,591	1,692.71

uopt data   OPGV data

Fig. 6.56. Column: d_c 320 μm, L 30 m, d_f 5.0 μm; helium carrier. Values assume ideality and are calculated for isothermal conditions. Values of H in mm, u_{opt} and OPGV in cm/sec. Values for "time" are values of t_R for that solute calculated at u_{opt} or OPGV.

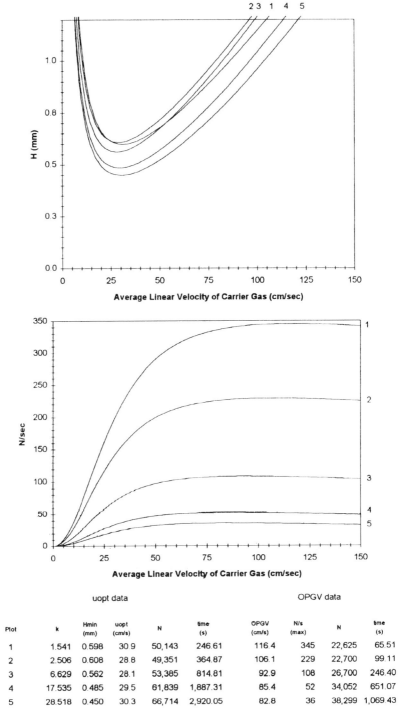

| | | uopt data | | | | | OPGV data | | |
Plot	k	Hmin (mm)	uopt (cm/s)	N	time (s)	OPGV (cm/s)	N/s (max)	N	time (s)
1	1.541	0.598	30.9	50,143	246.61	116.4	345	22,625	65.51
2	2.506	0.608	28.8	49,351	364.87	106.1	229	22,700	99.11
3	6.629	0.562	28.1	53,385	814.81	92.9	108	26,700	246.40
4	17.535	0.485	29.5	61,839	1,887.31	85.4	52	34,052	651.07
5	28.518	0.450	30.3	66,714	2,920.05	82.8	36	38,299	1,069.43

Fig. 6.57. Column: d_c 320 μm, L 30 m, d_f 5.0 μm; hydrogen carrier. Values assume ideality and are calculated for isothermal conditions. Values of H in mm, u_{opt} and OPGV in cm/sec. Values for "time" are values of t_R for that solute calculated at u_{opt} or OPGV.

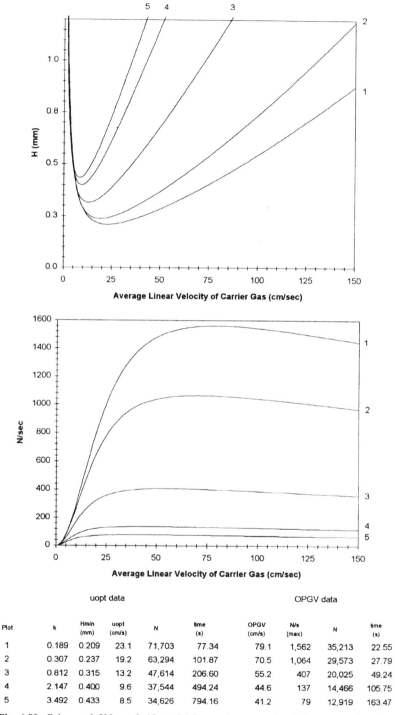

Fig. 6.58. Column: d_c 530 μm, L 15 m, d_f 1.0 μm; nitrogen carrier. Values assume ideality and are calculated for isothermal conditions. Values of H in mm, u_{opt} and OPGV in cm/sec. Values for "time" are values of t_R for that solute calculated at u_{opt} or OPGV.

Plot	k	Hmin (mm)	uopt (cm/s)	N	time (s)	OPGV (cm/s)	N/s (max)	N	time (s)
1	0.189	0.209	23.1	71,703	77.34	79.1	1,562	35,213	22.55
2	0.307	0.237	19.2	63,294	101.87	70.5	1,064	29,573	27.79
3	0.812	0.315	13.2	47,614	206.60	55.2	407	20,025	49.24
4	2.147	0.400	9.6	37,544	494.24	44.6	137	14,466	105.75
5	3.492	0.433	8.5	34,626	794.16	41.2	79	12,919	163.47

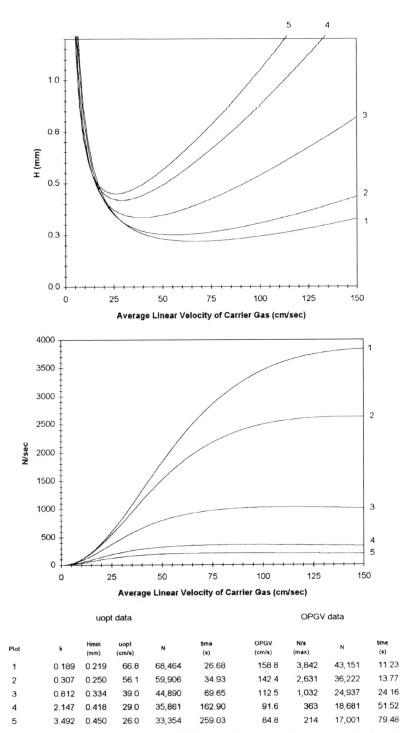

		uopt data				OPGV data			
Plot	k	Hmin (mm)	uopt (cm/s)	N	time (s)	OPGV (cm/s)	N/s (max)	N	time (s)
1	0.189	0.219	66.8	68,464	26.68	158.8	3,842	43,151	11.23
2	0.307	0.250	56.1	59,906	34.93	142.4	2,631	36,222	13.77
3	0.812	0.334	39.0	44,890	69.65	112.5	1,032	24,937	24.16
4	2.147	0.418	29.0	35,861	162.90	91.6	363	18,681	51.52
5	3.492	0.450	26.0	33,354	259.03	84.8	214	17,001	79.48

Fig. 6.59. Column: d_c 530 μm, L 15 m, d_f 1.0 μm; helium carrier. Values assume ideality and are calculated for isothermal conditions. Values of H in mm, u_{opt} and OPGV in cm/sec. Values for "time" are values of t_R for that solute calculated at u_{opt} or OPGV.

224

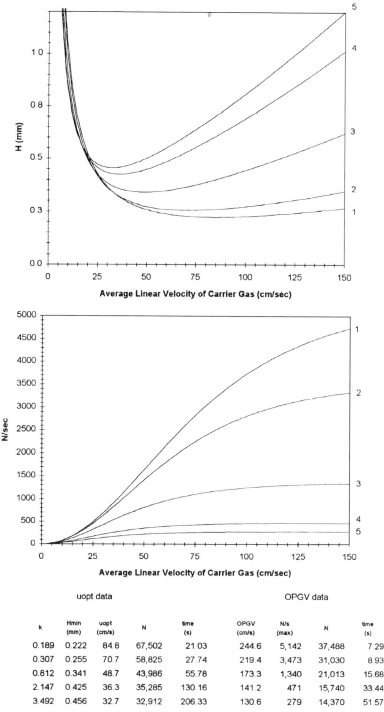

| | | uopt data | | | | | OPGV data | | |
Plot	k	Hmin (mm)	uopt (cm/s)	N	time (s)	OPGV (cm/s)	N/s (max)	N	time (s)
1	0.189	0.222	84.8	67,502	21.03	244.6	5,142	37,488	7.29
2	0.307	0.255	70.7	58,825	27.74	219.4	3,473	31,030	8.93
3	0.812	0.341	48.7	43,986	55.78	173.3	1,340	21,013	15.68
4	2.147	0.425	36.3	35,285	130.16	141.2	471	15,740	33.44
5	3.492	0.456	32.7	32,912	206.33	130.6	279	14,370	51.57

Fig. 6.60. Column: d_c 530 μm, L 15 m, d_f 1.0 μm; hydrogen carrier. Values assume ideality and are calculated for isothermal conditions. Values of H in mm, u_{opt} and OPGV in cm/sec. Values for "time" are values of t_R for that solute calculated at u_{opt} or OPGV.

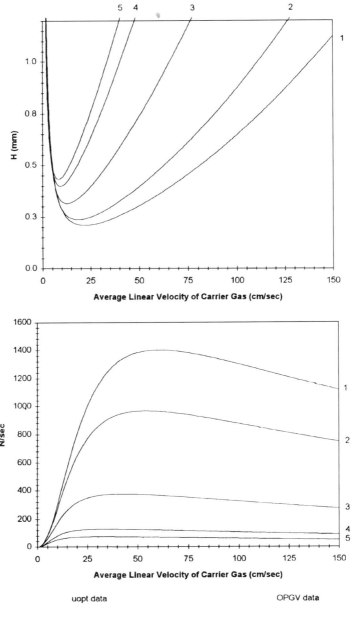

| | | uopt data | | | | OPGV data | | | |
Plot	k	Hmin (mm)	uopt (cm/s)	N	time (s)	OPGV (cm/s)	N/s (max)	N	time (s)
1	0.189	0.209	22.1	143,326	161.10	60.5	1,401	82,544	58.90
2	0.307	0.237	18.6	126,569	210.77	54.1	965	69,876	72.40
3	0.812	0.315	12.8	95,243	423.00	42.6	376	47,961	127.61
4	2.147	0.399	9.4	75,096	1,005.79	34.6	128	34,945	272.91
5	3.492	0.433	8.4	69,255	1,613.26	32.0	74	31,293	421.25

Fig. 6.61. Column: d_c 530 μm, L 30 m, d_f 1.0 μm; nitrogen carrier. Values assume ideality and are calculated for isothermal conditions. Values of H in mm, u_{opt} and OPGV in cm/sec. Values for "time" are values of t_R for that solute calculated at u_{opt} or OPGV.

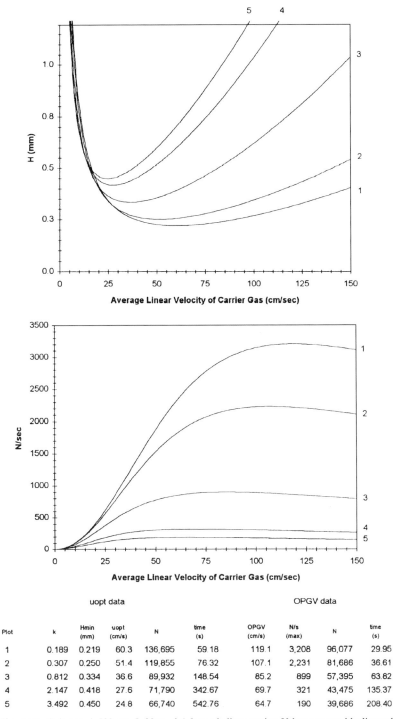

Plot	k	Hmin (mm)	uopt (cm/s)	N	time (s)	OPGV (cm/s)	N/s (max)	N	time (s)
1	0.189	0.219	60.3	136,695	59.18	119.1	3,208	96,077	29.95
2	0.307	0.250	51.4	119,855	76.32	107.1	2,231	81,686	36.61
3	0.812	0.334	36.6	89,932	148.54	85.2	899	57,395	63.82
4	2.147	0.418	27.6	71,790	342.67	69.7	321	43,475	135.37
5	3.492	0.450	24.8	66,740	542.76	64.7	190	39,686	208.40

Fig. 6.62. Column: d_c 530 μm, L 30 m, d_f 1.0 μm; helium carrier. Values assume ideality and are calculated for isothermal conditions. Values of H in mm, u_{opt} and OPGV in cm/sec. Values for "time" are values of t_R for that solute calculated at u_{opt} or OPGV.

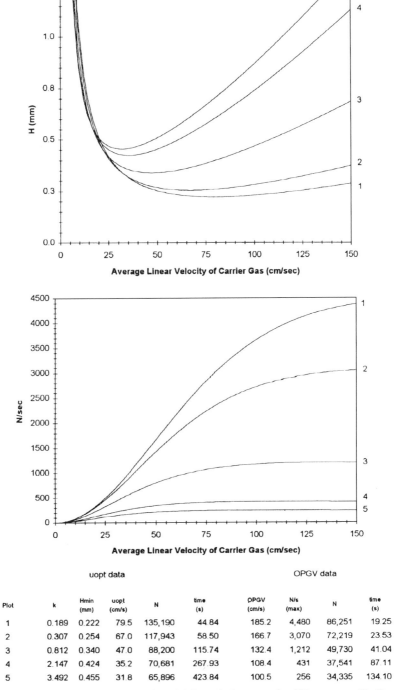

Plot	k	Hmin (mm)	uopt (cm/s)	N	time (s)	OPGV (cm/s)	N/s (max)	N	time (s)
			uopt data					OPGV data	
1	0.189	0.222	79.5	135,190	44.84	185.2	4,480	86,251	19.25
2	0.307	0.254	67.0	117,943	58.50	166.7	3,070	72,219	23.53
3	0.812	0.340	47.0	88,200	115.74	132.4	1,212	49,730	41.04
4	2.147	0.424	35.2	70,681	267.93	108.4	431	37,541	87.11
5	3.492	0.455	31.8	65,896	423.84	100.5	256	34,335	134.10

Fig. 6.63. Column: d_c 530 μm, L 30 m, d_f 1.0 μm; hydrogen carrier. Values assume ideality and are calculated for isothermal conditions. Values of H in mm, u_{opt} and OPGV in cm/sec. Values for "time" are values of t_R for that solute calculated at u_{opt} or OPGV.

228

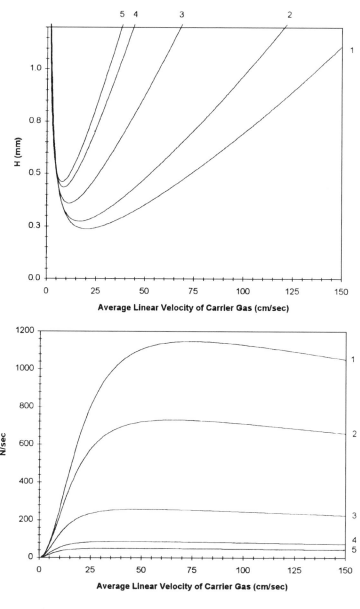

Plot	k	Hmin (mm)	uopt (cm/s)	N	time (s)	OPGV (cm/s)	N/s (max)	N	time (s)
1	0.283	0.238	20.4	63,129	94.23	74.4	1,146	29,631	25.86
2	0.460	0.274	16.7	54,780	130.75	65.6	729	24,306	33.36
3	1.216	0.360	11.5	41,616	288.17	51.4	258	16,674	64.68
4	3.217	0.435	8.8	34,445	720.75	42.6	87	12,858	148.46
5	5.233	0.461	8.0	32,536	1,171.46	39.9	51	11,860	234.54

(Headers: "uopt data" spans k, Hmin, uopt, N, time; "OPGV data" spans OPGV, N/s, N, time.)

Fig. 6.64. Column: d_c 530 μm, L 15 m, d_f 1.5 μm; nitrogen carrier. Values assume ideality and are calculated for isothermal conditions. Values of H in mm, u_{opt} and OPGV in cm/sec. Values for "time" are values of t_R for that solute calculated at u_{opt} or OPGV.

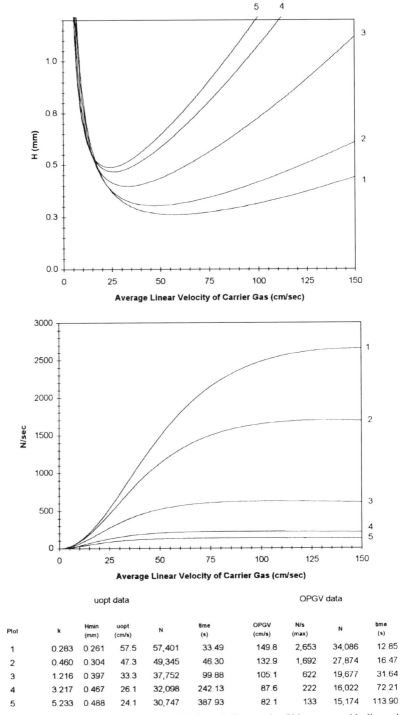

Plot	k	Hmin (mm)	uopt (cm/s)	N	time (s)	OPGV (cm/s)	N/s (max)	N	time (s)
		uopt data				OPGV data			
1	0.283	0.261	57.5	57,401	33.49	149.8	2,653	34,086	12.85
2	0.460	0.304	47.3	49,345	46.30	132.9	1,692	27,874	16.47
3	1.216	0.397	33.3	37,752	99.88	105.1	622	19,677	31.64
4	3.217	0.467	26.1	32,098	242.13	87.6	222	16,022	72.21
5	5.233	0.488	24.1	30,747	387.93	82.1	133	15,174	113.90

Fig. 6.65. Column: d_c 530 μm, L 15 m, d_f 1.5 μm; helium carrier. Values assume ideality and are calculated for isothermal conditions. Values of H in mm, u_{opt} and OPGV in cm/sec. Values for "time" are values of t_R for that solute calculated at u_{opt} or OPGV.

230

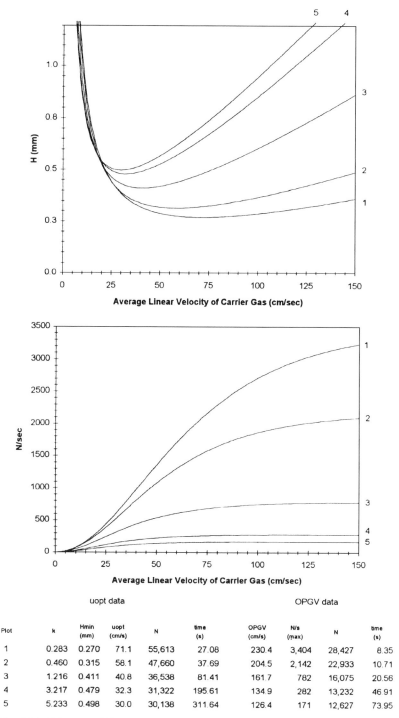

Plot	k	Hmin (mm)	uopt (cm/s)	N	time (s)	OPGV (cm/s)	N/s (max)	N	time (s)
1	0.283	0.270	71.1	55,613	27.08	230.4	3,404	28,427	8.35
2	0.460	0.315	58.1	47,660	37.69	204.5	2,142	22,933	10.71
3	1.216	0.411	40.8	36,538	81.41	161.7	782	16,075	20.56
4	3.217	0.479	32.3	31,322	195.61	134.9	282	13,232	46.91
5	5.233	0.498	30.0	30,138	311.64	126.4	171	12,627	73.95

Fig. 6.66. Column: d_c 530 μm, L 15 m, d_f 1.5 μm; hydrogen carrier. Values assume ideality and are calculated for isothermal conditions. Values of H in mm, u_{opt} and OPGV in cm/sec. Values for "time" are values of t_R for that solute calculated at u_{opt} or OPGV.

231

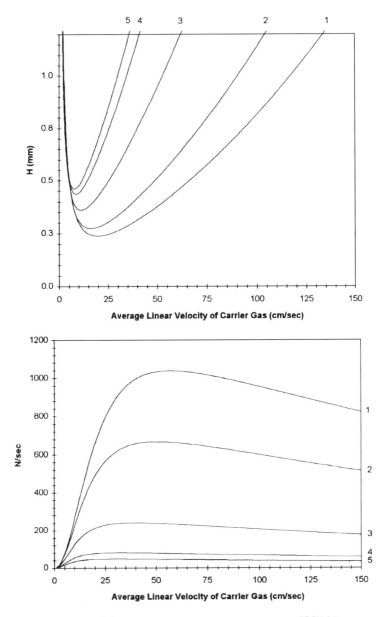

		uopt data				OPGV data			
Plot	k	Hmin (mm)	uopt (cm/s)	N	time (s)	OPGV (cm/s)	N/s (max)	N	time (s)
1	0.283	0.237	19.7	126,335	195.08	57.0	1,038	70,064	67.47
2	0.460	0.274	16.3	109,649	268.99	50.5	668	57,925	86.76
3	1.216	0.360	11.3	83,283	587.88	39.7	240	40,181	167.37
4	3.217	0.435	8.6	68,911	1,464.15	33.1	81	31,166	382.81
5	5.233	0.461	7.9	65,084	2,376.90	31.0	48	28,795	604.05

Fig. 6.67. Column: d_c 530 μm, L 30 m, d_f 1.5 μm; nitrogen carrier. Values assume ideality and are calculated for isothermal conditions. Values of H in mm, u_{opt} and OPGV in cm/sec. Values for "time" are values of t_R for that solute calculated at u_{opt} or OPGV.

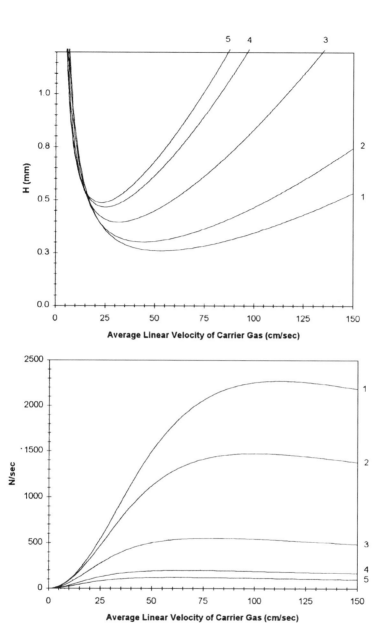

Plot	k	Hmin (mm)	uopt (cm/s)	N	time (s)	OPGV (cm/s)	N/s (max)	N	time (s)
		uopt data				OPGV data			
1	0.283	0.260	52.9	115,418	72.78	112.5	2,271	77,682	34.20
2	0.460	0.302	44.2	99,301	99.19	100.2	1,471	64,318	43.72
3	1.216	0.395	31.6	75,870	210.18	79.7	552	46,057	83.43
4	3.217	0.466	25.0	64,366	505.63	66.8	199	37,658	189.50
5	5.233	0.487	23.1	61,600	808.67	62.7	120	35,679	298.41

Fig. 6.68. Column: d_c 530 μm, L 30 m, d_f 1.5 μm; helium carrier. Values assume ideality and are calculated for isothermal conditions. Values of H in mm, u_{opt} and OPGV in cm/sec. Values for "time" are values of t_R for that solute calculated at u_{opt} or OPGV.

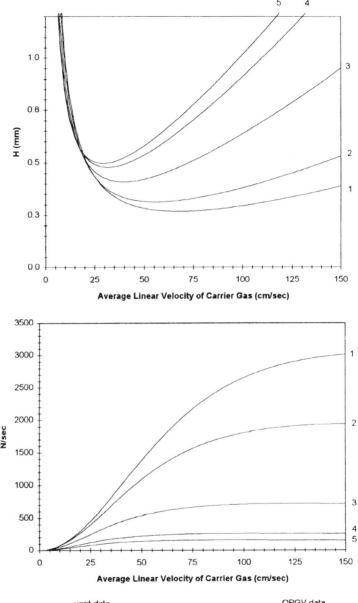

Plot	k	Hmin (mm)	uopt (cm/s)	N	time (s)	OPGV (cm/s)	N/s (max)	N	time (s)
			uopt data			OPGV data			
1	0.283	0.268	67.8	111,906	56.79	174.8	3,038	66,897	22.02
2	0.460	0.313	55.9	95,899	78.29	155.6	1,938	54,525	28.14
3	1.216	0.409	39.7	73,398	167.36	123.7	719	38,615	53.74
4	3.217	0.478	31.6	62,806	400.62	103.7	261	31,836	122.06
5	5.233	0.497	29.3	60,390	637.85	97.3	158	30,366	192.15

Fig. 6.69. Column: d_c 530 μm, L 30 m, d_f 1.5 μm; hydrogen carrier. Values assume ideality and are calculated for isothermal conditions. Values of H in mm, u_{opt} and OPGV in cm/sec. Values for "time" are values of t_R for that solute calculated at u_{opt} or OPGV.

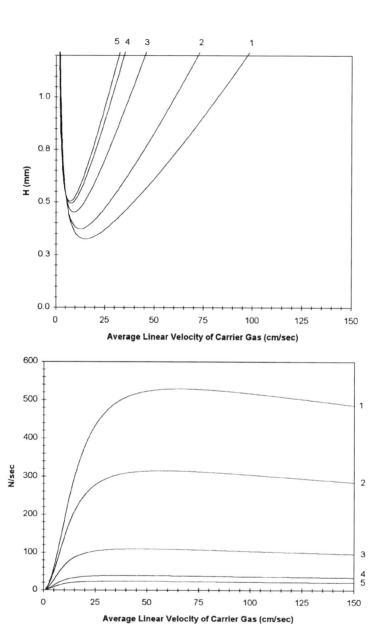

Plot	k	Hmin (mm)	uopt (cm/s)	N	time (s)	OPGV (cm/s)	N/s (max)	N	time (s)
1	0.564	0.325	15.1	46,183	155.09	65.8	529	18,863	35.67
2	0.917	0.372	12.5	40,359	230.77	57.8	314	15,646	49.78
3	2.426	0.451	9.3	33,290	553.97	46.5	109	12,087	110.43
4	6.416	0.493	7.8	30,414	1,432.02	40.3	39	10,726	275.77
5	10.436	0.503	7.3	29,814	2,341.87	38.4	23	10,429	446.68

Fig. 6.70. Column: d_c 530 μm, L 15 m, d_f 3.0 μm; nitrogen carrier. Values assume ideality and are calculated for isothermal conditions. Values of H in mm, u_{opt} and OPGV in cm/sec. Values for "time" are values of t_R for that solute calculated at u_{opt} or OPGV.

235

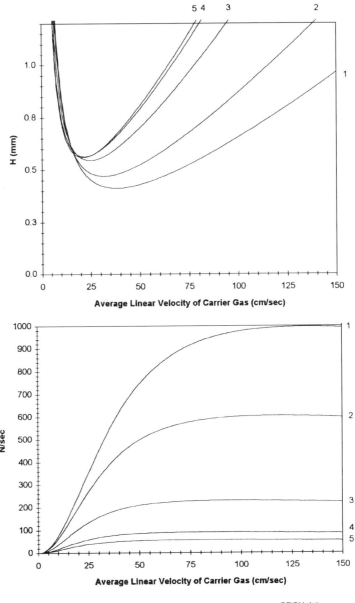

Plot	k	Hmin (mm)	uopt (cm/s)	N	time (s)	OPGV (cm/s)	N/s (max)	N	time (s)
		uopt data				OPGV data			
1	0.564	0.413	38.1	36,319	61.53	133.1	996	17,545	17.62
2	0.917	0.470	31.8	31,923	90.55	117.6	604	14,760	24.45
3	2.426	0.546	24.7	27,490	207.64	95.5	229	12,328	53.82
4	6.416	0.564	21.9	26,610	507.79	83.1	90	12,100	133.87
5	10.436	0.560	21.2	26,789	810.11	79.2	57	12,319	216.63

Fig. 6.71. Column: d_c 530 μm, L 15 m, d_f 3.0 μm; helium carrier. Values assume ideality and are calculated for isothermal conditions. Values of H in mm, u_{opt} and OPGV in cm/sec. Values for "time" are values of t_R for that solute calculated at u_{opt} or OPGV.

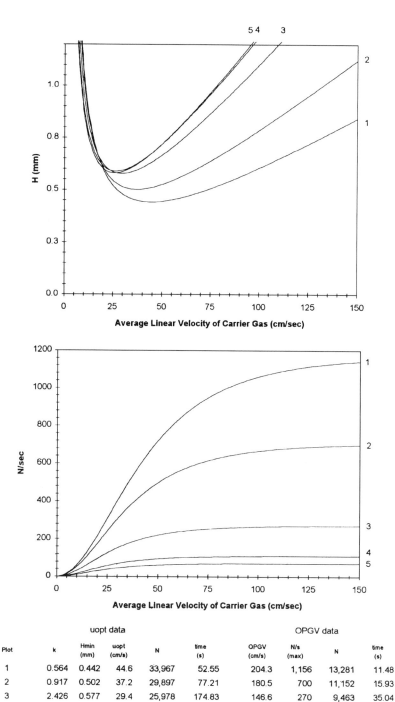

| | | uopt data | | | | | OPGV data | | |
Plot	k	Hmin (mm)	uopt (cm/s)	N	time (s)	OPGV (cm/s)	N/s (max)	N	time (s)
1	0.564	0.442	44.6	33,967	52.55	204.3	1,156	13,281	11.48
2	0.917	0.502	37.2	29,897	77.21	180.5	700	11,152	15.93
3	2.426	0.577	29.4	25,978	174.83	146.6	270	9,463	35.04
4	6.416	0.588	26.5	25,494	419.68	127.8	110	9,563	87.05
5	10.436	0.580	25.9	25,849	663.15	121.9	70	9,886	140.75

Fig. 6.72. Column: d_c 530 μm, L 15 m, d_f 3.0 μm; hydrogen carrier. Values assume ideality and are calculated for isothermal conditions. Values of H in mm, u_{opt} and OPGV in cm/sec. Values for "time" are values of t_R for that solute calculated at u_{opt} or OPGV.

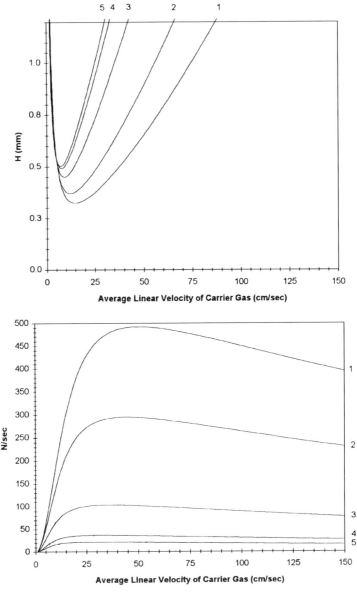

		uopt data					OPGV data		
Plot	k	Hmin (mm)	uopt (cm/s)	N	time (s)	OPGV (cm/s)	N/s (max)	N	time (s)
1	0.564	0.324	14.8	92,667	316.81	50.6	492	45,625	92.79
2	0.917	0.371	12.2	80,934	469.75	44.5	294	38,006	129.12
3	2.426	0.450	9.1	66,682	1,123.72	36.0	103	29,460	285.22
4	6.416	0.493	7.7	60,877	2,901.14	31.3	37	26,153	710.37
5	10.436	0.503	7.2	59,661	4,742.79	29.8	22	25,428	1,149.67

Fig. 6.73. Column: d_c 530 μm, L 30 m, d_f 3.0 μm; nitrogen carrier. Values assume ideality and are calculated for isothermal conditions. Values of H in mm, u_{opt} and OPGV in cm/sec. Values for "time" are values of t_R for that solute calculated at u_{opt} or OPGV.

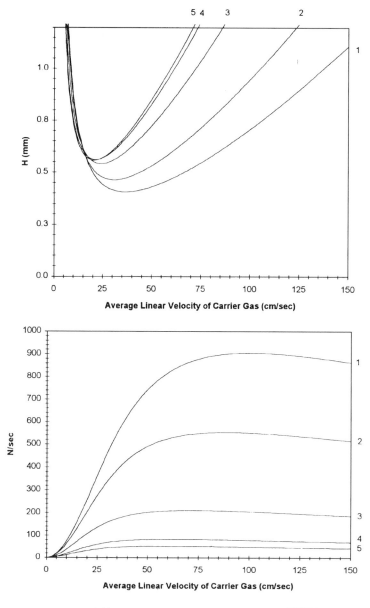

Fig. 6.74. Column: d_c 530 μm, L 30 m, d_f 3.0 μm; helium carrier. Values assume ideality and are calculated for isothermal conditions. Values of H in mm, u_{opt} and OPGV in cm/sec. Values for "time" are values of t_R for that solute calculated at u_{opt} or OPGV.

		uopt data				OPGV data			
Plot	k	Hmin (mm)	uopt (cm/s)	N	time (s)	OPGV (cm/s)	N/s (max)	N	time (s)
1	0.564	0.406	36.6	73,817	128.02	100.3	905	42,313	46.76
2	0.917	0.464	30.7	64,724	187.26	88.9	553	35,736	64.67
3	2.426	0.541	24.0	55,486	427.73	72.6	210	29,780	141.55
4	6.416	0.560	21.2	53,524	1,047.44	63.4	83	29,000	350.84
5	10.436	0.558	20.5	53,808	1,673.11	60.5	52	29,404	567.08

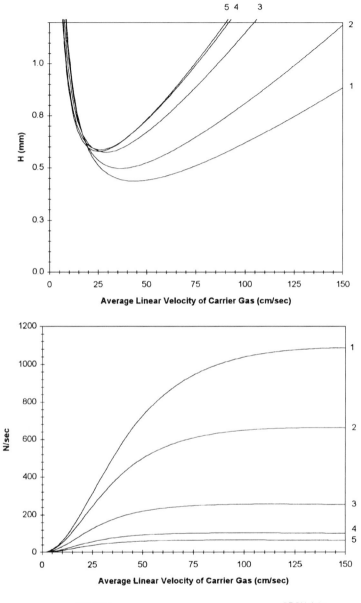

Fig. 6.75. Column: d_c 530 μm, L 30 m, d_f 3.0 μm; hydrogen carrier. Values assume ideality and are calculated for isothermal conditions. Values of H in mm, u_{opt} and OPGV in cm/sec. Values for "time" are values of t_R for that solute calculated at u_{opt} or OPGV.

| Plot | k | uopt data | | | | OPGV data | | | |
		Hmin (mm)	uopt (cm/s)	N	time (s)	OPGV (cm/s)	N/s (max)	N	time (s)
1	0.564	0.437	43.8	68,685	107.05	155.4	1,086	32,781	30.19
2	0.917	0.497	36.7	60,347	156.83	137.8	661	27,581	41.75
3	2.426	0.574	29.0	52,287	354.69	112.5	255	23,324	91.36
4	6.416	0.586	26.1	51,205	852.67	98.4	103	23,391	226.23
5	10.436	0.578	25.4	51,874	1,348.81	93.9	66	24,086	365.41

240

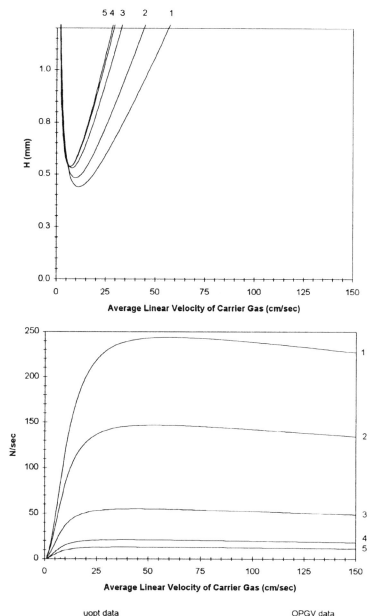

		uopt data				OPGV data			
Plot	k	Hmin (mm)	uopt (cm/s)	N	time (s)	OPGV (cm/s)	N/s (max)	N	time (s)
1	0.936	0.439	11.3	34,185	257.24	59.9	244	11,825	48.49
2	1.523	0.482	9.7	31,128	391.30	53.1	147	10,459	71.28
3	4.028	0.530	7.9	28,316	952.69	44.2	55	9,385	170.74
4	10.654	0.538	7.1	27,884	2,450.19	39.4	21	9,336	444.18
5	17.327	0.535	6.9	28,021	3,988.30	37.8	13	9,433	727.50

Fig. 6.76. Column: d_c 530 μm, L 15 m, d_f 5.0 μm; nitrogen carrier. Values assume ideality and are calculated for isothermal conditions. Values of H in mm, u_{opt} and OPGV in cm/sec. Values for "time" are values of t_R for that solute calculated at u_{opt} or OPGV.

241

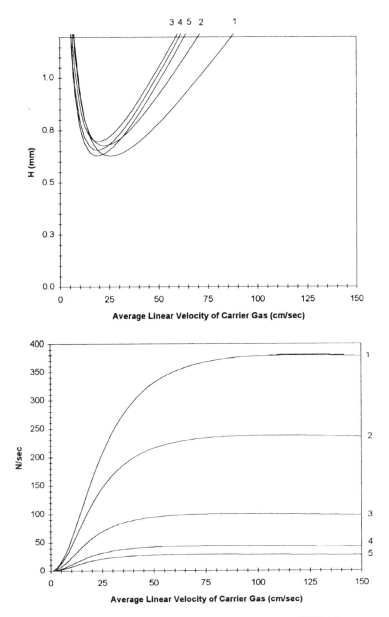

		uopt data				OPGV data			
Plot	k	Hmin (mm)	uopt (cm/s)	N	time (s)	OPGV (cm/s)	N/s (max)	N	time (s)
1	0.936	0.628	25.7	23,893	113.04	121.7	380	9,058	23.86
2	1.523	0.676	22.5	22,182	168.37	108.4	238	8,290	34.89
3	4.028	0.697	19.6	21,519	384.84	90.8	100	8,327	83.04
4	10.654	0.655	19.0	22,890	920.41	81.1	44	9,473	215.43
5	17.327	0.629	19.0	23,854	1,448.81	78.0	29	10,189	352.63

Fig. 6.77. Column: d_c 530 μm, L 15 m, d_f 5.0 μm; helium carrier. Values assume ideality and are calculated for isothermal conditions. Values of H in mm, u_{opt} and OPGV in cm/sec. Values for "time" are values of t_R for that solute calculated at u_{opt} or OPGV.

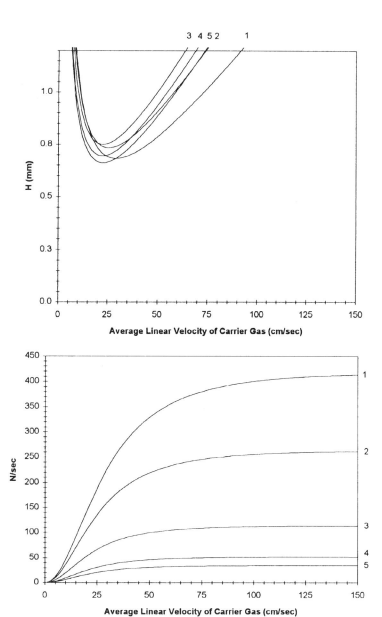

Plot	k	uopt data				OPGV data			
		Hmin (mm)	uopt (cm/s)	N	time (s)	OPGV (cm/s)	N/s (max)	N	time (s)
1	0.936	0.683	29.2	21,968	99.33	186.4	414	6,456	15.58
2	1.523	0.734	25.7	20,442	147.16	166.1	261	5,940	22.77
3	4.028	0.749	22.8	20,013	330.97	139.3	113	6,112	54.12
4	10.654	0.695	22.6	21,595	775.01	124.7	51	7,207	140.14
5	17.327	0.661	22.8	22,687	1,206.26	120.0	34	7,902	229.17

Fig. 6.78. Column: d_c 530 μm, L 15 m, d_f 5.0 μm; hydrogen carrier. Values assume ideality and are calculated for isothermal conditions. Values of H in mm, u_{opt} and OPGV in cm/sec. Values for "time" are values of t_R for that solute calculated at u_{opt} or OPGV.

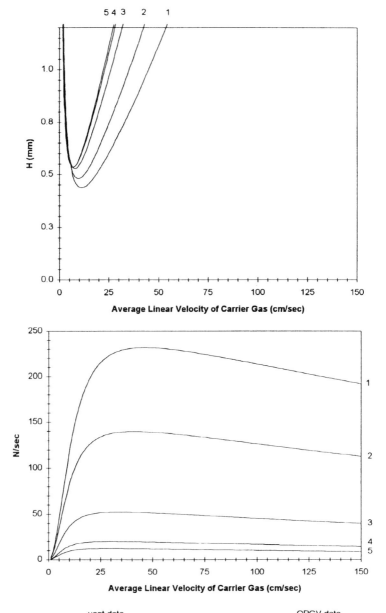

		uopt data				OPGV data			
Plot	k	Hmin (mm)	uopt (cm/s)	N	time (s)	OPGV (cm/s)	N/s (max)	N	time (s)
1	0.936	0.437	11.2	68,684	520.67	46.1	232	29,167	125.87
2	1.523	0.480	9.6	62,483	791.15	41.0	140	25,798	184.58
3	4.028	0.529	7.8	56,759	1,925.29	34.2	52	23,083	440.53
4	10.654	0.537	7.1	55,840	4,953.57	30.6	20	22,884	1,143.70
5	17.327	0.535	6.8	56,096	8,064.89	29.4	12	23,091	1,871.94

Fig. 6.79. Column: d_c 530 μm, L 30 m, d_f 5.0 μm; nitrogen carrier. Values assume ideality and are calculated for isothermal conditions. Values of H in mm, u_{opt} and OPGV in cm/sec. Values for "time" are values of t_R for that solute calculated at u_{opt} or OPGV.

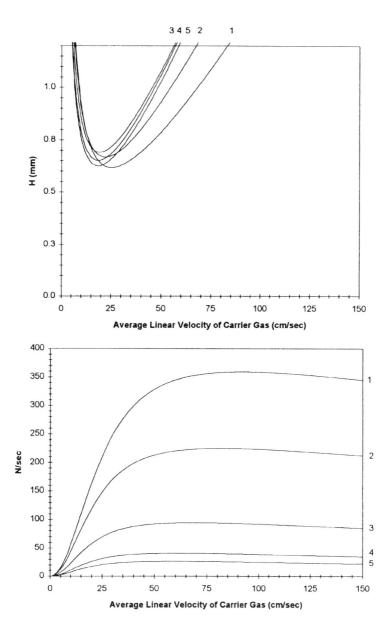

Plot	k	uopt data				OPGV data			
		Hmin (mm)	uopt (cm/s)	N	time (s)	OPGV (cm/s)	N/s (max)	N	time (s)
1	0.936	0.617	25.3	48,597	230.00	92.0	359	22,667	63.15
2	1.523	0.667	22.1	45,003	342.20	82.2	225	20,693	92.08
3	4.028	0.690	19.3	43,498	783.44	69.1	94	20,565	218.13
4	10.654	0.650	18.6	46,133	1,882.13	62.0	41	23,075	564.27
5	17.327	0.625	18.5	48,007	2,970.21	59.6	27	24,654	922.72

Fig. 6.80. Column: d_c 530 μm, L 30 m, d_f 5.0 μm; helium carrier. Values assume ideality and are calculated for isothermal conditions. Values of H in mm, u_{opt} and OPGV in cm/sec. Values for "time" are values of t_R for that solute calculated at u_{opt} or OPGV.

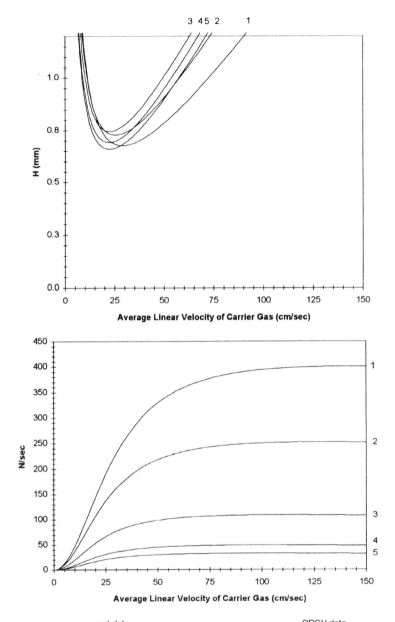

		uopt data				OPGV data			
Plot	k	Hmin (mm)	uopt (cm/s)	N	time (s)	OPGV (cm/s)	N/s (max)	N	time (s)
1	0.936	0.676	29.0	44,377	200.09	142.2	400	16,355	40.85
2	1.523	0.728	25.5	41,234	296.32	127.1	252	15,009	59.55
3	4.028	0.745	22.6	40,293	667.21	107.0	109	15,311	140.93
4	10.654	0.691	22.3	43,412	1,566.54	96.0	49	17,855	364.03
5	17.327	0.658	22.5	45,574	2,442.19	92.4	33	19,469	594.76

Fig. 6.81. Column: d_c 530 μm, L 30 m, d_f 5.0 μm; hydrogen carrier. Values assume ideality and are calculated for isothermal conditions. Values of H in mm, u_{opt} and OPGV in cm/sec. Values for "time" are values of t_R for that solute calculated at u_{opt} or OPGV.

Values of $\bar{u}_{opt}$ will be lower for columns with thicker films of stationary phase and/or for columns with films thicker than $\sim 0.5\,\mu$m of stationary phases that contain appreciable amounts of polar functional groups (D_S usually decreases with the degree of polarity).

The graphs presented for microbore columns are restricted to hydrogen carrier gas. Helium is an impractical carrier for these very fine-diameter columns. Graphs for the longer microbore columns will be of very limited utility in most cases because of the very high head pressures that would be required. Even a 10 m microbore column may demand head pressures of 60–70-psi hydrogen.

The plots in Figs. 6.4–6.81 all illustrate variation in H (upper plots) and N/sec (lower plots) as a function of $\bar{u}$ for the system specified in each individual figure caption. These plots were generated with a Microsoft Excel 4.0 spreadsheet linked to various macros written in Microsoft Visual Basic. In each case, these macros incorporate the formulae and equations detailed in Chapter 5.

To facilitate their utility for quick reference (i.e., not only as a way to estimate the effects of carrier gas velocity on specifically dimensioned column, but also to illustrate general trends that can be derived from the influence of column length, diameter, stationary phase film thickness, carrier gas viscosity, solute retention factor, and stationary phase diffusivity. Toward that goal, these plots have been organized so as to illustrate the behavior of the same five solutes under the same isothermal conditions on columns ranging from a 5 m $\times$ 0.05 mm, d_f 0.2-μm column to a 30 m $\times$ 0.53 mm, d_f 5.0-μm column. In all cases, it should be assumed that the stationary phase closely resembles a pure polydimethyl siloxane.

Those familiar with similar plots that appeared in earlier editions of this work will note that these new plots properly account for both the relationship between solute retention factor and temperature as well as variation in pressure at each point along the column.

6.10 Columns for Mass Spectrometry

The combination of gas chromatography with mass spectrometry is a powerful one. If compounds are to be characterized on the basis of their mass spectral fragmentation patterns, results are most useful if the spectrum is limited to ions derived from a single molecular species. Gas chromatography, as the most powerful separation technique, has a high potential for supplying the mass spectrometer with one molecular species at a time. In addition, the two modes of characterization (gas chromatographic retentions and mass spectral patterns) are truly orthogonal (i.e., they depend on entirely different molecular properties). This greatly increases the reliability of identifications that are based on both retention characteristics in a high-resolution system and matching fragmentation patterns.

In most applications, the mass spectrometer can be envisioned as a specialized detector, attached to the end of the chromatographic column. This detector is intolerant of background signal, and even moderate column bleed levels may require excessive signal attenuation. While bleed causes problems with the mass selective detector (MSD), the problems can be much more serious to the newer ion trap detector (ITD) because of their extreme bleed sensitivity. These complications triggered column manufacturers efforts that culminated in a new generation of low bleed columns (Chapter 4). One manufacturer adds the suffix "MS" to these part numbers to denote their special utility for mass spectrometry.

The mass spectrometer operates under high-vacuum conditions, and the gas flow that can be accepted is governed largely by the pumping capacity of that unit. The column may be connected through some type of separator or open split that diverts the major portion of the carrier gas, or low-flow (true capillary) columns can be plumbed directly into the ion source. The latter alternative requires sufficient pumping capacity on the part of the mass spectrometer to maintain the necessary vacuum while accepting the total carrier flow through the column. This, in turn, requires that the pressure drop through the column be sufficient to ensure both good chromatography in the column and adequate vacuum in the mass spectrometer.

The pressure drop through a 530-μm-diameter column is usually too low to permit it to be connected directly into the ion source of a conventional mass spectrometer. As is the case with packed columns, some type of pressure restriction or interface will be required. In general, the pressure drop required for direct interfacing can be provided by least 20 m and preferably 30 m of a 320-μm column, 15 m and preferably 25–30 m of a 0.25-μm column, or shorter lengths of smaller-diameter columns.

Direct connection of the column into the high-vacuum region of the mass spectrometer also requires vacuum tight connections. The porosity of standard graphite ferrules is slight and they are rarely faulted for general chromatographic purposes, where their superior compressibility is used to advantage. That porosity is sufficient to cause problems under high-vacuum conditions, and most authorities prefer graphitized Vespel ferrules for the high-vacuum column connection. The limited compressibility of these ferrules demands more precise tolerances between the ferrule hole and the column outer diameter and may require sizing the ferrule hole with jeweler's drills.

The zirconia ceramic connector described in Section 6.7 offers several advantages. Not only does it exhibit a much lower leak rate than other modes of column–mass spectrometer interfacing; it facilitates column changing. The "Connex" connector is usually affixed to the inlet end of the fused silica transfer line that terminates in the analyzer section of the spectrometer. To change columns, the filament is switched off, and as the column is disconnected, a connector containing an undrilled zirconia ferrule is placed on the open end of the transfer line, which has been left in position. This can be done very quickly, and the mass spectrometer, is exposed to air for perhaps 0.25 sec. After the new column is positioned, the zirconia "plug" is removed, and the zirconia ferrule attached to the new column is snapped into position. Again, this requires but a fraction of a second. The mass spectrometer usually regains full vacuum in approximately 25 min, when the filament can be reenergized and the unit is back on line.

References

1. K. Yabumoto, D. K. Ingraham, and W. Jennings, *J. High Res. Chromatogr.* **3**:248 (1980).
2. W. Jennings, *Gas Chromatography with Glass Capillary Columns*, 2nd ed. Academic Press, New York, 1980.
3. W. Jennings and T. Shibamoto, *Qualitative Analysis of Flavor and Fragrance Volatiles by Glass Capillary Gas Chromatography*. Academic Press, New York, 1980.
4. M. F. Mehran, W. J. Cooper, R. Lautamo, R. R. Freemen, and W. Jennings, *J. High Res. Chromatogr.* **8**:715 (1985).
5. D. F. Ingraham, C. F. Shoemaker, and W. Jennings, *J. Chromatogr.* **239**:39 (1982).
6. G. Takeoka, H. M. Richard, M. F. Mehran, and W. Jennings, *J. High Res. Chromatogr.* **6**:145 (1983).
7. M. F. Mehran, W. J. Cooper, and W. Jennings, *J. High Resol. Chromatogr.* **7**:215 (1984).
8. M. F. Mehran, M. G. Nickelsen, N. Golkar, and W. J. Cooper, *J. High Res. Chromatogr.* **13**:429 (1990).
9. G. Takeoka and W. Jennings, *J. Chromatogr. Sci.* **22**:177 (1984).
10. M. F. Mehran, W. J. Cooper, M. Mehran, and W. Jennings, *J. Chromatogr. Sci.* **24**:142 (1986).
11. W. Jennings and G. Takeoka, in *Analysis of Volatiles: Methods and Applications* (P. Schreier, ed.), p. 63. de Gruyter, Berlin, 1984.
12. W. Jennings, *Am. Lab.* **16**:14 (1984).
13. A. E. Coleman, *J. Chromatogr. Sci.* **13**:198 (1973).
14. K. Grob and G. Grob, *J. High Res. Chromatogr.* **5**:349 (1982).
15. G. Schomburg, R. Dielmann, H. Borowski, and H. Husmann, *J. Chromatogr.* **167**:337 (1978).
16. V. Paramasigamani and W. Aue, *J. Chromatogr.* **168**:202 (1979).
17. W. Jennings, *J. Chromatogr. Sci.* **21**:337 (1983).

CHAPTER 7

INSTRUMENT CONVERSION AND ADAPTATION

7.1 General Considerations

Some chromatographic instruments still in use today were originally supplied as packed column instruments and remain dedicated to packed column use. In most cases, it is a fairly simple matter to convert such an instrument to open tubular columns. In years past, such conversions necessitated the purchase of optional capillary accessories from the instrument manufacturer and/or considerable ingenuity on the part of the researcher. The variety of relatively inexpensive items available today greatly simplify retrofitting the packed column instrument. The performance of even an older instrument will be much improved by conversion to open tubular columns, provided the response of the electronics is sufficiently fast [1].

7.2 Oven Considerations

The fused silica open tubular column is characterized by a very low thermal mass and quickly responds to radiant heat. The heating coils of most modern instruments are mounted behind a barrier that prevents radiant heat from reaching the column. In some other instruments, a line of sight exists between the column and the oven heater; this can cause anomalous chromatographic behavior.

If the oven temperature selector is abruptly increased by 100°C, the heating element is fully energized. However, under normal operation, including isothermal and routine programmed conditions, the heating element flickers on and off in response to feedback from the sensor-controlled circuitry. Where a line of sight exists between the heating element and the column, the column is exposed to short bursts of radiant heat, to which it quickly responds. When the column mounting geometry is such that the column is uniformly exposed to these bursts of radiant heat, chromatographing bands are repetitively accelerated and decelerated as the heating element flickers. It is more common, however, that the exposure is not uniform: portions of the column are shielded from the heating element, sometimes by other portions of the column, and sometimes by parts of the column cage or mounting hardware. Under these conditions, the front of a chromatographing band can be exposed to a burst of radiant heat, while the rear of that band is shielded. This quickly degrades the band, resulting in split and/or malformed peaks (i.e., the "Christmas tree" effect [2, 3]). The problem is easily demonstrable, and has been rectified by placing a barrier of aluminum foil to block the heater-to-column line of sight and shield the column from radiant heat. The barrier should be placed where it does not interfere with the flow of oven air. Some prefer to loosely shroud the column with foil [4]. Foil barriers are neither necessary nor desirable in other cases.

Some attention should also be directed to suspension of the column within the oven. Column-mounting hardware should support the column cage and should not come into contact with the column. Neither should any portion of the column be allowed to come in contact with oven walls. Except for those portions attached to the inlet and to the detector, the low thermal mass column should be able to follow the temperature of the oven air without restriction. It is also important to column longevity that the column-mounting hardware is located in a position that permits unstrained connections to the injector and detector (see Chapters 2 and 6).

7.3   Carrier Gas Considerations

Many of the advantages offered by modern open tubular columns are attributable to improved methods of tubing deactivation and to the high-purity crosslinked, surface-bonded polymers with which the tubing is coated. The lifetime of any column, however, is shortened by abuse, and subjecting the stationary phase to attack by gasborne impurities is one form of abuse. In general, most stationary phases can be adversely affected by many carrier gasborne impurities; chromatographic performance and column longevity can be affected by the carrier gas quality. While the silarylene siloxane polymers described in Chapter 4 exhibit improved resistance to thermal and oxidative degradation, these, too, will benefit from the use of higher-quality carrier gas. To ensure optimum performance and extended column life, instrument conversion should also involve a critical evaluation of the gas lines and gas purification devices.

Because of these considerations, the premium grades of carrier gas ("five-nines," or "99.999%") are usually less expensive in the long run. Stainless-steel-diaphragm regulators and flow controllers are strongly preferred. Oxygen contamination of the carrier gas by diffusion through neoprene diaphragms and O-rings has been documented many times. A minimum number of fittings should be used, and all fittings and lines should be metal. Even at room temperature, oxygen diffuses through most plastics, including Teflon, and the rate of diffusion increases with temperature.

The superiority of hydrogen as a carrier gas was discussed in Chapter 5. The dependability of hydrogen generators has improved greatly in recent years, and these offer a convenient source of high-purity hydrogen. One cautionary note: in an effort to measure the outputs of a hydrogen generator, a skilled analyst recently used two flowmeters, one attached to the O_2 outlet, one attached to the H_2 outlet; the flow meter attached to the O_2 side exploded.

To form an explosive mixture with air, the hydrogen concentration must be within the range of 4%–10% by volume. Because it was the oxygen side that exploded and considering that these concentration limits would be lower if the dilutent were oxygen rather than air, the most probable explanation is that the procedure forced a slight degree of mixing of the separated gases in the generator, and a static charge led to ignition on the O_2 side; the H_2 side was probably well above the explosive range (i.e., too "rich"). Fortunately, there no injuries resulted in this particular incident.

The other choice is to use bottled gas. While the user should recognize the potential for explosion, hydrogen can be used with a high degree of safety. The best way to avoid reaching explosive concentrations of hydrogen is to limit the amount of hydrogen that can be discharged into any confined space. If hydrogen is supplied by means of a pressure-controlled line, an unregulated volume of hydrogen can be discharged through a leak or a broken column; this can result in a dangerous situation. The safety of any hydrogen-supplied installation can be improved by employing separate mass flow controllers at the outlet of the hydrogen tank, and for each and every subsequent hydrogen demand function (Fig 7.1).

In the case of bottled gas, a two-stage stainless-steel-diaphragm pressure regulator should be installed on the hydrogen cylinder, followed by a moisture trap and a high-capacity oxygen scrubber (both mounted vertically), in that order. Some oxygen scrubbers also react with water, so the gas should be dried before it is conducted through the oxygen scrubber, and thence through a "master flow controller" (stainless steel diaphragm) to a low-pressure hydrogen supply line. The toggle valves, through which each line supplying a separate hydrogen demand accesses the manifold, permit shutting down the hydrogen supply to a given detector or a given injector without disturbing the flow controller settings.

Separate stainless-steel-diaphragm mass flow controllers are positioned along the supply line to direct the reduced-pressure hydrogen to each separate demand point. That for a flame ionization detector might be set to deliver 30 cm^3/min; that supplying an on column injector attached to a 0.32-mm capillary might be set to 3 cm^3/min. The required amounts of hydrogen at these various points are

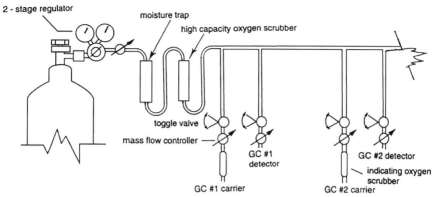

Fig. 7.1. Schematic of a mass-flow-controlled hydrogen system, supplying several instruments with hydrogen carrier and detector hydrogen. As described in more detail in the text, when each hydrogen outlet is limited by a mass flow controller, as contrasted to a simple pressurized line, the possibility of discharging an unrestrained stream of hydrogen into a confined space is limited. Hydrogen is a superior carrier gas, but should always be handled with respect.

then totaled, and the "master" flow controller set slightly above this value. If, for example, the system is supplying three FIDs at 30 cm³/min, two split injectors at 100 cm³/min, and one on-column injector at 3 cm³/min, the total demand is 293 cc/min. The master flow controller is then set to 300 cm³/min. It plays no role in gas regulation, unless the main supply line is ruptured. At that point, it would become limiting, and discharge of hydrogen gas into the laboratory atmosphere would be limited to 300 cm³/min. In operation, the hydrogen discharge through any particular line is limited to the setting of the mass flow controller supplying that line, and the possibility that the hydrogen concentration in the space where that line terminates could attain a concentration of 4% even in the event of a broken column (which would discharge into the oven), has been greatly diminished.

Traps and scrubbers should always be mounted *vertically* rather than *horizontally*. Packings in the traps have a tendency to settle, and as packings settle, a packing-free flow path exists across the top of the horizontally mounted trap. The path of least resistance lies above the packing, and the gas passes through without being affected. With vertically mounted traps, the gas flow must continue to percolate through the packing, even if settling is severe.

Carrier gas lines should be metal rather than plastic. One model of retractable on-column injector requires some flexibility on the carrier gas line and is normally supplied with a length of heavy-walled Teflon. On GC/MS installations where that has been replaced with a length of loosely coiled 1/16-in. stainless steel tubing, a marked reduction in oxygen contamination has been noted.

The degree to which oxygen causes column deterioration is influenced by the type of stationary phase and the temperatures to which the column is exposed.

These interrelationships were touched upon in Chapter 4, and are discussed further in Chapter 10. For systems that are susceptible, some insurance against oxygen entry through leaky fittings is desirable. Additional indicating oxygen scrubbers should also be installed immediately before each inlet system to minimize the amount of oxygen coming in contact with the column. Because many parts of the system (including septa, flow controller O-ring seals, pressure-regulator diaphragms, elastomeric lines) are permeable to oxygen, a completely oxygen-free system is rarely possible.

7.4   Packed–Large-Diameter Open Tubular Column Conversion

Using simple reducing adapters, the 530-μm large-diameter open tubular column can be directly substituted for the packed column, because it must merely be attached directly to the conventional packed column injector, it can operate under carrier gas flows as high as 30 cm^3/min, and it can use the same detector without modification. This is the simplest and most direct conversion. The much higher gas flow volumes that are often used with these large-diameter columns can duplicate those used by the packed column and impose no additional demands in terms of operator technique or training. The benefits of this conversion include instrumental simplicity and user-friendliness of the converted instrument.

The larger-diameter column, of course, encompasses a larger volume of mobile (gas) phase per unit of column length, and to maintain a more nearly normal phase ratio, thicker films of stationary phase are required. This thicker film further benefits the perceived inertness of the columns. In addition, the columns are normally used with conventional packed column injectors and operated without splitting or purging; the entire sample is passed through the column to the detector. Both of these factors—the greater inertness of the column and the capability of accepting direct undiverted injections—increase the quantitative reliability of the system.

Large-diameter open tubular columns have special utility in purge-and-trap systems. Higher-volume gas flows reduce the times required to purge the sample to the trap, and to desorb that trapped sample to the column. It is often desirable to use the same carrier source during sample purge, desorption, and analysis. This is facilitated by using a high flow (i.e., Megabore) column.

Large-diameter columns offer another advantage over conventional capillaries: their greater circumference translates to a larger surface area (ergo more stationary phase) per unit length, even at constant film thickness. If the phase ratio (β) is held constant, the amount of stationary phase is even larger, since the film thickness must then be increased in proportion to the increase in column radius [see Section 1.13 and Eq. (1.15)]. More stationary phase increases the sample accepting capacity of the column, and the dynamic range of the system.

On the other hand, large-diameter columns offer a lower N/m ratio, and should not be considered as true capillary columns. Most analysts would be better advised to begin with the new generation of 0.45-mm-ID columns. These offer all the advantages of the 0.53-mm columns, and because their N/m ratios are ~20% higher, their separation potential is higher, and analysis times can be shorter.

The packed column chromatographer works with a very forgiving system, but pays a high price: inferior separations, lowered sensitivities, longer analysis times, and a less inert system. But the first injections made by even an experienced packed column chromatographer into a capillary system can be traumatic. True capillary columns (e.g., ID 180–320 μm) are more demanding, both in terms of instrument design, and in terms of operator technique. An initial conversion to one of the larger-diameter open tubular columns offers a convenient path to modernization, because these columns can be just as forgiving as the packed column. When the analyst feels comfortable with either of the large-diameter columns operating in high flow mode, he or she should gradually increase its separation potential by reducing the carrier flow. Lower flows place more emphasis on the injection efficiency and other aspects of operator technique. As the flow drops below 10 mL/min, makeup gas should be added to the detector, and at a flow of ~3–5 mL/min, the column is approaching its optimum flows (see Chapter 6).

It is now time to consider downsizing to 0.32 mm, 0.25 mm, or smaller diameter columns. While the user friendliness decreases, the increase in the N/m ratio permits the analyst to generate superior separations in equivalent times, or equivalent separations in much shorter times. The future lies in shorter columns of smaller diameter.

Inlet Conversion

Conversion kits are available from several supply houses. The simpler kits provide a glass injector liner and adapters that reduce the column connections on the packed column injector and detector to 1/16-in. fittings. Even the simplest conversion kits are usually satisfactory if these columns are limited to high-flow, low-resolution mode operation and if the size of the injected sample is restricted to about 1 μL. Proper adapter design, which is discussed in Chapter 3, becomes increasingly important with larger sample injections and/or higher resolution (i.e., low-flow) operation. Figure 7.2 shows an example of a top-sealing adapter kit, and Fig. 7.3 illustrates one suitable for bottom-sealing packed column injectors terminating in a 1/4-in. Swagelok fitting. For the analysis of sensitive compounds, it can be advantageous to deactivate the sample-contacting glass liner. A suitable deactivation procedure is described in Section 7.7.

Detector Conversion

Most adapters for the detector side are similar to a reducing tee, the sidearm of which is normally capped when the column is operated in high-flow mode. With FID, the outlet end of the column should terminate inside the flame jet so it is swept by the combustion hydrogen (Fig. 7.4 [6]). Some flame jets have internal

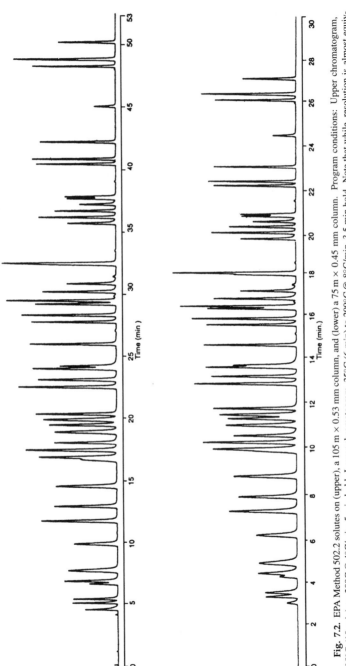

Fig. 7.2. EPA Method 502.2 solutes on (upper), a 105 m × 0.53 mm column, and (lower) a 75 m × 0.45 mm column. Program conditions: Upper chromatogram, 35°C (10 min) to 200C @ 4°C/min, 5 min hold. Lower chromatogram, 35°C (6 min) to 200°C @ 8°C/min, 3.5 min hold. Note that while resolution is almost equivalent, the upper run required 51 minutes, while the lower run required just over half that time, or 27.5 minutes.

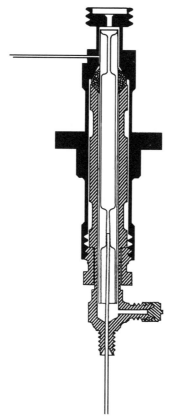

Fig. 7.3. An adapter for installing open tubular columns on one model of a top-sealing specially threaded injector.

diameters that are too small to accept large-diameter open tubular columns. These require a "butt-type" junction, which is usually suitable in high-flow mode but may lead to peak broadening in low-flow mode, and active sites within the flame jet may cause problems. Several jet options are available for some popular instruments and can be provided by the instrument supplier on inquiry.

Makeup Gas Considerations

The large-diameter open tubular column can be operated without makeup gas by capping the adapter sidearm (Fig. 7.5). Alternatively, that port can be used for the addition of make-up gas. Makeup gas can serve two useful purposes:

1. In cases where solutes are eluted under conditions of low volumetric flow, closely eluting bands have opportunities to remix before they reach the zone of

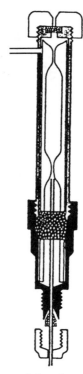

Fig. 7.4. An adapter for installing open tubular columns on one design of a bottom sealing injector.

detection, and they also stay too long in the detector. This results in broadened peaks and loss of resolution. These problems are minimized in a well-designed FID by housing the end of the column within the flame jet. Eluting solutes are discharged into the stream of combustion hydrogen, ensuring their rapid transport to and from the zone of detection and producing clean, sharp peaks from short chromatographic bands.

2. Makeup gas may permit the detector to be operated in a plateau region of high sensitivity [7]. Most detectors are designed to operate under packed column flow conditions, and sensitivity can be seriously eroded by lower carrier gas flows or by changes in the ratios of the detector gases (for FID, this would be the ratio

Fig. 7.5. Effect of make-up gas on the performance of large diameter open tubular columns [6]. Column, 30 m × 0.53 mm, with helium flow at (left to right) 10, 8, 6, 4, and 2 mL/min; upper chromatograms without make-up gas. Nitrogen make-up at 30 mL/min was added for the bottom chromatograms. See text for discussion.

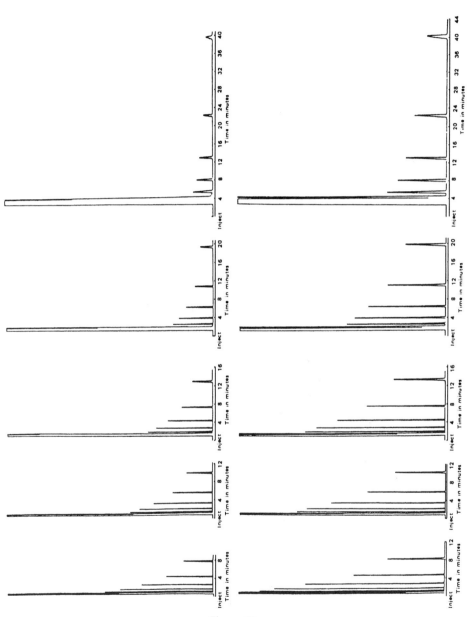

Figure 7.5.

carrier : combustion hydrogen : air). Most FIDs exhibit reasonably flat response (i.e., a plateau) and optimum sensitivity at gas flow ratios approximating 1:1:10 for nitrogen, hydrogen, and air, respectively. In addition, the total flow volume must be commensurate with the size of the jet orifice: the linear flow velocity through the orifice must be sufficient to maintain the flame, but with a given orifice size, higher flow volumes result in linear velocities so high that the flame is self-extinguishing; i.e., the gas velocity is so high that the flame "blows out."

The first makeup gas benefit described above is rarely applicable to large diameter open tubular columns, but advantages are derived from the second. As shown in Fig. 7.6 [6], makeup gas has little or no effect on peak shape, even in low-flow mode, provided the outlet end of the column is housed within the flame jet. It can, however, produce enhanced sensitivity. Makeup gas is discussed in greater detail in a later section.

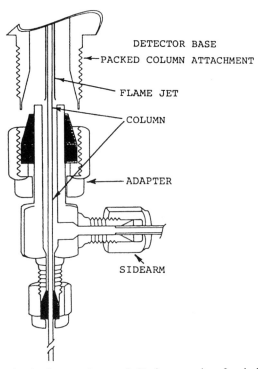

Fig. 7.6. Schematic of a detector adapter, suitable for conversion of packed column detectors to open tubular operation. The column should normally terminate within the stem of the flame jet. The side arm (which provides the entry point for make-up gas) is often capped off when operating in high flow mode (carrier gas at 10 mL/min or higher).

7.5   Packed–Capillary Column Conversion

Many aspects of this subject have been considered previously [1]. In general, the conversion will require the addition of a capillary-compatible inlet and addition of makeup gas at the detector. The latter point was discussed above, and the various types of capillary-compatible inlet are compared in Chapter 3. Kits that permit retrofitting of split, splitless, and on-column inlets to almost any packed column instrument are available from several supply houses.

7.6   Makeup Gas Considerations

Because the C_M term of the van Deemter equation is unimportant for the packed column, nitrogen is often used as the mobile phase in packed column gas chromatography. The FID gases under these conditions are (approximately) 30 cm^3/min nitrogen carrier, 30 cm^3/min combustion hydrogen, and 300 cm^3/min air. In the open tubular column, it is the C_M term that is limiting and hydrogen and helium become the superior carriers. By using hydrogen or helium as carrier while employing nitrogen as a makeup gas, detector sensitivity can be maintained while enjoying superior open tubular gas chromatography. The type and amount of makeup gas should be selected with a view to the recommendations of the detector manufacturer (use as makeup what they recommend as packed column carrier) and Table 7.1.

7.7   Inlet Deactivation

The column is usually blamed for problems that are sometimes evidenced by active solutes, including solute tailing (usually attributable to a reversible adsorption) and partial or total loss of the solute (irreversible adsorption). Where this is, in fact,

TABLE 7.1

Makup Gas Recommendations

Eletector	Makeup	Total flow makeup + column (cm^3/min)
Flame ionization	Nitrogen	30–40
Thermal conductivity	Same as carrier	<5[a]
Electron capture	Nitrogen/argon + methane	30–60[a]
Thermionic	Helium	10–30
Flame photometric	Helium	30–60
Hall	Helium	20–40

[a]Determined by detector volume; consult detector manual.

the fault of the column, corrective measures can sometimes be taken. These are considered in Chapter 10. Frequently, however, the problem does not lie in the column but is associated with active sites in the inlet.

The vaporizing forms of injection include direct flash vaporization and hot on-column injection with the large-diameter (e.g., Megabore) open tubular columns and split, splitless, and programmed temperature vaporization with smaller diameter (capillary) open tubular columns. The vaporization chamber is generally glass and should be readily removable to facilitate cleaning out the injection residues. Materials allowed to remain in the heated inlet continue to oxidize and generally become acidic. Not only do these residues react with solutes of later injections, but their continued degradation can generate baseline problems and eventually "poison" the column (see Chapter 10). Inlets are probably best cleaned by immersion in concentrated HNO_3, rinsed in deionized water, and deactivated. Dichromate cleaning solutions should be avoided, as the high concentration of chromium ion may complicate later steps.

Even a clean inlet can cause activity problems. Conventional glass contains both appreciable amounts of metal oxides and an abundance of surface silanols (Chapter 2). Those constructed of quartz contain less metal ion, but it is sufficient to cause problems. Left untreated, inlets from either material can generate chromatograms testifying to excessive activity, even with the most inert fused silica column. The body of knowledge accumulated from problems encountered (and eventually solved) with conventional glass capillary columns can now stand us in good stead. Metal ions in the glass surface can be removed by a leaching process. Glass wool and glass beads that may be used in the linear should also be treated. The benefits of using presilylated glass or quartz wool are lost when these materials are subsequently folded and pushed into the liner. As the individual fibers break, new active sites are exposed. A better procedure is subject the entire prepacked liner (containing glass beads, industrial diamonds, and/or glass wool) to the deactivation process.

The reagents used in deactivation are dangerous, and should be handled with great care. The following procedure usually achieves a high degree of deactivation, while minimizing the risks of exposure [8]. Liners are most appropriately deactivated in batches. Several liners in an appropriately sized beaker are totally immersed in 25% HCl or HNO_3, and held overnight at room temperature or 3–4 hr at a temperature of 60–65°C. The process should be repeated if the acid becomes markedly discolored. The acid is decanted, and the beaker of liners is thoroughly rinsed several times to remove all traces of acid, and the metal ions that this treatment has leached from the glass surface. It is then placed in a drying oven at ~150°C. The glass has a dense population of surface silanol groups. Appropriately positioned silanols hydrogen bond to each other, and others hydrate. At temperatures in the vicinity of 180°C, hydrogen-bonded silanols dehydroxylate to formed strained siloxane bridges. These structures resist deactivation by silylation,

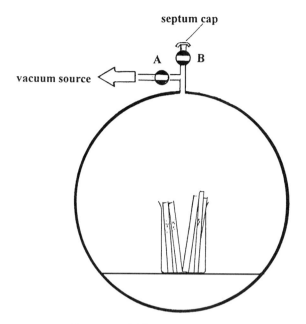

septum cap

vacuum source

A B

1) Immerse assembled liners in 25% HNO₃ overnight;

2) Rinse thoroughly in deionized water;

3) Stand liners in a beaker in vacuum oven, and hold overnight
@ 200°C under 25" vacuum *(ca. 125 mm Hg);*

4) Close valve "A", open valve "B" and inject 10μL neat
HMDS, DMCS, or 50/50 mixture;

5) Close valve "B", open valve "A", and turn oven heat off;
open when cooled to room temperature.

Fig. 7.7. Schematic of a vacuum oven, adapted for use in small part (injector liner) silylation [8]. See text for details of adaptation and operation.

but on exposure to water or water vapor, revert to active silanols. The goal in the drying step is to produce glass surfaces that are thoroughly *dehydrated*, but to *avoid dehydroxylation* [9].

The beaker of liners is now placed in a vacuum oven whose vacuum make–break connection has been adapted to include a septum-sealed inlet (Fig. 7.7). The oven is held at 150°C and a vacuum equivalent to 25 ins. of mercury (∼115 mm Hg or 0.15 torr) overnight. Valve A is then closed, valve B opened, and 10 μL of freshly opened dimethylchlorosilane (DCMS) or hexamethyl disilazane (HMDS) or a 50/50 mixture of the two (*pure reagent; no solvent*) is injected through the septum. Valve B is then closed, the oven temperature increased to 200°C, where a second silylanizing injection is made. Valve A is then opened, and the oven heater turned off. After the oven has cooled, the deactivated liners are stored in sealed containers until needed.

Immediately after each such use, the syringe should be rinsed with methylene chloride or other solvent that will not react with the silylating reagents, and discharged into a waste beaker containing (e.g.), methanol or other material that will dissipate the reagent. Otherwise, water vapor in the air can react with the silylating reagent in the syringe and freeze the plunger to the syringe barrel.

References

1. W. Jennings, *Gas Chromatography with Glass Capillary Columns*, 2nd ed. Academic Press, New York, 1980.
2. F. Monari and S. Trestianu, *Proc. 5th Int. Symp. Capillary Chromatogr.*, Riva del Garda, Huethig, Heidelberg, p. 327 (1983).
3. F. J. Schwende and D. D. Gleason, *J. High Res. Chromatogr.* **8**:29 (1985).
4. S. A. Mooney, *J. High Res. Chromatogr.* **5**:507 (1982).
5. E. Guthrie (J & W Scientific, Inc.), personal communication (1985).
6. M. F. Mehran, *J. High Res. Chromatogr.* **9**:272 (1986).
7. M. M. Thomason, W. Bertsch, P. Apps, and V. Pretorius, *J. High Res. Chromatogr.* **5**:690 (1982).
8. J. A. Settlage, personal communication; see also *J. High Res. Chromatogr.* **17**:739 (1994).
9. W. Jennings, *Comparisons of Fused Silica and Other Glass Columns in Gas Chromatography*. Huethig, Heidelberg, 1983.

CHAPTER 8

SPECIAL ANALYTICAL TECHNIQUES

8.1 General Considerations

Gas chromatography offers an extremely high potential for separating the components of volatile mixtures, but there are occasions when the achievement of a particular goal can be facilitated by modifying the method of sample injection or other parameter(s). This section is not intended as a comprehensive treatment of all the ingenious and specialized methods that have been employed to achieve better results for any given gas chromatographic problems. Instead, it presents a highly selective review of a few of the methods that have been used.

Sometimes it is simply the complexity of the sample that interferes with the isolation and detection of specific components. Heart-cutting techniques, in which a selected group of solutes from one column is redirected to a dissimilar column, can sometimes be used to achieve the necessary separation.

Where the goal is the detection of minor components, injection of a massive volume of sample may be required. Large-volume injections can lead to lengthened sample bands at the beginning of the column, with adverse effects on the lengths of eluted solute bands, on peak widths, and on sensitivities. These problems can be especially troublesome in purge and trap methodologies, which usually require some means of refocusing or shortening the sample band deposited on the column. If it assumed that the injection technique has been optimized, the routes to improved separation can usually be reduced to methods of increasing N, α, or k [Eq. (1.23)]. Temperature and column phase ratios are the most general means of manipulating k and were discussed in Chapter 5. Methods of increasing N and α, methods of achieving shorter bands from massive gas injections, and a few selected applications of heart cutting are considered below.

8.2   Flow Stream Switching

Some approaches to be considered here require redirecting the flow stream before or during chromatography. Both mechanical valves and "valveless" pressure switching devices have been employed for this purpose. The use of fluidic logic elements would seem an ideal way to approach flow stream switching and has been explored by several workers [1–4]. Figure 8.1 illustrates a switch that utilizes two fluidic logic Coanda wall attachment elements. The entering flow stream follows one of the two available flow routes. The flow dynamics then generate a reduced pressure between the flow stream and the wall that holds the flow stream to this route. By delivering a pressure pulse (i.e., a short burst of nitrogen) to the control point at that port, pressure is restored, and the flow stream switches to the alternate route.

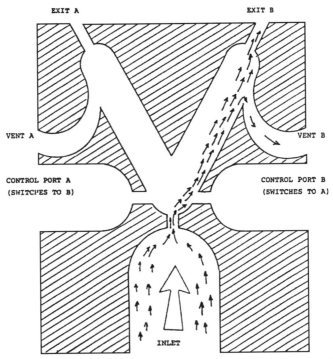

Fig. 8.1. A flow switch utilizing two Coanda fluidic logic devices [1–4]. See text for details of operation.

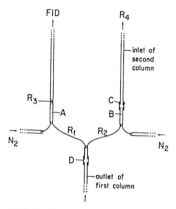

Fig. 8.2. A typical Deans switch [5, 6]. In normal operation, the effluent from column 1 proceeds along the leg R_1 to the FID. A nitrogen pulse at control point "A" switches the column effluent to the second column via R_2. A pulse at control point "B" redirects the stream to R_1 and FID [15]. Reprinted with permission of Huethig Publishing, Heidelberg.

The reduced pressure now generated along the second wall will hold the flow stream in this second position until it is canceled by a pressure pulse at that control port. Commercial fluidic switches are available, but in models that are suitable only with flow volumes appreciably larger than those used with all but the largest open tubular columns. Both fluidic switches and the Deans switch (discussed below) are influenced by differential changes or restrictions in the pressure drops through the alternative flow paths. Differential restrictions in the two flow paths (that can be occasioned by temperature programming) can cause inadvertent switching.

Figure 8.2 illustrates a typical Deans switch [5, 6], which utilizes an imposed pressure at the appropriate point to direct the flow stream through the opposite of the two alternate paths. Both the fluidic switch and the Deans switch can be constructed so that only inert materials are in the sample flow stream. On the other hand, temperature shifts (including those encountered in programmed operation) may change the relative pressures in the different legs and inadvertently trigger switching.

Many of the disadvantages cited earlier for mechanical valves [7, 8] have been overcome. It is, of course, essential that the flow passages through these devices be consistent with those of the column. Excessive volumes associated with the switch would contribute to remixing of partially separated components and to band lengthening. Figure 8.3 illustrates a zero-dead-volume void-filling column-to-valve-port connection that is available in several styles.

Mechanical valves are quite positive in their switching action, and some models now have internal passages that are dimensionally consistent with and designed for use with fused silica columns. With each pass through the valve, samples come in contact with both the valve body (usually stainless steel) and the rotor material

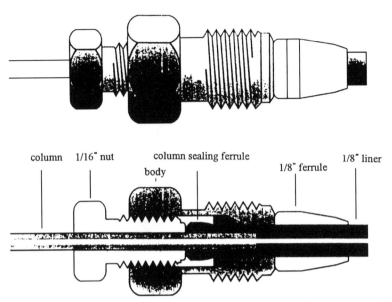

Fig. 8.3. A zero dead volume void-filling column-to-valve-port connection. Various models are available for a variety of applications. Courtesy of Valco, Inc.

(generally a Teflon-filled polyimide). Valve residence times are usually on the or-der of milliseconds, and suitability of these valves for a large number of test solutes has been demonstrated [8]. Nevertheless, they cannot be considered totally inert toward all solutes. Among the other disadvantages that have been found for me-chanical valves are cold-trapping of higher-molecular-weight solutes. Such valves usually have an appreciable thermal mass, and so the temperature of the valve body usually lags behind that of the oven unless the valve is supplied auxiliary heat [8].

8.3   Multidimensional Chromatography

The potential of multidimensional separations has been analyzed in some detail [9]. Forms of multidimensional separations have long been used in gas chromatog-raphy. Many such applications have been well reviewed [9, 10]. The term has oc-casionally been applied to systems where the effluent from a single column is split to two detectors. Usually at least one of these detectors is specific or semispecific. While this generates complementary information, the additional dimension is in the detection rather than the chromatography. Bertsch [10] reserved the term "two-dimensional gas chromatography" for systems containing two columns of different selectivities operated under conditions that eliminate ambiguities in correlating the solutes eluting from the two columns. The flexible fused silica column has made it very simple, for example, to connect columns coated with dissimilar stationary phases to a single inlet so that the sample is simultaneously injected on the different columns, each of which is connected to a dedicated detector. Additional information obtained in such systems is of limited value unless it can be unequivocally established that a particular solute is peak x on column A and peak y on column B (e.g., by coinjection of known standards). In general, the utility of this approach varies inversely with the complexity of the sample. It is difficult to extract useful information from the fact that a mixture of 500 solutes exhibits a certain elution pattern on one column and another elution pattern on a different column.

Where peaks from the one column can be correlated with specific peaks from the second column, two-dimensional systems can be employed to establish precise retention indices on two dissimilar stationary phases and have been used for the assignment of specific identifications. It is important to recognize that retention behavior establishes only that a solute cannot be any number of compounds whose retention characteristics on that column under those conditions are known to be demonstrably different. Retentions, in other words, can never prove what a solute is, but only what it is not. If a particular compound is suspected, retention behavior can establish that if it is present in a given sample it must be a particular peak or a component of that peak.

Assignments based on the agreement of gas chromatographic retentions *and* mass spectral fragmentation patterns have a high degree of validity because the two probes are nonredundant [12, 13], that is, they are based on entirely different solute properties. Assignments based on the agreement of retention characteristics on two dissimilar columns are based on the volatilities of the solute *as affected by its interaction with the different stationary phases*. While superior to retention data from a single column, the method is much less definitive than one based on matching two nonredundant modes of solute characterization.

While two-dimensional chromatography can be performed with relative simplicity in an off-line mode (e.g., [14]), it has been argued that an on-line approach generally produces more reliable results. The latter almost always involves flow stream switching [6].

In flow stream switching, a discrete portion of the total chromatographing band (which can be restricted to a single peak) is diverted from one column onto a second column (heart cutting). Both Deans switches [11] and mechanical valves, discussed in the preceding section, have been used for this purpose. Figure 8.4 illustrates flow switching with an eight port-valve fitted with a sampling loop, designed for application in flavor studies [8]. In the first step (valve positioned as shown at the bottom), effluent from column A passes through the nondestructive PID and the sample loop to the atmosphere, where it can be subjected to sensory (sniff) monitoring. As the valve is turned (top figure), the contents in the sample loop are switched to a second dissimilar column.

Werkhoff [15] used multidimensional GC to resolve a single peak from a raspberry-flavor chromatogram on one column to a second chiral-specific column, which separated the enantiomers (Fig. 8.5). Glausch *et al.* [16] separated

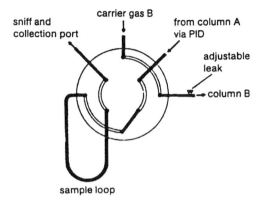

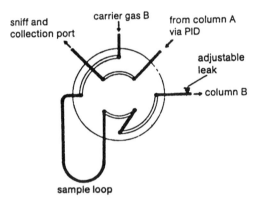

Fig. 8.4. Flow switching with an eight port valve used in flavor studies. In the upper position, column B is furnished with carrier gas in a stand-by mode, while effluent from column A proceeds from a non-destructive detector through a small sample loop to a sniff and collection port. As an interesting aroma appears, the valve is switched to the positions indicated by the lower figure, and the contents of column B are swept to column B for analysis on a second column. The adjustable leak is a splitter to shorten the sample band and sharpen peaks [8].

the enantiomers of PCB 149, by directing a cut from an apolar DB-5 column to a chiral-specific column (Fig. 8.6). Sandra reasoned that the precise location and duration of the heart-cut window were crucial to MDGC. He and his coworkers proposed real-time control of column switching [17].

Trapping in an open tubular column, that was then switched on line in MDGC, has also been explored [18]. Another novel application involved the detection of stilbene hormones in corned beef by a multidimensional approach combining HPLC with MDGC [19].

Dual-oven instruments permit the two columns to be operated under different temperature profiles, and can offer some distinct advantages [8, 20].

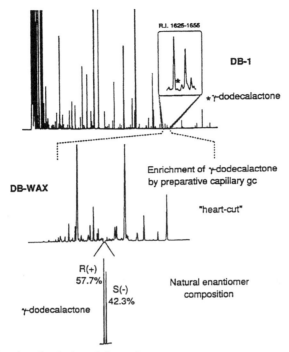

Fig. 8.5. Diversion of a single peak from a DB-WAX column to a second chiral column, which resolved the enantiomers [15].

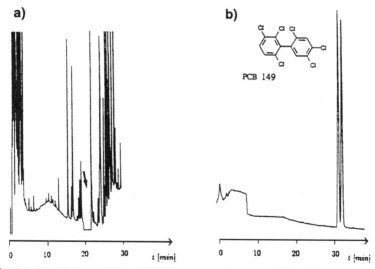

Fig. 8.6. Polychlorinated biphenyl (PCB) congeners from the technical mixture Chlophen A60. The enantiomers comprising isomer 149 were resolved by a heart cut to a second chiral column [16].

8.4   Recycle Chromatography

Some separations could be improved by employing a separation system that developed very large numbers of theoretical plates. Columns of smaller-diameter or longer columns are sometimes used in attempts to achieve this goal (see Chapter 5). Several disadvantages, attributable to the higher pressure drops and steeper van Deemter curves, accompany either of these routes.

In studies of the optimum and optimum practical gas velocities of segments of different length from the same glass capillary column, Yabumoto and VandenHeuvel [21] concluded that the OPGV varied inversely with the length of the column. The primary disadvantage of the short column is that it can possess only a limited number of theoretical plates. Recycling a partially resolved fraction through a short column would offer all of the advantages of the short column while subjecting the solutes to much higher numbers of theoretical plates. Several workers have explored this concept [22, 23].

Recycling requires transport of the chromatographing band from the column outlet back to the column inlet and is best accomplished by mounting a flow-switching device centrally in the column, that is, utilizing two short column segments (Fig. 8.7). Imperfections or excess volume in the switching device would cause longer, more dilute bands and encourage remixing of partially separated solutes. A recycle unit employing a prototype mechanical valve connected to two 20-m columns generated over 2,000,000 theoretical plates on a low-k solute (butane) at the optimum carrier gas velocity [24]. In essence, the solutes were

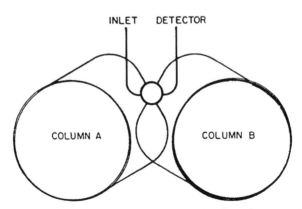

Fig. 8.7. Schematic of a recycle unit [24]. By repetitively shunting a chromatographing band from the outlet of one short column segment to the inlet of a second short column segment, it is possible to generate very large numbers of theoretical plates in comparatively short times. See text for details. Schematics for incorporating the unit into a configuration amenable to heart cutting from an analytical column, and the use of two recycle units containing dissimilar stationary phases for selectivity tuning (discussed below) have also been suggested [25].

chromatographed through 400 m of column, but under conditions commensurate with those encountered in a 40-m column. This required 20 "repass cycles" and a total analysis time of 16 min. When operated at OPGV, a second unit developed over 4000 theoretical plates per second.

Only a limited section of the chromatographing band can be contained within a column segment; the distance between any two components x and y is

$$d_{xy} = n_{cyc}[(k_y - k_x)/(k_y + 1)]L_{seg} \qquad (8.1)$$

where n_{cyc} is the number of full cycles, L_{seg} the length of one segment ($L_{seg1} = L_{seg2}$), and k_x and k_y represent the retention factors of the solutes of the two components [14]. Equation (8.1) can also take the form

$$d_{xy}/L_{seg} = n_{cyc}[(k_y - k_x)/(k_y + 1)] \qquad (8.2)$$

The limit of the recycle system is reached when $d_{xy} = L_{seg}$; if

$$n_{cyc} > (k_y + 1)/(k_y - k_x) \qquad (8.3)$$

some of the chromatographing components must exceed the envelope of the system. Because there is no reason to recycle solutes unless k_x is very nearly equal to k_y, this is not a serious limitation [25].

A more serious problem in recycle chromatography lies in the coordination of the timing of the valve activation with the chromatography. In the studies cited above, the length of time required for passage of the chromatographing band through one column segment was estimated from an initial injection, and an electronic timer was set to switch the valve pneumatically at that interval. Any error in that set switching interval is cumulative and is ultimately evidenced when a valve switch clips one end from the band and directs it instead to the detector. The ideal moment for valve activation on each pass could be better determined by a nondestructive in-line detector installed between the outlet end of each segment and the valve. Unfortunately, few, if any, nondestructive detectors are suitable for in-line operation at the flow volumes commensurate with small-diameter open tubular columns. The 530-μm column is sufficiently large to accommodate a microthermistor bead, and this may permit such an approach.

Recycle chromatography offers a means of applying a greater number of theoretical plates to unresolved solutes, but there is no advantage to recycling solutes that can be resolved by more ordinary means. The recycle unit should utilize isothermal conditions, even though programmed temperatures are preferable for most complex mixtures. Housing the recycle and normal chromatographic units in separating controlled ovens would permit improved flexibility and utility [25]. A band that would otherwise produce an unresolved or multicomponent peak could be redirected to isothermal recycle while the rest of the chromatogram proceeded in a separate unit that could be programmed in the normal manner.

8.5 Specifically Designed Stationary Phases

Some solute mixtures that resist separation on any known stationary phase have been subjected to multidimensional separations, while other efforts in this area have led to the synthesis of well-defined stationary phases, precisely designed to maximize the separation factors of specific groups of solutes.

The first efforts in this direction employed binary stationary phase mixtures and, although often credited to later authors, can be traced to work of Maier and Karpathy [26]. Their approach can be illustrated by considering a simple hypothetical system consisting of three components—1, 2, and 3—which yield only two peaks when chromatographed on either of two different stationary phases, A and B. On stationary phase A the two peaks consist of 3 and 1 + 2, whereas on stationary phase B they consist of 2 and 1 + 3 (Fig. 8.8, top).

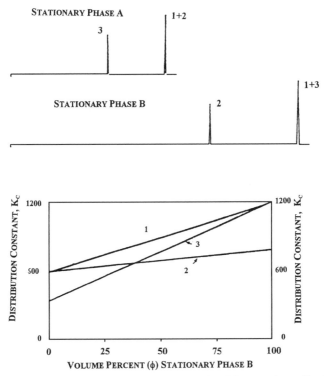

Fig. 8.8. Simplified explanation of the Maier and Karpathy [26] approach to utilization of binary mixtures of stationary phases. Distribution constants (K_c) for each solute are calculated from the isothermal chromatogram on stationary phase A, and plotted on the left ordinate of the bottom graph. Similarly, K_cs of each solute on the isothermal chromatogram (*same temperature*) on stationary phase B; these are plotted on the right ordinate of the bottom graph. K_c for each of these solutes in any specified mixture of A + B can now be predicted from a line connecting the two values for that solute.

Maier and Karpathy established that the distribution constant of a solute in a mixture of two stationary phases amounted to the distribution constant in one phase times the fractional volume of the mixture occupied by that phase, plus the distribution constant in the other phase times the volume fraction of that phase:

$$K_{c(A+B)} = K_{c(A)}\phi_{(A)} + K_{c(B)}\phi_{(B)} \tag{8.4}$$

where $\phi_{(A)}$ and $\phi_{(B)}$ represent the volume fractions of A and B in the binary mixture, respectively. The relationship $K_c = \beta k$ [Eq. (1.14)] permits calculation of distribution constants on columns of known β, or, if β is constant, solute retention factors can be used instead. These concepts were applied to the simple hypothetical system presented above in constructing the graph in Fig. 8.8 (bottom), where the retention factors of each solute in stationary phase A (0 vol% B) are on the left ordinate and values in stationary phase B (100% B) are on the right ordinate.

Equation (8.4) established that the K_c (or at constant β, the k) of a solute in a binary mixture is a linear function of the volume fraction of either stationary phase. Hence the lines in Fig. 8.8 (bottom) represent the retention factors of the indicated solute in any binary mixture of A and B ranging from 100% A + 0% B to 0% A + 100% B. This graph makes it apparent that solutes 2 and 3 will exhibit the same K_c (and coelute) in a mixture containing 37% stationary phase B in A. The differentiation of all solute retention factors will be greatest (and solutes will be most widely dispersed) at mixture compositions where the lines are most widely spaced. One such point occurs at 22.5 vol% B, where the solute elution order will be 3, 2, and 1 (order of the K_c values). Another is at 55 vol% B, where the solute elution order will be 2, 3, and 1. Maier and Karpathy also suggested that the desired proportions of the two stationary phases could be achieved by (1) use of a packing coated with the proper mixture of A and B; (2) blending in a single column the correct proportions of packings that had been separately coated, one with A and one with B; or (3) series coupling of predetermined lengths of two columns, one containing stationary phase A and the other stationary phase B.

Laub and Purnell [27, 28] reasoned that the composition of the ideal mixture could be more precisely defined by plotting the separation factors of each solute pair as a function of the volume percent of one stationary phase in that binary mixture. The lines formed by connecting those points produce a "window diagram" (Fig. 8.9), and the highest "window" occurs at the optimum stationary phase mixture, where the separation factors of all solutes are maximized.

Ingraham et al. [29] used a computerized version of this approach to establish that a mixture of 5 vol% Carbowax 20M in polydimethylsiloxane would yield the best separation of alcoholic fermentation products whose differentiation had previously required two separations on different columns. These same concepts were then applied to another mixture, whose components were also reportedly inseparable on any single column [30]. Figure 8.10 shows a window diagram generated by plotting the separation factors of every possible solute pair of that

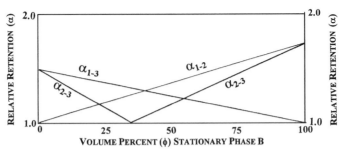

Fig. 8.9. Window diagram for the hypothetical mixture examined in Fig. 8.8 [27, 28]. There are two "windows," one at 22.5, the other at 55 volume percent stationary phase B in the binary mixture A + B. See text for discussion.

mixture as functions of the volume percent of DB-1701 (right axis) in DB-1 (left axis). Windows appear at 25 and at 66 vol% DB-1701. Because the solutes are more retained in DB-1701, solute retention factors would be larger in the latter window and the magnitude of the $[(k + 1)/k]^2$ multiplier of Eq. (1.22) would be significantly smaller (see Table 1.3). A 66-vol% DB-1701 column could therefore achieve the separation with fewer theoretical plates than a 25-vol% DB-1701 column. The appropriate lengths of capillaries coated with the two stationary phases were then coupled to achieve the separation.

It was also noted that because of the diffusivity and velocity gradients engendered by the pressure drop through the coupled columns, the segment at the inlet

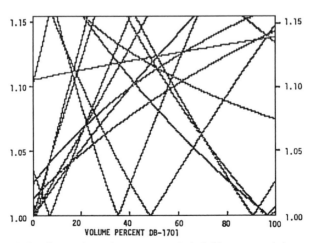

Fig. 8.10. Window diagram for a mixture composed of ethyl hexanoate, ethyl octanoate, geranial, geraniol, limonene, linalyl acetate, myrcene, neral, nonanol, octanal, octyl acetate, and terpinene-4-ol [30]. See text for discussion.

end influenced the separation more (and the outlet end influenced it less) than would be predicted on the basis of their respective lengths. In other words, the results generated by a coupled column containing equal-length segments separately coated with stationary phases A and B are influenced by the velocity and diffusivity gradients through the column. In a practical sense, this is often determined by which segment is used at the inlet end.

A graphical approach to determining column length fractions required to produce the desired results was employed by Takeoka *et al.* [30]. Although this worked well, Purnell and Williams [33] argued that it was possible to calculate precise correction factors that would compensate for the pressure-drop-engendered deviations. However, they offered no experimental data to compare with their calculated values, and Krupcik *et al.* [34], who also tried calculated corrections, reported that the calculated values did not agree with the deviations that they encountered experimentally.

Mehran *et al.* applied the coupled-column approach to separate a series of priority pollutants that had resisted separation on any single column [35], and suggested that the pressure-drop-engendered deviations could be eliminated by synthesis of a new stationary phase containing the indicated concentrations of the functional groups in the two separate stationary phases. This approach led to the subsequent synthesis of a new stationary phase, DB-1301, specifically designed for the separation of that series of pollutants [36].

8.6   Selectivity Tuning

The basic concept of selectivity tuning can be demonstrated by coupling two geometrically similar columns containing two dissimilar stationary phases, A and B. The chromatogram generated by a heterogeneous solute mixture differs, depending on whether A or B is used as the inlet end; this phenomenon was referred to above. The separation is invariably influenced more by the inlet end, and less by the outlet end, than would be predicted on the basis of their length fractions [30].

A number of workers have explored effects engendered by coupled columns operated at different temperatures and/or pressures (e.g., [37–39]), and others (e.g., [40–42]) have studied the separation effects exercised by carrier flow (i.e., residence time) in the different column segments.

Hinshaw and Ettre [41] used pneumatics similar to those shown in Fig. 8.11 to demonstrate the effects of what they appropriately termed "relative retentivity" in a coupled column system. Figure 8.12 shows results generated by these authors on a lime oil sample.

Selectivity tuning has also been widely used in LC (e.g., [43]). In view of the simplicity of the adaptations required for the chromatographic equipment and the potential for selectively adjusting the relative retentions of problematic solutes, it seems surprising that these methods have not been more widely adopted in the field.

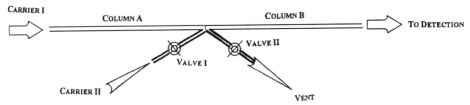

Fig. 8.11. The concept of "selectivity tuning," adapted from Hinshaw and Ettre [40, 41]. Columns A and B are coupled together, with a "Y" providing a port at the point of junction. To gain a higher contribution from column A (and a lesser contribution from column B), the micro-metering valve I is opened, introducing a controlled amount of additional carrier gas (carrier II is at a slightly higher pressure than carrier I). This increases the residence time in (and the contribution from) column A, and decreases the time in (and the contribution from) column B. The degree to which these contributions are affected is controlled by valve I. To gain a higher contribution from column B (and a lesser contribution from column A), valve I is closed, and valve II is opened. This decreases the time in (and the contribution from) column A, and increases the time in (and the contribution from) column B. Again, the magnitude of these effects is controlled by valve II.

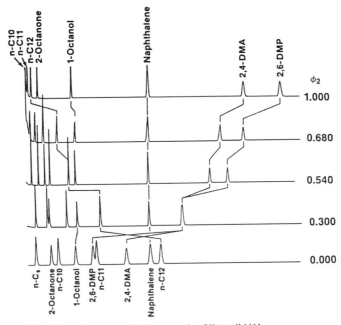

Fig. 8.12. Selectivity tuning, as applied to a sample of lime oil [41].

8.7   Vapor Samples and Headspace Injections

Headspace injections not only have an appealing simplicity, but also allow the investigator to avoid sample preparation procedures that may engender qualitative and quantitative changes in the composition of the sample [44] and changes that can occur in the interval between sample preparation and sample analysis, even under conditions of low-temperature storage [45]. Minimization of artifact formation, and simplification of the analytical process are strong incentives for the use of direct sample injections whenever possible.

On the other hand, headspace injections are normally dominated by the more volatile components of the sample, and the detection of higher-boiling and/or trace components has been difficult. Techniques such as purge and trap [46] were developed largely to permit prechromatographic isolation of detectable quantities of trace components from volumes of vapor too massive for their direct injection. Although these methods extend the utility of vapor phase analysis, problems related to inadequate retention of some components by the trapping substrate ("breakthrough," usually of the more volatile constituents) and inadequate recovery of other components from the trapping substrate (sometimes resulting from posttrapping degradation on the substrate) have also been reported (see, e.g., [44]).

The primary benefit of purge-and-trap techniques is that larger volumes of vapor can be sampled. The limits of detection can be considerably extended, provided the volatiles entrained in that sample can be recovered from the gaseous matrix and introduced onto the chromatographic column as a short starting band. Toward the latter goal, the carrier gas stream is normally employed to discharge the contents of the loaded trap into the column. The trap is usually heated during this step, while the column (or a portion thereof) is chilled.

By using a syringe with a very fine needle, it is also possible to eliminate the intermediate trapping and discharge steps and deposit the vapor sample directly inside the chilled column [47]. Fused silica needle syringes have been used for this purpose quite successfully. The method has also been used for the detection of halocarbons in water, with detection limits reportedly extending into the parts-per-trillion range [48]. Work now in progress indicates that by injecting headspace samples as large as 10 cm^3, sensitivities for even higher-boiling and strongly polar solutes in aqueous samples can be greatly enhanced. Solutes contained in the vapor sample must be given sufficient opportunity to cold trap as a focused band in the first chilled section of the column. To avoid an elongated sample band and broader peaks, lower injection speeds should be employed for samples dominated by lower-boiling solutes, or where the column is chilled with carbon dioxide as opposed to liquid nitrogen. The speed of injection should also be decreased as the sample size is increased.

A major limitation of these procedures is that they require considerable manual manipulation and dexterity. Encouraging results have been obtained in preliminary

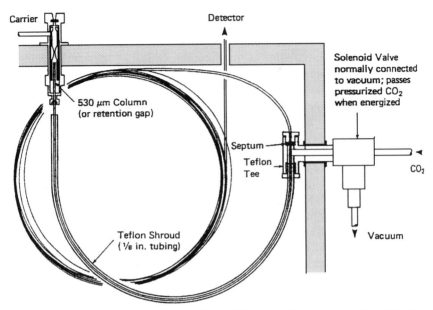

Fig. 8.13. Schematic of a device for achieving band focusing with on column headspace injections. The coolant is introduced counter to the carrier flow, while heat is introduced in the same direction as carrier flow. This achieves band focusing in both the trapping and the "injection" steps.

attempts to simplify the introduction and removal of coolant in such a way as to enhance shortened bands (Fig. 8.13), which may make possible full automation of this procedure for vapor phase sampling.

8.8   Fast Analysis

Segments of the chromatographic community currently exhibit a rekindled interest in faster analyses. Much of this interest is found in commercial environmental analytical laboratories, where large numbers of samples are subjected to essentially the same analysis. Partly in response, at least one equipment manufacturer has exploited the concept of using shorter, smaller-diameter columns, and utilizing computer programs to predict temperature and pressure programs to enhance results and minimize time.

As pointed out by Gaspar [49], by exploiting column parameters and working conditions, it is possible to perform fast analyses on conventional equipment. Extracolumn factors play major roles in ultrafast analyses (i.e., analytical runs of a few seconds), and their correction usually mandates specially designed instrumentation. Assuming short columns coated with thin films of high-diffusivity stationary phase and hydrogen carrier gas, Gaspar argues that

$$\bar{u}_{\text{opt}} = 8 D_M f(k)/d_c \qquad (8.5)$$

and

$$H_{\min} = d_c/f(k) \qquad (8.6)$$

These expressions assume that "band broadening" occurs only in the column. The validity of this assumption is higher with long columns and low linear gas velocities. High-speed analyses lean toward short columns and higher gas velocities, and extracolumn contributions become increasingly important.

Gaspar *et al.* [50] modified the Golay equation to include the effects of extracolumn contributions, as

$$H = B/\bar{u} + C\bar{u} + D\bar{u}^2 \qquad (8.7)$$

where D includes contributions from the injector, detector, connections, electrometer, and recorder or data-handling equipment:

$$D = \sigma_{\text{EC}}^2/[(k+1)^2 L] \qquad (8.8)$$

where σ_{EC}^2 is the sum of extracolumn variances expressed in time units.

The time necessary to generate one theoretical plate has been expressed [50, 51] as

$$T_H = \sigma^2/t_R \qquad (8.9)$$

From this, they derive Eq. (8.10):

$$T_H = (k+1)[(B/\bar{u}^2) + C + D\bar{u}] \qquad (8.10)$$

from which they derive the minimum T_H at an optimum velocity expressed as

$$\bar{u}' = (2B/D)^{1/3} \qquad (8.11)$$

This is the velocity at which the tangent, drawn through the origin, touches the van Deemter (or Golay) $H/\bar{u}$ curve. For all practical purposes, this is the same OPGV discussed in Chapters 1 and 5.

The value of H at this velocity can be calculated as

$$H' + \frac{3}{2}(2B)^{2/3} D^{1/3} + (C/D^{1/3})(2B)^{1/3} \qquad (8.12)$$

This corresponds to the optimum practical gas velocity, OPGV [51].

Gaspar also presented equations for approximating the optimum column length, and the optimum practical gas velocity. The former took the form

$$L' = 1.92 \left[(\sigma_{\text{EC}}^2 B^2 N_{\text{req}}^3)/(K+1)^2 \right]^{1/4} \qquad (8.13)$$

and the latter

$$u' = 1.56\{[B^2 N_{req}(k+1)^2]/\sigma_{EC}^2\}^{1/4} \tag{8.14}$$

To minimize analysis times, Gaspar derives the expression shown as

$$t_R = (L/\bar{u})(k+1) \tag{8.15}$$

Analysis time, of course, is dictated by t_R of the last component. Analysis time, then, varies directly with the column length and inversely with the carrier gas velocity. The wisdom of first determining the minimum length of column required [Eq. (8.13)] and the optimum gas velocity [Eq. (8.14)] now becomes evident. This can also take the form

$$t_R = N_{req}(k+1)H/\bar{u} \tag{8.16}$$

which again is entirely logical; analysis time varies directly with the number of required theoretical plates, and the magnitude of H (small values of H indicate a more efficient system, developing a given number of plates in a shorter column), and indirectly with the gas velocity.

Leclercq and Cramers combined the Golay–Giddings and Poiseuille equations to derive expressions for calculating maximum plate numbers and minimum time conditions for given columns at fixed but selectable (including vacuum) outlet pressures [52]. Cramers [53] concluded that the required analysis time for a given resolution in a low-pressure-drop system (short or large diameter column) can be expressed as

$$t_{R(p_i/p_o)=1} = \frac{4}{3}\{[(1+k)(11k^2+6k+1)]/k^2\}R^2 r_c^2[\alpha^2/(\alpha-1)^2](p_o/pD_{m,1}) \tag{8.17}$$

where R is the resolution between two sequentially eluting peaks, r_c is the column radius, and $p_1 D_{m,1}$ represents the solute-carrier gas diffusion coefficient at unit pressure p_1. Hence, for low-pressure-drop columns, t_R is proportional to $(r_c)^2$.

In a high-pressure-drop system (long column, small diameter column),

$$t_{R(p_i/p_o \to \infty)}$$
$$= 24\{[(1+k)(11k^2+6k+1)]/k^3\}[\alpha^3/(\alpha-1)^3]R^3 r_c[2\eta/pD_{m,1}]^{1/2} \tag{8.18}$$

where η is the dynamic viscosity of the carrier gas. The influence of column radius has declined, while the importance of stationary phase selectivity (α) has increased.

These considerations indicate that the retention time is a complex function of stationary phase selectivity, column length and radius, and the thickness of the stationary phase film. Varying these parameters can affect analysis time, and can also affect inlet pressure, which may necessitate other instrumental requirements. The dimensions of the column ultimately selected can also affect the maximum sample capacity and the minimum amount that can be detected with reliability.

References

1. R. L. Wade and S. P. Cram, *Anal. Chem.* **44**:131 (1972).
2. S. P. Cram and S. N. Chesler, *J. Chromatogr.* **99**:267 (1974).
3. G. Gaspar, P. Arpino, and G. Guiochon, *J. Chromatogr. Sci.* **15**:256 (1977).
4. R. Annino, M. F. Gonnord, and G. Guiochon, *Anal. Chem.* **51**:379 (1979).
5. D. R. Deans, *Chromatographia* **1**:18 (1968).
6. G. Schomburg, F. Weeke, F. Mueller, and M. Oreans, *Chromatographia* **16**:87 (1983).
7. R. J. Miller, S. D. Sterns, and R. R. Freeman, *J. High Res. Chromatogr.* **2**:55 (1979).
8. W. Jennings, S. G. Wyllie, and S. Alves, *Chromatographia* **10**:426 (1977).
9. J. C. Giddings, in *Multidimensional Chromatography: Techniques and Applications* (H. J. Cortes, ed.), pp. 1–27. Dekker, New York, Basel, 1990.
10. W. Bertsch, in *Multidimensional Chromatography: Techniques and Applications* (H. J. Cortes, ed.), pp. 74–144. Dekker, New York, Basel, 1990.
11. W. Bertsch, *J. High Res. Chromatogr.* **1**:187,289 (1978).
12. D. R. Deans, *J. Chromatogr.* **203**:19 (1981).
13. D. H. Freeman, *Anal. Chem.* **53**:2 (1981).
14. J. C. Giddings, *Anal. Chem.* **39**:1027 (1967).
15. P. Werkhoff, S. Brennecke, and W. Bretschneider, *Chem. Mikrobiol. Technol. Lebensm.* **15**:47 (1993).
16. A. Glausch, G. J. Nicholson, M. Fluck, and V. Schurig, *J. High Res. Chromatogr.* **17**:347 (1994).
17. K. Himberg, E. Sippola, F. David, and P. Sandra, *J. High Res. Chromatogr.* **16**:645 (1993).
18. H. G. J. Mol, H. G. Janssen, and C. A. Cramers, *J. High Res. Chromatogr.* **16**:413 (1993).
19. C. G. Chappell, C. S. Creaser, and M. J. Shepherd, *J. High Res. Chromatogr.* **16**:479 (1993).
20. B. M. Gordon, C. E. Rix, and M. F. Borgerding, *J. Chromatogr. Sci.* **23**:1 (1985).
21. K. Yabumoto and W. J. A. VandenHeuvel, *J. Chromatogr.* **140**:197 (1977).
22. D. Dedford (Phillips Petroleum Co.), U.S. Patent 3,455,090 (1969).
23. A. M. Ried, *J. Chromatogr. Sci.* **14**:203 (1976).
24. W. Jennings, J. A. Settlage, R. J. Miller, and O. G. Raabe, *J. Chromatogr.* **186**:189 (1979).
25. W. Jennings, J. A. Settlage, and R. J. Miller, *J. High Res. Chromatogr.* **2**:441 (1979).
26. H. J. Maier and O. C. Karpathy, *J. Chromatogr.* **8**:308 (1962).
27. R. J. Laub and J. H. Purnell, *J. Chromatogr.* **112**:71 (1975).
28. R. J. Laub and J. H. Purnell, *Anal. Chem.* **48**:799 (1976).
29. D. F. Ingraham, C. F. Shoemaker, and W. Jennings, *J. Chromatogr.* **239**:39 (1982).
30. G. Takeoka, H. M. Richard, M. Mehran, and W. Jennings, *J. High Res. Chromatogr.* **6**:145 (1983).
31. J. C. Giddings, *Anal. Chem.* **36**:741 (1964).
32. J. C. Sternberg, *Anal. Chem.* **36**:921 (1964).
33. J. H. Purnell and P. S. Williams, *J. High Res. Chromatogr.* **6**:799 (1953).
34. J. Krupcik, G. Guiochon, and J. M. Schmitter, *J. Chromatogr.* **213**:189 (1981).
35. M. F. Mehran, W. J. Cooper, and W. Jennings, *J. High Res. Chromatogr.* **7**:215 (1984).
36. M. F. Mehran, W. J. Cooper, R. Lautamo, R. R. Freeman, and W. Jennings, *J. High Res. Chromatogr.* **8**:715 (1985).
37. T. S. Buys and T. W. Smuts, *J. High Res. Chromatogr.* **4**:102 (1981).
38. T. S. Buys and T. W. Smuts, *J. High Res. Chromatogr.* **4**:317 (1981).

39. R. E. Kaiser, R. I. Rieder, L. Leming, L. Blomberg, and P. Kusz, *J. High Res. Chromatogr.* **8**:580 (1985).

40. J. V. Hinshaw and L. S. Ettre, *Chromatographia* **21**:561 (1986).

41. J. V. Hinshaw and L. S. Ettre, *Chromatographia* **21**:669 (1986).

42. J. R. Jones and J. H. Purnell, *Anal. Chem.* **62**:2300 (1990).

43. T. Welsch, U. Dornberger, and D. Lerche, *J. High Res. Chromatogr.* **16**:18 (1993).

44. W. Jennings and A. Rapp, *Sample Preparation for Gas Chromatographic Analysis*. Huethig, Heidelberg, 1983.

45. G. Takeoka, M. Guentert. S. L. Smith, and W. Jennings, *J. Agric. Food Chem.* **34**:576 (1986).

46. T. A. Bellar and J. J. Lichtenberg, *J. Am. Water Works Assoc.* **66**:739 (1974).

47. G. Takeoka and W. Jennings, *J. Chromatogr. Sci.* **22**:177 (1984).

48. M. F. Mehran, W. J. Cooper, M. Mehran, and W. Jennings, *J. Chromatogr. Sci.* **24**:142 (1986).

49. G. Gaspar, *J. Chromatogr.* **556**:331 (1991).

50. G. Gaspar, R. Annino, C. Vidal-Madjar, and G. Guiochon, *Anal. Chem.* **50**:1512 (1978).

51. G. Gaspar, C. Vidal-Madjar, and G. Guiochon, *Chromatographia* **15**:125 (1982).

52. P. A. Leclercq and C. A. Cramers *J. High Res. Chromatogr.* **8**:764 (1985).

53. C. A. Cramers, *J. High Res. Chromatogr.* **9**:676 (1986).

CHAPTER 9

SELECTED APPLICATIONS

9.1 General Considerations

Any attempt to create a compilation of selected applications intended as a guideline must be selective. As a consequence, that compilation can almost always be faulted by individual chromatographers for the omission of examples that they consider critical to their specific fields. On the other hand, some of the included examples can be criticized because developments in gas chromatography continue to occur at a rapid pace; in many cases it is now possible to generate results that are superior to the examples used as illustrations. Developments that have made those improvements possible include (1) better methods of sample pretreatment and storage; (2) improvements in commercial instrumentation, including superior inlet devices that take into account our newer understandings of the injection process; and (3) the availability of better-deactivated columns in a wider range of diameters and coated with a greater variety of stationary phase types and film thicknesses. Some of the chromatograms used as examples in this chapter could now be improved on because of developments of this type that have occurred since those chromatograms were originally obtained. However, although it is true that today's analyst can usually reap the advantages of at least another year or two of developmental improvements, the examples cited can be useful as points of departure.

Application guidelines for stationary phase selection used in catalogs of suppliers such as Supelco, Chrompak, J&W Scientific, and others can also be useful. Intercomparisons are complicated by the fact that columns coated with a particular stationary phase carry different names and codes depending on the manufacturer. In actuality, those columns may in fact be different; most manufacturers purchase stationary phases per se, some refine the purchased preparations, and others synthesize their stationary phases in-house. Qualitative differences are common between different lots of tubing and can also occur in different sections of the same lot. Variations in the deactivation processes, in coating procedures, and in the level of quality control all influence the ultimate behavior of the finished column. Stationary phase citations used here are not intended as endorsements of any particular brand; stationary phase "equivalents" are indicated in suppliers' catalogs and the reference of Yancey [1]. At the same time, it must be recognized that columns containing stationary phases of a given "type" but produced by different manufacturers are not necessarily equivalent. Most stationary phases are obtained from polymer suppliers whose chromatographic sales represent a minuscule portion of a market that is primarily directed to lubricants or detergent chemicals; a column manufacturer resorting to in-house synthesis of stationary phases takes this more expensive route in an effort to increase column quality and reproducibility. In addition, there are a few cases where no equivalent phases exist at the time of this writing.

Another approach to the classification of applications in gas chromatography is to divide pertinent data in the literature into subject matter areas; although some overlap is inevitable, a more serious degree of redundancy can be avoided by assignments to the four major categories: (1) food, flavor, and fragrance; (2) petroleum and industrial chemical; (3) environmental; and (4) biological and medical. Natural products and pheromones are most logically included in the first group, as are chemicals of major significance in the food processing industries (e.g., monitoring of ethylene dibromide and nitrosamines in foods). Fatty acids have been considered under both the food and biological headings. Analyses concerned with synthetic fuels, coal, and shale are considered under the second section. Examples of forensic applications such as the detection of accelerants in arson residues appear in the second section, while measurements of blood alcohol and drugs of abuse are included in the fourth (biological) section. The third section considers the detection of contaminants in air, water, and soil, ranging from halomethanes in water to dioxins in soil. Pesticide analyses appear in both the first and fourth sections; pesticide contamination is usually a matter of environmental concern, but different procedures may be required for the detection of residual pesticides in foods. Separation of enantiomers can be of interest in all of the above fields; the protein amino acids are considered under food, while the other restricted examples cited here were arbitrarily assigned to the biological area.

9.2   Food, Flavor, and Fragrance Applications

Column recommendations for food, flavor, and fragrance applications can be gleaned from suppliers' catalogs or the primary literature in the application area of interest. Most of the objective approaches in flavor research employ gas chromatography, but sensory correlations are required before any flavor significance can be assigned to gas chromatographic patterns. The GC sniffport has become an essential tool for the flavorist and perfumist to directly correlate olfactory response with the response of the chromatographic detector. Sniffport applications are best accomplished with a low-dead-volume "Y" fitting inserted on the end of the column with two short pieces of deactivated tubing, of equal length, connecting the "Y" to the detector and the sniffport. Sniffports with humidified makeup gas provide the best prevention of the fatigue associated with long hours of sniffing hot GC effluent. High capacity columns, with 0.53-mm IDs and thick films are preferred to get the highest concentration of analytes to the nose. Figure 9.1 shows a typical sniffport result. The analyses may be complicated by the fact that the volatiles of interest usually occur at very low concentrations, and they are often dispersed through a matrix containing materials that would, if introduced onto the column, shorten its useful life to a considerable degree; a sample preparation step may be essential. Chromatograms can be extremely complex and are often influenced by the sample preparation methods employed [2].

Figure 9.2 shows chromatograms of black pepper produced by (top to bottom) headspace injection, simultaneous steam distillation/extraction, and Soxhlet extractions with dichloromethane and liquid carbon dioxide [3]. The headspace injection accentuates the extremely volatile constituents and discriminates against higher-boiling components; Soxhlet extraction discriminates against the more volatile solutes and accentuates the higher-boiling substances; results obtained by simultaneous steam distillation/extraction lie between the two. Soxhlet extraction with liquid carbon dioxide is especially useful for fragile essences that could be damaged by heat; Fig. 9.3 illustrates the results of a flower extraction.

The analysis of headspace components from natural systems can be done directly if there is a sufficient concentration of the analytes of interest in the vapor phase. If this critical concentration is not present, a concentration step is necessary. Dynamically passing headspace vapors through absorbent trapping materials in the most common method of headspace sampling. Tenax is the most popular absorbent in these applications. Activated charcoal is also a powerful absorbent because of its thermal and chemical stability and it is not deactivated by water, is gaining popularity as an absorbent in the food and flavor area. Kaiser applied activated charcoal traps to the analysis of flower headspace components in orchids [4]. The components from a single flower were trapped (see Fig. 9.4) onto specially prepared charcoal traps ranging in size from 1 to 5 mg. These were then extracted with 10–50 μL of carbon disulfide prior to analysis. A portion of the resulting chromatogram of *Cattleya labiata* is shown in Fig. 9.5. To obtain complete recovery of the more polar components, extraction with 10–50 μL of ethanol was also required.

Coleman has applied purge-and-trap sampling to the headspace analysis of tobacco and peanuts [5]. The automation of the method is an effective way to characterize new tobacco varieties. The volatile components of sour orange, *Citrus*

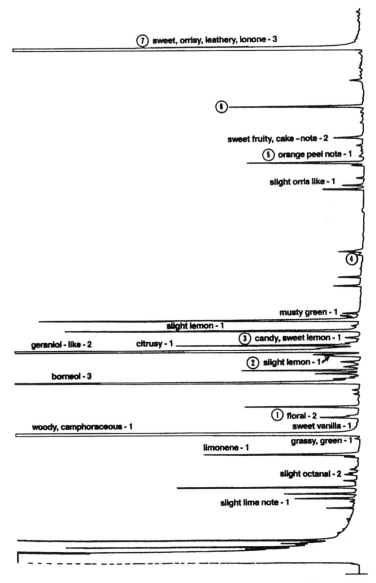

Fig. 9.1. Sample sniff-port chromatogram of lemon oxygenates. Phrases like woody-camphoraceous, sweet-fruity, musty-green and floral are from the trained evaluators' vocabulary of descriptors. Notations are comments of one panel member. The oil was analyzed on a 0.53 mm I.D. 5% phenyl methylsiloxane column. Adapted from [2], and reprinted with permission.

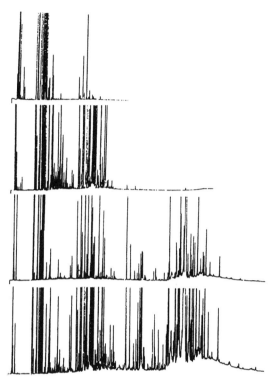

Fig. 9.2. Chromatograms of black pepper volatiles produced by (top to bottom) (1) direct injection of 200 μL of headspace vapor, (2) simultaneous steam distillation-extraction (Nickerson Likens), and Soxhlet extractions with (3) dichloromethane, and (4) high pressure liquid carbon dioxide. Column, 30 m × 0.25 mm, 0.25 μm film of polydimethylsiloxane; 40–280°C @ 4°/min, headspace injection on-column, all other injections split 1 : 100. From [3], and reprinted by permission of the American Chemical Society.

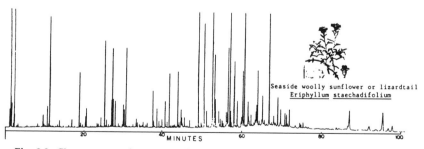

Seaside woolly sunflower or lizardtail
Eriphyllum staechadifolium

Fig. 9.3. Chromatogram of an essence of *Eriophyllum staechadifolium* (seaside woolly sunflower, also known as lizardtail), produced by high pressure liquid carbon dioxide Soxhlet extraction. Chromatographic conditions as in Fig. 9.2. From [3], and reprinted by permission of the American Chemical Society.

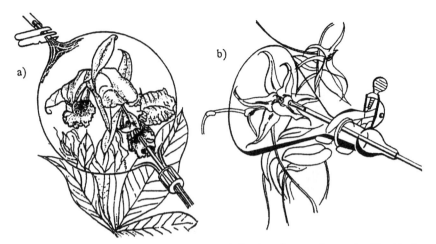

Fig. 9.4. Two methods of trapping headspace volatiles from flowers. A pump is utilized to draw the headspace volatiles through the trapping material. The trapping vessel is located to maximize contribution from the flower as opposed to other plant tissues, taking care to avoid damage to the flower or the surrounding plant. From [4], reprinted with permission.

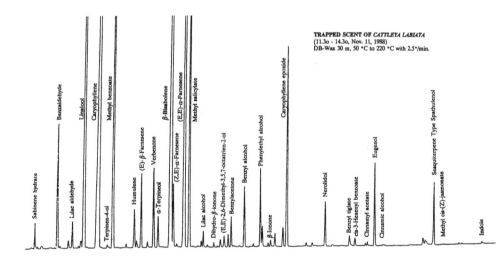

Fig. 9.5. A portion of the trapped headspace (scent) from an orchid, *Cattleya labiata*. The sample was obtained by CS_2 desorption of the charcoal trap, Fig. 9.4. *Adapted from Ref. 4 with permission.*

sudachi, have also been investigated by purge-and-trap GC, and the recoveries were compared with simultaneous distillation extraction (see Fig. 9.6). Two trapping materials, Porapak Q and Tenax GC, were compared. The Porapak Q trap allowed for the identification of three new low-boiling components, isobutyraldehyde, 1-propanol, and (Z)-2-pentanol, and seven new terpenes [6].

Solid phase microextraction (SPME), has also been used to effectively sample the volatiles of hops [7] and cinnamon [8]. In SPME a stationary phase coated fiber is inserted into a container of sample volatiles and allowed to equilibrate. The fiber is then exposed directly to the inside of the injector for desorption, and the chromatographic process is begun (Fig. 9.7) [9]. Volatile and semivolatile components of cinnamon, concentrated by SPME, are shown in Fig. 9.8.

Gas chromatography is widely used in the analysis of the essential oils, where the complexity of the sample can offer a very real challenge. Many of these commercially important products are fragile. Chromatographic separations are employed to quantitate specific compounds that may be indicative of positive or negative quality notes in the oil, and to detect the clandestine addition of addends such as antioxidants or less expensive oil components used as diluents. Lemon oil, which has a long history as an important oil of commerce, can serve as an example. An early form of lemon oil adulteration was the addition of turpentine [10], but modern analytical methods have led instead to dilution with less expensive but normal constituents of lemon oil (e.g., benzyl alcohol [11, 12]). Hence gas chromatography is used not only to establish that the content of such materials is within normal ranges but also to measure other components whose concentrations are known to increase as the oil is abused (e.g., *p*-cymene, myrcene), to quantitate the more desirable constituents (e.g., citral), and to detect preservatives such as butylated hydroxytoluene and butylated hydroxyanisole that are illegal additives in some countries (Fig. 9.9 [13]).

The complexity of the oil is such that the complete separation on any one column has yet to be demonstrated. One of the major commercial laboratories performs uses high-methyl polymethylsiloxane columns for lemon oil analysis. It is possible to achieve a more complete separation on some other stationary phases, but columns coated with these more polar phases exhibit shorter lifetimes. The longer-life methyl silicone column separates all critical components. The same reasoning can be applied to the analysis of a number of other complex mixtures; Fig. 9.10 shows a chromatogram of a peppermint oil.

The characterization of essential oil and flavor components has traditionally been accomplished by a comparison of the retention indices generated on dissimilar stationary phases, usually a methyl silicone and a polyethylene glycol (wax). Several compilations of retention indices are available [14–16], but most of the major flavor houses have constructed proprietary databases. The information is often augmented with mass spectral and/or infrared characteristics. In difficult cases, high resolution Fourier transform nuclear magnetic resonance (FT-NMR)

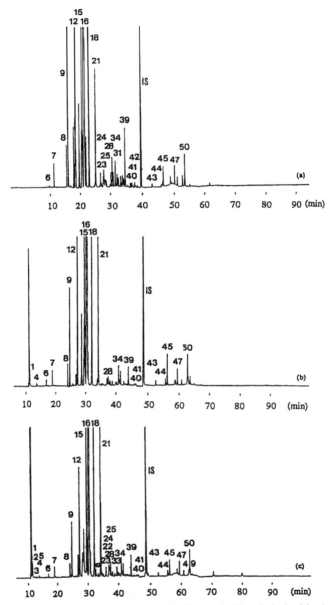

Fig. 9.6. Chromatograms of the volatiles profile from sudachi peel obtained by (top), the SDE method, (center) purge and trap sampling with a Tenax™ GC trap, and (bottom) a Porapak™ Q trap. The column was a 50 m × 0.25 mm I.D. OV-101 with a 0.2 μm film thickness. Adapted from [6] and reprinted with permission.

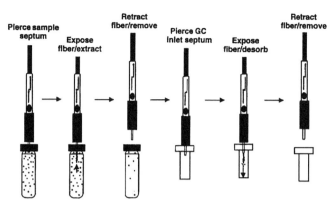

Fig. 9.7. Steps in solid phase micro extraction (SPME). (a) with the fiber retracted, the outer metal sheath punctures the vial, and the stationary phase fiber is then inserted into the sample. In (b), analytes are absorbed onto the fiber, (c) the unit is withdrawn from the vial, and (d) the sheath inserted through the GC septum. In (e), the fiber is extended and the heat of the injector desorbs the analytes. The unit is then withdrawn from the injector. *Courtesy of Supelco Inc.*

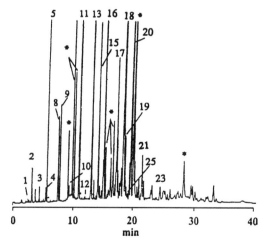

Fig. 9.8. Gas chromatographic profile of headspace volatile compounds sampled by SPME from cinnamon. Extracted compounds range from β-pinene (1) to cinnamic acid (23). From [8], and reprinted with permission.

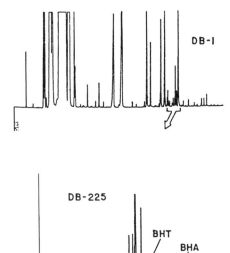

Fig. 9.9. Heart cutting applied to the detection of butylated hydroxy toluene (BHT) and butylated hydroxyanisole (BHA) in a commercial lemon oil. Retentions of the two antioxidants on each of the two dissimilar columns was first established by injections of the pure materials. A narrow "window" that would contain those solutes was then redirected from the 30 m × 0.32 mm dimethylpolysiloxane column to the 30 m × 0.32 mm 50%-dimethyl-50%-phenylcyanopropyl polysiloxane column. From [13], and reprinted with permission of Preston Publications.

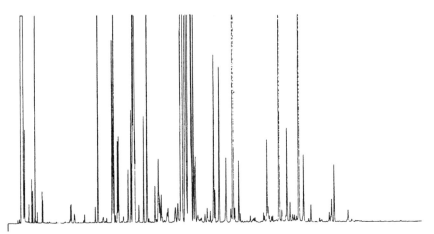

Fig. 9.10. Peppermint oil on a 30 m × 0.32 mm 5%-diphenyl-95%-dimethyl polysiloxane column. On-column injection of 0.5 μL, 40°C (1 min), 60°/min to 60°C, 3°C/min to 250°C.

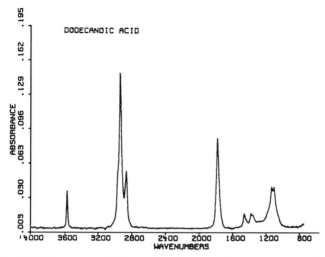

Fig. 9.11. GC/FT-IR spectrum of dodecanoic acid separated from a cognac sample. From [17], reprinted with permission.

has been applied to trapped GC effluents, sometimes collected from multiple injections. The application of GC/FT-IR spectroscopy has also been utilized. The quality of gas phase spectra is usually superior to condensed phase spectra, and can provide structural information that is not readily apparent from MS and NMR data. The GC/FT-IR spectrum of dodecanoic acid, separated from a cognac sample, is shown in Fig. 9.11 [17]. Commercial libraries of spectral data are available to aid such comparisons.

In their studies of the volatile constituents of black tea, Mick and Schreier [18] subjected tea leaf infusions to vacuum steam distillation, solvent extraction, and fractionation on silica gel. Figure 9.12 illustrates the complexity of three such fractions. An extensive GC data set, 44 teas analyzed on a wax column, was subjected to cluster analysis in attempts to correlate the degree of fermentation and the flavor profile of tea [19]. The antimicrobial activity of green tea volatiles have also been studied [20].

The study of thermally generated flavors, especially as derived from cooked meat continues to be important. Studies of the thermal decomposition of thiamin (vitamin B_1) have explored the generation of sulfur-containing volatiles. Figure 9.13 shows the GC-FPD (sulfur-specific detector) chromatograms of heat-treated thiamine solutions at pH 1.5, 7, and 9.5. The samples were obtained by simultaneous distillation/extraction of previously heated aqueous solutions. Over 90 volatile sulfur compounds were characterized from these mixtures [21].

Figure 9.14 shows a chromatogram typical of that produced by headspace injection of vapors overlying ripening banana [22]. Injections of large volumes of

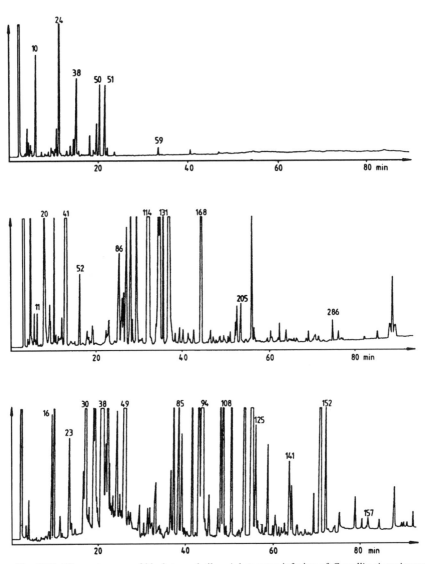

Fig. 9.12. Chromatograms of black tea volatiles. A hot water infusion of *Camellia sinensis* was subjected to steam distillation, and the concentrated pentane-dichloromethane extract fractionated by gradient elution from a silica gel column. The fraction shown in the top chromatogram was least polar and eluted with pentane; the fractions shown in the center and bottom chromatograms eluted with increasing concentrations of ethyl ether. Column, 30 m × 0.32 mm, polyethylene glycol, 5 min at 60°C, 2°/min to 240°C. From [18], reproduced with permission of the American Chemical Society.

Fig. 9.13. Volatile extracts of thermally treated Thiamin solutions at (top), pH 1.5, (center) pH 7.0, and (bottom) pH 9.5. GC-FPD (sulfur compounds) on a 60 m × 0.32 mm I.D. dimethylpolysiloxane column with a 0.25 μm film. Oven, 60 to 220°C at 3°C/min.

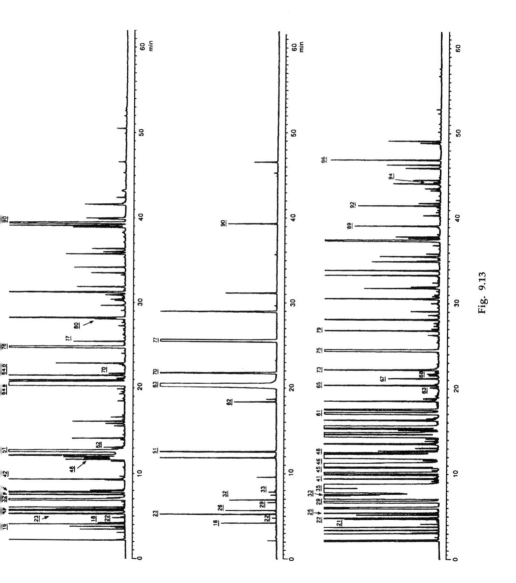

Fig. 9.13

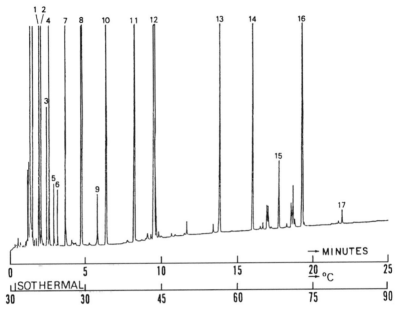

Fig. 9.14. Chromatogram of headspace volatiles from ripening banana [22].

headspace require a refocusing mechanism prior to the beginning of the chromato-graphic process. This can be accomplished with a commercial cryogenic focusing module. If only a few samples are involved, the first coil of the column can be submerged in an appropriate coolant (e.g., liquid nitrogen) during the injection. When the coolant is removed, the column assumes oven temperature and the chro-matographic separation begins. The volume of these vapor injections are limited by the formation of ice plugs in the column from the humid samples.

Injection of 500 μL of a wine headspace produced the chromatogram shown in Fig. 9.15 [23]; the first major peak is probably acetaldehyde, followed by methyl acetate. Note that the PEG-type stationary phase does not normally differentiate 2-methylbutanol from 3-methylbutanol; a high-methyl polysiloxane-type stationary phase achieves slightly better separation of these two solutes. With wine as with other materials, chromatograms of liquid–liquid extraction essences are usually more complex than those produced by headspace injection (Fig. 9.16) [23].

Solid phase extraction (SPE) is a useful tool to prepare dirty or complex sam-ples for analysis. The C_{18}-based packings effectively extract organic compounds from aqueous matricies such as wine. This approach was utilized to analyze *trans*-resveratrol, suggested to be an active principle in the protection of atherosclero-sis and coronary heart disease, in red wines. Figure 9.17 shows the selected ion monitoring (SIM) chromatogram of a red wine ethyl acetate eluent from a C_{18}

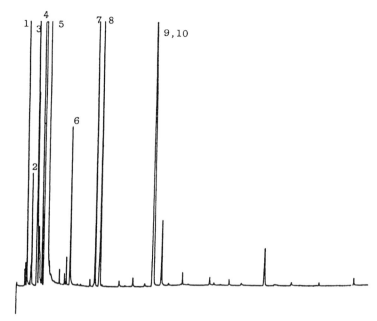

Fig. 9.15. Chromatogram of 500 μL of Muscat wine headspace on a 30 m × 0.32 mm DB-WAX column. Column, 30°C (2 min) to 40°C @ 2°/min, to 170°C @ 4°/min. Peak assignments, 1) acetaldehyde; 2) methanol; 3) ethyl acetate; 4) ethanol; 5) propanol; 6) 2-methyl-1-propanol; 7) butanol; 8) 2-pentanol; 9) 2-methyl-1-butanol; 10) 3-methyl-1-butanol [23].

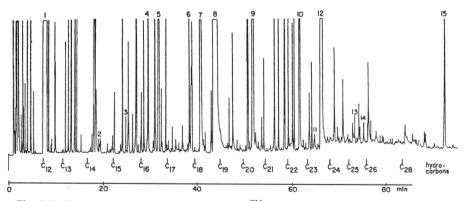

Fig. 9.16. Chromatogram of a dichloromethane-Freon™ extract of a California Riesling on a 30 m × 0.32 mm DB-WAX™ column. Temperature, 40°C (1 min) to 200°C @ 2°/min. Retentions of the n-paraffin hydrocarbons are indicated above the abscissa. Peak assignments, 1) 2-methyl-1-butanol + 3-methyl-1-butanol; 2) furfural; 3) linalool; 4) ethyl decanoate; 5) diethyl succinate (artifact?); 6) 2-phenethyl acetate; 7) hexanoic acid; 8) 2-phenethanol; 9) octanoic acid; 10) decanoic acid; 11) dihy-droxyhexanoic acid gamma lactone; 12) monoethylsuccinate; 13) N-2-phenethylacetamide; 14) ethyl pyroglutamate; 15) 2-(4-hydroxyphenyl) ethanol [23].

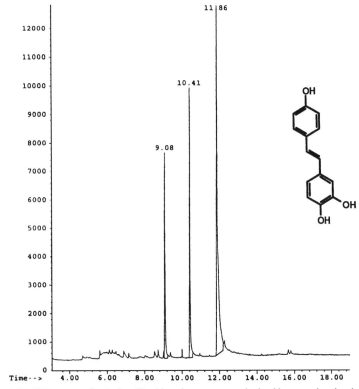

Fig. 9.17. SPE extract of red wine. Total ion chromatogram obtained by summing the abundances of ions 227, 228 and 229. The third peak is *trans*-resveratrol. The column was a 30 m × 0.25 mm I.D., 5%-phenyl-95%-methyl polysiloxane with a 0.25 μm film thickness. *Adapted from Ref. 24 with permission.*

SPE cartridge. The third peak corresponds to a trans-resveratrol concentration of 2.2 mg/L in the wine [24]. For more polar solutes such as catechin and epi-catechin, the recovery is improved by removing ethanol from the wine prior to extraction [25].

The application of chiral separations to the characterization of essential oils and foods is an important area of research and testing. This stems from the need to authenticate materials, both raw ingredients and finished food products, to insure that they are free from adulteration and support the claims of labeling and advertising. The increasingly high cost of rare essential oils has made them vulnerable to augmentation by oils and oil fractions of lesser value. The ratio of the enantiomers of specific components is a viable way to authenticate the purity and origin of an oil. Chiral separations of many natural molecules have been published [26–28]. Chiral GC authentications of natural products are best implemented when they

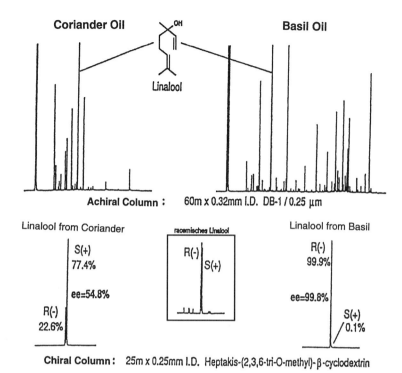

Fig. 9.18. A multidimensional chiral separation of linalool. Linalool is separated from the Coriander and Basil oils on an achiral methylpolysiloxane column, and rechromatographed on the chiral column noted. The enantiomeric ratios of linalool in the two oils are different and characteristic of each oil. From [31], reprinted with permission.

are coupled with multidimensional chromatography. A nonchiral column, usually with a methylsiloxane or wax stationary phase, provides the initial separation of the complex mixture. The area of interest is cut from this analysis and rechromatographed on a column with chiral selectivity. Figure 9.18 demonstrates the multidimensional chiral characterization of the linalool in both coriander and basil oils [29–31]. The chromatogram produced by a racemic mixture of derivatized protein amino acids on a Chirasil-Val column is shown in Fig. 9.19 [32].

The analysis of lipids, in particular triglycerides and fatty-acid methyl esters (FAMEs) therefrom, is an active area of research and application. These substances make important contributions to human health and well-being. The gas chromatography of lipids was reviewed in 1989 [33]. Because of their low volatilities, short columns coated with thinner films (of ten of apolar stationary phases) are generally used for triglyceride analysis. The stationary phase, the column deactivation treatment, and the column material per se must be able to withstand the

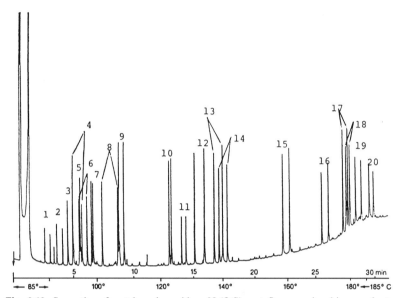

Fig. 9.19. Separation of protein amino acids as N-(O,S)-pentafluoropropionyl isopropyl esters on Chirasil-Val™ 8. Solutes: 1), alanine; 2) valine; 3) threonine; 4) α-isoleucine; 5) glycine; 5) isoleucine; 7) proline; 8) leucine; 9) serine; 10) aspartic acid; 11) cysteine; 12) methionine; 13) phenylalanine; 14) glutamine; 15) tyrosine; 16) ornithine; 17) lysine; 18) histadine; 19) argenine; 20) tryptophane. In each case, the D isomer elutes first. From [32].

higher temperatures that may be required for these analyses. High-phenyl siloxane phases achieve improved resolution of the higher-molecular-weight triglycerides. As an example, differences in the hard- and soft-melt fractions of milk fat were characterized by gas chromatography prior to foaming studies (Fig. 9.20) [34]. Milk fat has been reported to contain over 1000 different triglycerides [35].

The analysis of FAMEs is traditionally accomplished with PEG phases. Methods to calculate the effective chain lengths based on retention time are both accurate and reproducible [36]. The recognized importance of the *cis* and *trans* relationships, and the location of the unsaturation is a challenging GC application. Encouraging results are currently obtained with highly polar cyanopropyl siloxane phases [37–39]. Figure 9.21 shows the separation $C_{18:0}$, $C_{18:1}$, and $C_{18:2}$ isomers on siloxane phases with 100% and 95% cyanopropyl loading, and 100% polycyanoethylsiloxane (OV-275). The highly polar columns exploit differences in the separation factors (α) of the analytes, but theoretical plates (N) are also important to their resolution. Long ($\geq$ 100-m) columns are often used for these separations.

Gas chromatographic analysis of the mycotoxins has been handicapped by decomposition of the derivatives during analysis. Kientz and Verweij studied the efficacy of several derivatizing reagents for the type A and type B trichothecenes

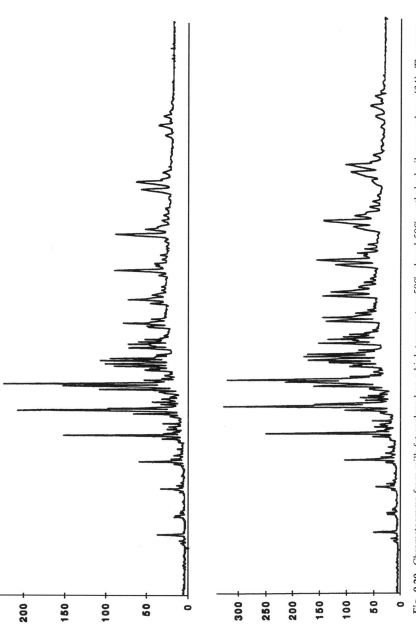

Fig. 9.20. Chromatograms from milk fat analyzed on a high temperature 50%-phenyl-50%-methylpolysiloxane column [34]. The upper temperature limit on the program used was 355°C. The bottom chromatogram is the total milk fat fraction and the top chromatogram is the soft melt fraction.

1. C18:0
2. C18:1 *trans* isomers
3. C18:1 *cis* and *trans* isomers
4. C18:1 *cis* isomers
5. C18:2

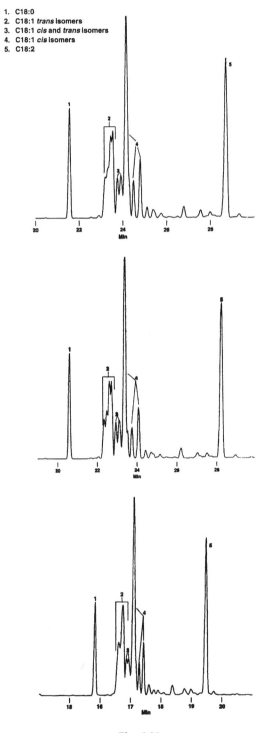

Fig. 9.21

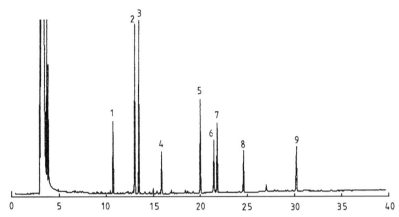

Fig. 9.22. Mycotoxins (trichothenes) derivatized with trifluoracetamide in the presence of sodium bicarbonate [40].

and reported that such decompositions were largely attributable to the presence of excess reagent during the analysis [40]. Figure 9.22 shows a chromatogram of mycotoxins derivatized with the volatile trifluoroacetic anhydride in the presence of sodium bicarbonate.

Headspace sampling can be very effective in the determination of residual solvents in foodstuffs and has been used to detect solvent residues in decaffeinated instant coffee [41]. The application of SPME to the determination of halogenated volatiles in the headspace of a variety of beverages has been reported [42].

Some food packaging (and handling) materials possess a potential for contamination that must also be addressed. That potential may be related to trace amounts of residual solvents used in the manufacture of food packaging films, to the presence of extractable plasticizers (e.g., 2-ethylhexanol in plastic hose and tubing), or to degradation products from the packaging materials. Plastic bottles of polyethylene terephthalate, intended for use as beverage containers and stored capped to preserve their cleanliness, have been found to contain significant quantities of acetaldehyde [43]. A related problem occurs in the use of plastic ware with convenience foods. Some frozen "TV dinners" are purchased "microwave-ready" with the edible materials artfully arranged on plastic dishes. Heat is transferred to the plastic in contact with the heated food, which would be expected to encourage migration of residual monomers and catalysts from the plastic into the foodstuff [44].

Fig. 9.21. A comparison of stationary phase selectivity in the analysis of the isomers of C: 18 fatty acid methyl esters: (top) cyanopropylpolysiloxane (SP™2560); center, 90%-dicyanopropyl-10%-cyanopropylphenyl-polysiloxane (SP-2380); bottom, 100% cyanoethylpolysiloxane (OV-275). All columns were 100 m × 0.25 mm I.D. with a 0.20 μm film. The oven temperature was 175°C in all cases. *Figure courtesy of L. Sidisky, Supelco. Reprinted with permission from Supelco Inc.*

The application of GC to the analysis of pesticide and industrial chemical residues in food products has gained heightened importance because of consumers demands for foods that are characterized as "natural." The difficulty of this work is in the extraction and concentration of analytes that are present in trace levels. Attention to the injection methods employed, and the use of highly inert columns and selective detectors are sometimes necessary in the analysis of low levels of these "active" analytes. This area has been the subject of several recent reviews [45]. Ethylene dibromide (EDB) has been widely used to inhibit postprocessing infestation by weevils and other insects in packaged foods such as grain products and cake mixes. It was reasoned that the material would be expelled from the food-stuff during subsequent cooking, but the demonstrated carcinogenicity of higher levels of EDB triggered concern and a number of legislative acts banned the use of that chemical. Because of the complexity of most samples isolated from food, unambiguous separation of something like EDB can be a problem. Section 8.5 describes techniques employed to design stationary phases for specific applications based on maximizing the relative retentions of all solutes and how those techniques were applied to the synthesis of a new stationary phase tailored to halocarbon separations (see also [46]). The DB-624 column, described in Section 9.5, employs a thicker film of this stationary phase to achieve separation of a series of volatile

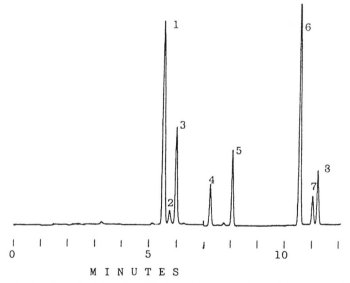

Fig. 9.23. Detection of ethylene bromide (EDB). Column, 30 m × 0.53 mm DB-624; 35 to 120°C @ 5°/min, ECD. Solutes, 1) chloroform; 2) 1,1,1-trichloroethane; 3) carbon tetrachloride; 4) trichloroethylene; 5) bromodichloromethane; 6) tetrachloroethylene; 7) chlorodibromomethane; 8) ethylene dibromide. Courtesy of J&M Scientific.

priority pollutants by optimization of both relative retentions (α) and partition ratios (k). This column is also capable of achieving the separation of ethylene dibromide from several solutes with which it is prone to coelute (Fig. 9.23).

9.3   Petroleum- and Chemical-Related Applications

Petroleum-related analyses range from the separation of light gas mixtures to the characterization of heavy crude oils to oil shales and materials produced in coal liquefaction. Particularly with the naturally occurring materials, the dominant solutes are hydrocarbons.

Differentiation of the paraffinic, olefinic, and aromatic hydrocarbons are normal goals, and detection of sulfur- and/or nitrogen-containing components is sometimes required. Hydrocarbon preparations derived from shale oil or coal liquefaction can be extremely complex and can also include a variety of other functional groups. Multidimensional chromatography may be required for the analysis of these complex mixtures; it is also useful for the estimation of certain octane-boosting additives in automotive fuels. The evaluation of oxygenated compounds in finished products has also become increasingly important (e.g., [13]).

Packed columns are still used for some separations of light hydrocarbons, fixed gases, low-molecular-weight sulfur- and nitrogen-containing compounds, and low-molecular-weight alcohols and halocarbons. Packings for these applications usually consist of molecular sieve, alumina, or one of the porous polymers (e.g., Chromosorb 104, Chromosorb 106, GS-Q, Porapak N).

Interactions between the stationary phase and hydrocarbon-type solutes are generally limited to dispersion forces (Chapter 4). The polymethylsiloxane stationary phases, whose interaction potentials are limited to dispersion, are well suited to the separation of these compounds. At low temperatures, polymethylsiloxanes (like other substances) lose their fluid characteristics and behave more like conventional solids. Crosslinking slows this low-temperature transition, and higher diffusivities persist for some time at subambient temperatures. Held at low temperatures for extended periods, they eventually exhibit solidlike properties, and chromatography suffers. A brief exposure to higher temperature restores fluidlike diffusivities. Mooney et al. [47] demonstrated the separation of low-molecular-weight hydrocarbons in a valved inlet system, using a relatively thick-film column at subambient conditions (Fig. 9.24).

Thicker-film columns that have become available since that time have proved especially useful for the analysis of these low-boiling mixtures. Figure 9.25 illustrates a chromatogram of natural gas obtained from the North Sea area, chromatographed on a "standard" film capillary and programmed from -50 to $200°C$ [48]. In the refinery gas separation shown in Fig. 9.26, the very thick film combined with a relatively low starting temperature shifted the solute partition ratios enough to achieve the resolution of methane, ethylene, and ethane at $0°C$.

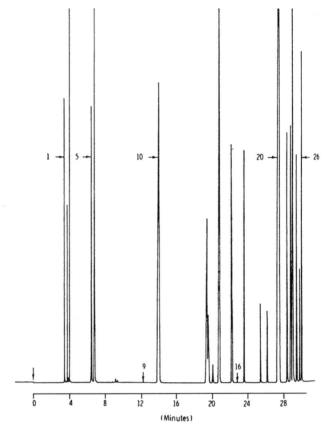

Fig. 9.24. Separation of C-1 through C-5 hydrocarbons [47]. Column, 60 m × 0.25 mm, 1.0 μm film of dimethylpolysiloxane, −40°C (15 min) to 100°C @ 10°C/min. Solutes: 1) methane; 2) ethylene; 3) acetylene; 4) ethane; 5) propylene; 6) propane; 7) propadiene; 8) methylacetylene; 9) cyclopropane; 10) *iso*butane; 11) *iso*butylene; 12) 1-butene; 13) 1,3-butadiene; 14) *n*-butane; 15) *trans*-2-butene; 16) methylcyclopropane; 17) *cis*-2-butene; 18) 1,2-butadiene; 19) 3-methyl-1-butene; 20), *n*-pentane; 21) 1-pentene; 22) methyl-1-butene; 23) *n*-pentane; 24) *trans*-2-butene; 25) *cis*-2-pentene; 26) methyl-2-butene. Reprinted with permission of Huethig Publishing Co.

Separations have been shown on porous layer open tubular (PLOT) columns coated with alumina (usually as Al_2O_3/KCl). Figure 9.27 shows chromatograms generated by low-molecular-weight hydrocarbons on a fused silica PLOT column of this type; the increased retentions permit separation of these low-molecular-weight solutes at appreciably higher temperatures, obviating the need for subambient operation. The application of this phase to the analysis of ethylene is shown in Fig. 9.28.

The selectivities of these Al_2O_3 PLOT columns can be dramatically changed by using different salt solutions in their manufacture; for example, the Al_2O_3/Na_2SO_4

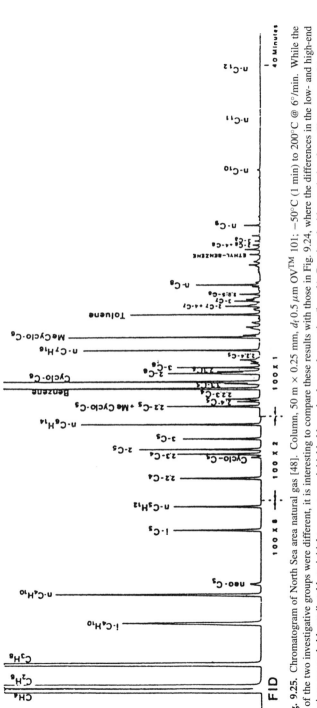

Fig. 9.25. Chromatogram of North Sea area natural gas [48]. Column, 50 m × 0.25 mm, d_f 0.5 μm OV™ 101; −50°C (1 min) to 200°C @ 6°/min. While the goals of the two investigative groups were different, it is interesting to compare these results with those in Fig. 9.24, where the differences in the low- and high-end separations are probably attributable to initial temperatures, initial holds, and column phase ratios (i.e., d_f). Reprinted with permission from *Chromatographia*.

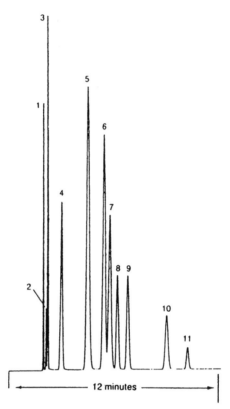

Fig. 9.26. Refinery gas on a thick-film column. Column, 30 m × 0.32 mm, d_f 5.0 μm dimethyl polysiloxane. One mL gas injected directly on column; 0°C (3 min) to 50°C @ 10°/min. Solutes: 1) methane; 2) ethylene; 3) ethane; 4) propane; 5) *iso*butane; 6) 1-butene; 7) *n*-butane; 8) *trans*-2-butene; 9) *cis*-2-butene; 10) *iso*pentane; 11) *n*-pentane. Courtesy of J&W Scientific.

PLOT column exhibits a different selectivity than the KCl version. PLOT columns are also available with porous layers made of molecular sieve material or synthetic polymers similar to the Porapak columns. Figure 9.29 shows the application of a molecular sieve PLOT column to permanent gas analysis. Ingenious multicolumn valved inlet systems, with potential for backflushing and/or redirecting individual fractions to open tubular, micropacked, and PLOT columns, have been designed (e.g., [48]) and are capable of achieving separations that are difficult, if not impossible, to duplicate with conventional equipment and/or conventional open tubular columns.

Figure 9.30 shows a chromatogram typical of a naphtha separation. The complexity of the sample requires extremely efficient columns or sophisticated valve-switching applications to garner usable information from the analysis. This subject

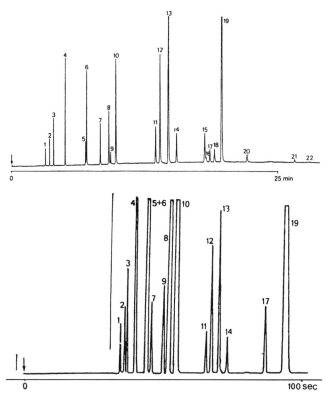

Fig. 9.27. Separation of the C-1 through C-5 hydrocarbons on a 50 m × 0.32 mm fused silica PLOT Al$_2$O$_3$/KCL column. Top, nitrogen carrier at 1 atm, 70 to 200°C @ 3°/min. Solutes: 1) methane; 2) ethane; 3) ethene; 4) propane; 5) cyclopropane; 6) propene; 7) ethyne; 8) *iso*butane; 9) propadiene; 10) *n*-butane; 11) *trans*-2-butene; 12) 1-butene; 13) *iso*butene; 14) *cis*-2-butene; 15) *iso*pentane; 16) 1,2-butadiene; 17) propyne; 18) *n*-pentane; 19) 1,3-butadiene; 20) 3-methyl-1-butene; 21) vinylacetylene; 22) ethylacetylene. Bottom: High speed analysis of C-1 to C-4 hydrocarbons; column as above, hydrogen carrier @ 3 atm, *iso*thermal 130°C. Solutes: 1) methane; 2) ethane; 3) ethene; 4) propane; 5) cyclopropane; 6) propene; 7) ethyne; 8) *iso*butane; 9) propadiene; 10) *n*-butane; 11) *trans*-2-butene; 12) 1-butene; 13) *iso*butene; 14) *cis*-2-butene; 15) *iso*pentane; 16) 1,2-butadiene; 17) propyne; 18) *n*-pentane; 19) 1,3-butadiene. Reprinted with permission of Chrompak, USA.

has recently been reviewed [49]. The financial and regulatory requirements associated with the production and formulation of gasoline for internal-combustion engines have placed added demands on analysts working in this field. The evolution of column technology and regulatory requirements at this writing has made this a very active field of research. The so-called PIANO (paraffins, isoparaffins, aromatics, naphthenes and olefins) methods are done in a variety of ways with varying complexity and reliability [49]. ASTM methods for detailed hydrocarbon analysis using a single column focused on resolving and characterizing

Response(mV)

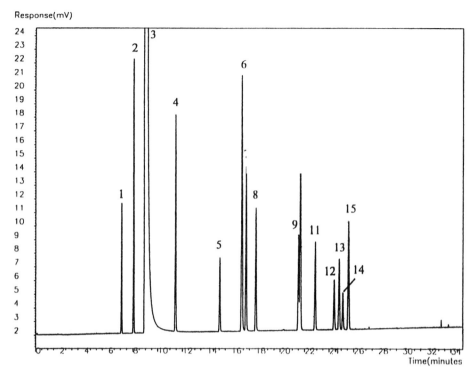

Fig. 9.28. Ethylene separated from impurities on a 50 meter Al$_2$O$_3$PLOT column with KCl deactivation coupled to a 30 meter thick film (5 μm) methyl silicone column. Both columns have a 0.53 mm I.D. The methyl silicone column improves the separation of methyl acetylene, iso-pentane and n-pentane. The components are: (1) methane, (2) ethane, (3) ethylene, (4) propane, (5) propylene, (6) acetylene, (7) isobutane, (8) n-butane, (9) trans-2-butene, (10) 1-butene, (11) cis-2-butene, (12) isopentane, (13) methylacetylene, (14) n-pentane, (15) 1,3-butadiene.

all of the components in gasoline and reformates are currently under revision. The requirements of separating the aromatic components from the aliphatics is best accomplished with the addition of a short length of 5% phenyl–95% methyl polysiloxane column onto the front of a much longer (e.g., 100-m) methylsiloxane column. The length of the precolumn is determined by optimizing the separation of several critical pairs (e.g., 2,2,3-trimethylpentane from toluene) (Fig. 9.31) [50]. A 450-m column, 0.20 mm ID and capable of generating over 1.3 million effective theoretical plates, has been used in the analysis of gasoline [51]. Some 970 components were differentiated in the 650-min. (10.8-hr!) run time.

The analysis of aromatics for regulatory purposes can be accomplished with a 60-m-thick film methylsiloxane column and MS detection in the selected ion monitoring mode (see Fig. 9.32). The addition of oxygenated compounds to gasoline at

1. Neon
2. Oxygen
3. Nitrogen
4. Methane
5. Carbon monoxide

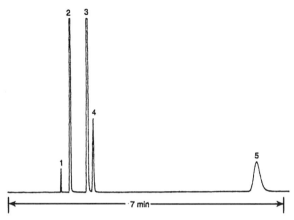

Fig. 9.29. Separation of permanent gases at 50°C on a 30 m × .53 mm I.D., molecular sieve PLOT column.

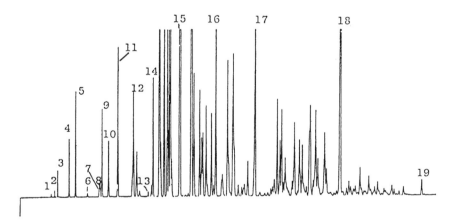

Fig. 9.30. Chromatogram produced by injection of one mL neat naphtha on a 30 m × 0.32 mm column, d_f 1.0 μm dimethylpolysiloxane. Hydrogen carrier @ 38 cm/sec (100°C); 28°C (4 min) to 100°C @ 2.5°/min. Analysis time 28 min. Solutes (attenuation ×1): 1) propane; 2) methylpropane; 3) n-butane; 4) (attenuation to ×4) 2-methylbutane; 5) n-pentane; 6) 2,2-dimethylbutane; 7) cyclopentane; 8) 2,3-dimethylbutane; 9) 2-methylpentane; 10) 3-methylpentane; 11) (attenuation to ×8) n-hexane; 12) methylcyclopentane; 13) (attenuation to ×16) benzene; 14) cyclohexane; 15) n-heptane; 16) toluene; 17) n-octane; 18) n-nonane; 19) n-decane. Courtesy of J&W Scientific.

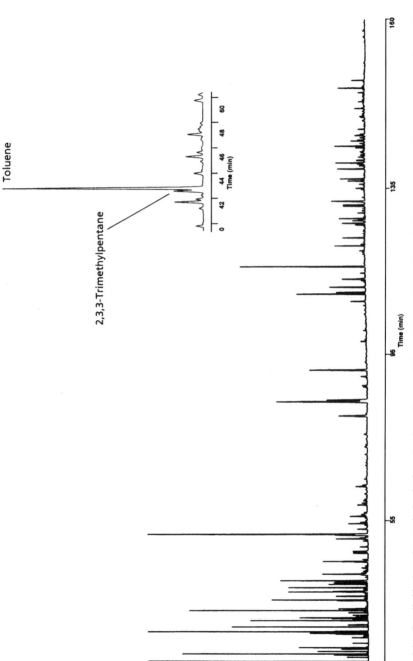

Fig. 9.31. Detailed hydrocarbon analysis of reformulated gasoline on a 100 m × 0.25 mm I.D. methylpolysiloxane column with a short piece (*ca.* 3 meters) of a 5% phenyl column attached to the front end [50]. The resolution of 2,3,3-trimethylpentane from toluene is highlighted. Courtesy of J&W Scientific.

1. Methyl-tert-butylether	12. p-Xylene	23. 1,4-Diethylbenzene
2. n-Hexane	13. o-Xylene	24. n-Butylbenzene (valley)
3. Benzene-d6 (ISTD)	14. n-Propylbenzene	25. 1,2-Diethylbenzene
4. Benzene	15. 1-Methyl-3-ethylbenzene	26. 1,2,4,5-Tetramethylbenzene
5. Isooctane	16. 1-Methyl-4-ethylbenzene	27. 1,2,3,5-Tetramethylbenzene
6. n-Heptane	17. 1,3,5-Trimethylbenzene	28. Naphthalene-d8 (ISTD)
7. Toluene	18. 1-Methyl-2-ethylbenzene	29. Naphthalene
8. n-Octane	19. 1,2,4-Trimethylbenzene	30. n-Dodecane
9. Ethylbenzene-d10 (ISTD)	20. n-Decane	31. Pentamethylbenzene
10. Ethylbenzene	21. 1,2,3-Trimethylbenzene	32. 2-Methylnaphthalene
11. m-Xylene	22. Indan	33. 1-Methylnaphthalene

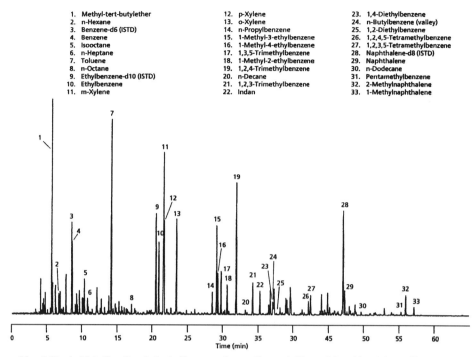

Fig. 9.32. A thick film dimethylpolysiloxane column, 60 m × 0.25 mm I.D. with a 1.0 μm film, coupled with MS detection used to quantitate aromatic compounds in gasoline. Deuterated benzene (d6), ethylbenzene (d10) and naphthalene (d8) served as internal standards. Courtesy of J&W Scientific.

required levels has necessitated optimization of these methods. Both single-column and valve-switching methods exist. These determinations have greatly benefited from the development of oxygen specific detectors, O-FID, atomic emission, and IR detection (Fig. 9.33). Selective detectors have also been applied to the detection of sulfur compounds in fuels (Fig. 9.34) [52–56].

Separations of weathered gasolines have significance for arson analysis, and methods of isolating samples from arson residues have been described [57]. The application of solid phase microextraction (SPME) to arson analysis has been reported [58]. Figure 9.35 compares the recovery of lighter fluid by SPME and ASTM method E1412-91. The latter uses activated charcoal strips and CS_2 desorption. The SPME method was determined to be simpler and faster and did not involve extraction solvents.

The components of crude oil extend into the higher-molecular-weight range and require the use of less retentive (e.g., thinner-film) columns, capable of higher-temperature operation. Beyond generating a "composite picture" [59], such separations are employed as "fingerprints" to establish the source of oil spills, in the

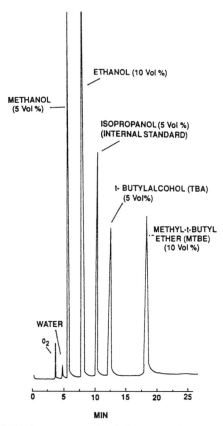

Fig. 9.33. The GC-OFID chromatogram of typical oxygenated compounds found in reformulated gasoline. The separation was performed on a 30 m × 0.32 mm I.D. methylpolysiloxane column with a 3 μm film thickness [53].

dating of crudes by measurement of the C_{17}/pristane and C_{18}/phytane ratios and in characterizing steroidal hydrocarbons that are common constituents of crude oils and ancient sedimentary rocks (Fig. 9.36) [60–62]. The latter can be of geological interest; a sample displaying a high content of the C_{29} steranes (e.g., isomers of 24-ethylcholestane) as compared to the C_{27} steranes would suggest

Fig. 9.34. Chromatograms of gasoline with a sulfur chemiluminescence detector, SCD [55]. Because the SCD uses an FID for SO production, both hydrocarbon (top) and sulfur (bottom) chromatograms are obtained from a single sample and column. The sulfur chromatogram is free from hydrocarbon interference. The sample was analyzed on a 30 m × 0.32 mm I.D. dimethylpolysiloxane column with a 4.0 μm film thickness. Adapted from Ref. 56.

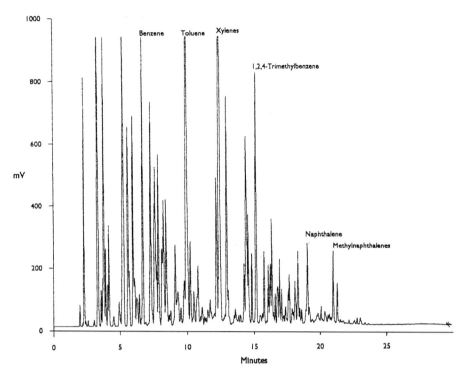

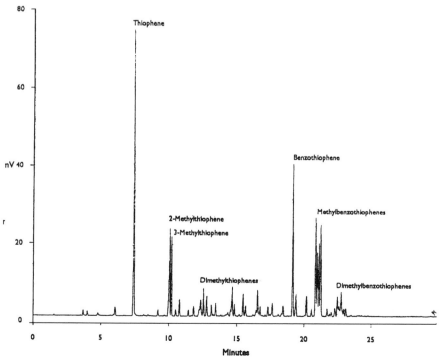

Fig. 9.34

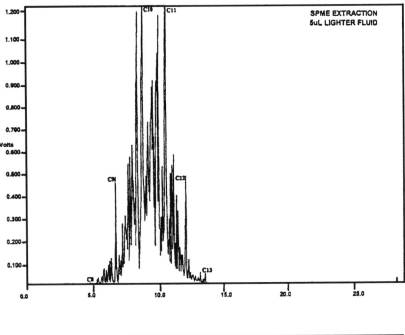

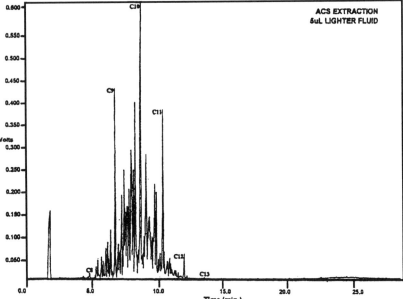

Fig. 9.35. Comparison of a lighter fluid extract obtained by SPME (top) and ASTM method E 1412-91 (bottom). FID fingerprint data obtained in this manner is used in arson investigations. From [58], and reprinted with permission of the Huethig Publishing Company.

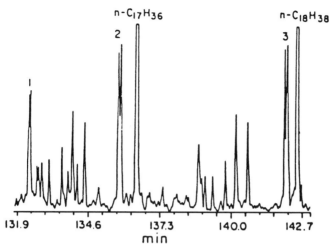

Fig. 9.36. Section from a chromatogram of a petroleum sample showing the separation of nor-pristane (1), pristane (2), and phytane (3) [62]. Column, 140 m × 0.27 mm, coated with $C_{87}H_{176}$ and programmed from 100 to 230°C at 0.8°/min. Temperature range of the portion illustrated was 132–143°C.

that the precursor organic matter contained land-derived biochemical compounds rather than those derived from marine organisms [63]. Capillary gas chromatography/mass spectrometry can also be employed to characterize shales and fossils isolated therefrom [64]. Figure 9.37 shows a chromatogram of the extract of a fossil wood found in Posidonomia shale [63].

The upper operating temperature limit of the gas chromatographic column would seem to impose some restrictions on the analysis of crude oils and other samples that range into very high-molecular-weight (and low-volatility) solutes. With the fused silica column, the protective outer polyimide coating is most temperature-limiting and begins to show signs of degradation at about 340°C; the rate of degradation increases rapidly as the temperature is increased. Higher-temperature polyimides extend this range somewhat, and conventional glass capillaries avoid the constraint of an outer coating material. The latter option, however, suffers from high activity, among other shortcomings (Chapter 2).

Unflawed fused silica tubing is extremely resistant to breakage; however, some flaws persist and carry forward from the blank to the drawn tubing (Chapter 2), and others originate if the tubing is exposed to water vapor [65]. Under ideal conditions, metal coatings that are completely impervious to water can be applied to fused silica, and flaw-free fibers or tubing so protected would be highly resistant to mechanical breakage. Under conditions of temperature cycling, the difference in thermal expansion of the metal and silica could lead to complications. Polyimide

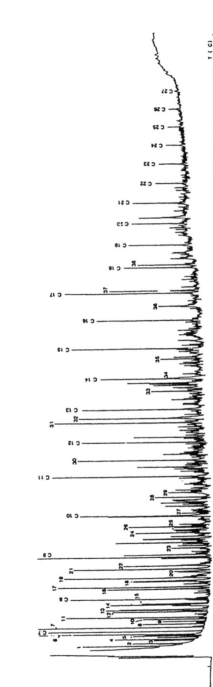

Fig. 9.37. An extract of fossilized wood on a 30 m × 0.25 mm dimethylpolysiloxane column [64]. Temperature, 20°C (3 min), 30°C (min) to 300°C @ 40°/min. Paraffinic hydrocarbons indicated by "C" prefix. Other indicated solutes: 1) methylcyclopentane; 2) cyclohexane; 3) benzene; 4) 2-methylhexane; 5,6) dimethylcyclopentane isomers; 7) methylcyclohexane; 8) ethylcyclopentane; 9,10) trimethylcyclopentane isomers; 11) toluene; 12) 2-methylheptane; 13) dimethylcyclohexane; 14) methylethylcyclopentane; 15) dimethylcyclohexane; 16) ethylcyclohexane; 17) trimethylcyclohexane; 18) ethylbenzene; 19) m- and p-xylenes; 20) alkyl-substituted cyclohexane; 21) o-xylene; 22) methylethylcyclohexane; 23) propylcyclohexane; 24) propylbenzene; 25) methylethylbenzene; 26) butylbenzene; 27, 28) alkyl benzenes; 29) naphthalene; 30) tetramethylbenzene; 31) 2-methylnaphthalene; 32) 1-methylnaphthalene; 33, 34) dimethylnaphthalene isomers; 35) C-16 isoprenoid; 36) C-18 isoprenoid; 37) 2,6,10,14-tetramethylpentadecane (pristane); 38) 2,6,10,14-tetra methylhexadecane (phytane).

coatings are highly resistant but not completely impervious to water. However, it is also probable that during its application, the polyimide (as contrasted with metal) is better able to flood and seal existing flaws that would otherwise grow at a stress-proportional rate and eventually lead to breakage. Because commercial fused silica tubing does have surface flaws, the polyimide-coated tubing is generally stronger and more durable than the aluminum-coated tubing commonly available today.

In addition, the thermal stability of the polyimide coating is exceeded only slightly by that of the stationary phase, which imposes the next limitation on some (e.g., hydrocarbon) high-temperature analysis. The lability of most organic compounds at these higher temperatures is even more ominous; many chemical bonds (e.g., C—O, C—N, C—S) do not have the thermal stabilities necessary to survive exposure to these extreme temperatures. Hence the gas chromatographic separation may begin with well-defined solutes and end with their pyrolysis products. The quantitative and qualitative validities of such analyses are improved by employing chromatographic techniques not contingent on heat-induced volatilities of low-volatility thermally labile solutes (e.g., liquid chromatography, supercritical fluid chromatography).

Metal wall coated open tubular columns with greatly improved deactivation procedures have recently become available. These columns suffer none of the high-temperature limitations of fused silica or aluminum clad columns and can be prepared with bonded stationary phases. Short metal columns with thin films of bonded methyl siloxane stationary phase have been shown to be excellent for the high-temperature simulated distillation of heavy crude oil (see Fig. 9.38) [66].

Such samples have a boiling range from -44 to $1355°F$. PTV injection into a $-30°C$ column is required to separate the CS_2 solvent from the C_6 hydrocarbons. Thicker film columns ($3 \mu m$) are used for lower-temperature simulated distillations (Fig. 9.39) [67]. While these thick film-columns will effectively resolve the lower-boiling components from the solvent, the bleed produced by the thick stationary phase limits the upper range to C_{60} at approximately $350°C$. Thin-film metal capillaries allow recognition of C_{110} at $450°C$. Figure 9.40 is a reference high-temperature crude oil analyzed in this manner. Figure 9.41 shows the calibration curve used to calculate the boiling-point range of the crude oil; the column used measured only those components in the range bp $156–1355°F$.

In spite of the considerable progress that has been made in column deactivation, difficulties with extremely active solutes are not uncommon. Because thicker stationary phase films tend to mask active sites on the tubing surface, the very thick-film columns are usually perceived as being more inert than fused silica capillaries of "conventional" film thickness. They can be used to advantage in the analysis of some active solutes, provided those solutes are sufficiently low-boiling that the resulting increase in k does not hinder their elution. As an example, Fig. 9.42 shows a series of short-chain alcohols on a thick-film column.

Industrial chemical applications outside the petroleum field are as diverse as the number of products and processes that exist for their manufacture and use.

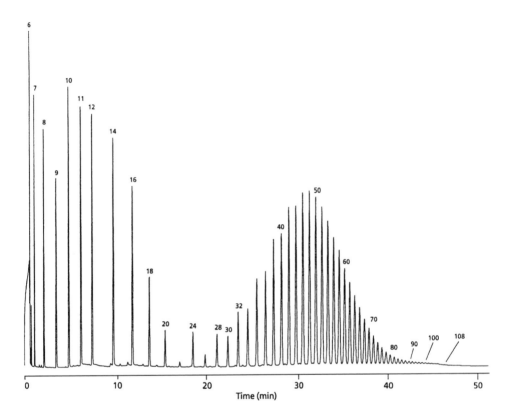

Fig. 9.38. Standard n-alkanes chromatographed on a 5 m × 0.53 mm I.D. metal capillary column with a 0.15 μm film of bonded methylpolysiloxane [66]. The oven was programmed from −30 to 430°C at 10°C per minute. The sample was in CS_2 solution. The PTV injector used was programmed from 55 to 450°C at 1°C per second.

Gas chromatography plays a vital role in the analysis of both major and minor components in finished products and raw materials. The analytical range is from the determination of ethylene oxide in DMF (Fig. 9.43 [68]), to the evaluation of nonionic surfactants (Fig. 9.44 [69]).

Solvent assay can be a challenging problem, because a massive amount of the principal constituent must be introduced into the column if the detector is to respond to the minor constituents; thus, the dynamic range of the system must be sufficiently large to accommodate massive quantitative differences in component concentrations. Figure 9.45 shows an analysis of "99.7% pure benzene" on a 0.53-mm-ID column coated with a 5-μm film of stationary phase. The low end of the dynamic range of a system is dictated by the level of detector "noise," while

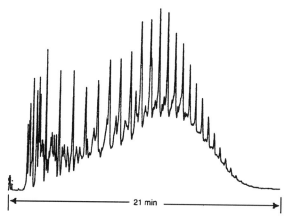

Fig. 9.39. A reference gas oil chromatographed according to ASTM D2887E. The 10 m × 0.53 mm I.D., 3 μ md_f column was programmed only to 350°C, limiting the range to C_{60}. The sample was introduced by Megabore direct injection.

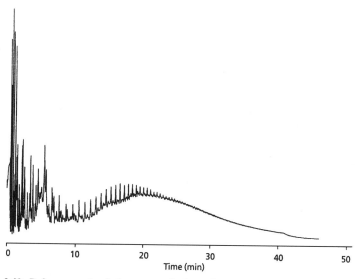

Fig. 9.40. Reference crude oil chromatographed on a 5 m × 0.53 mm I.D. metal capillary column with a 0.15 μm film of bonded methylpolysiloxane. The oven was programmed from −30 to 430°C at 10°C per minute. The sample was in CS_2 solution. The PTV injector used was programmed from 55 to 450°C at 1°C per second.

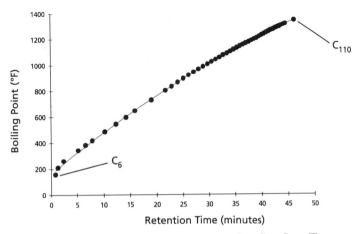

Fig. 9.41. Boiling point versus retention curve for n-alkanes from C_6 to C_{110}. These curves provide the calibration for the evaluation of crude oils such as the one shown in Fig. 9.40. Calculations from these data are automated.

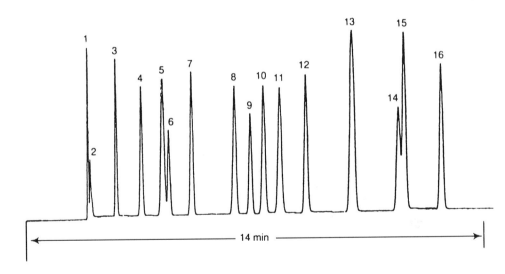

Fig. 9.42. Short chain alcohols on a 30 m × 0.32 mm column, 5.0 μm film of dimethylpolysiloxane. Split injection, 30°C (0.5 min) to 110°C @ 5°/min. Solutes, 1) methanol; 2) acetaldehyde; 3) ethanol; 4) 2-propanol; 5) 2-methyl-2-propanol; 6) methyl acetate; 7) 1-propanol; 8) 2-butanol; 9) ethyl acetate; 10) 2-methyl-1-propanol; 11) 2-methyl-2-butanol; 12) 1-butanol; 13) 2-pentanol + 3-pentanol; 14) 3-methyl-1-butanol; 15) 2-methyl-1-butanol; 16) 1-pentanol.

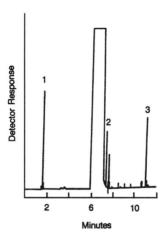

Fig. 9.43. Trace ethylene oxide impurity in DMF solution chromatographed on a 30 m × 0.25 mm I.D. bonded wax column with a 0.25 μm film [68]. Compounds: (1) ethylene oxide, (2) 2-chloroethanol, and (3) ethylene glycol.

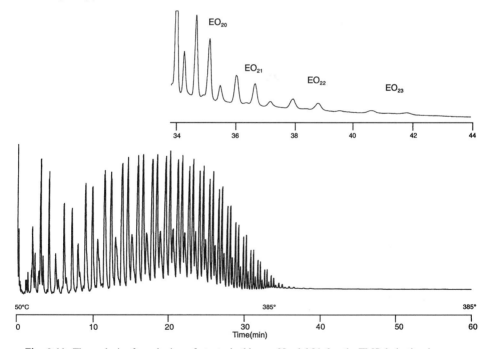

Fig. 9.44. The analysis of non-ionic surfactants, in this case Neodol 91-6 as the TMS derivative, is possible only with short columns and thin stationary phase films. In this case a 3 m × 0.25 mm I.D. length of DB-XLB with a 0.1 μm film was temperature programmed from 50 to 385°C at 10°C per minute. On-column or PTV injection is necessary for such low volatility samples.

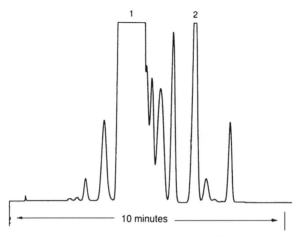

Fig. 9.45. Direct injection of 1.0 μL of a "99.7 % benzene". Column, 30 m × 0.53 mm, 5.0 μm dimethylpolysiloxane; 45°C (8 min) to 100°C @ 5°/min. Peak (1) is benzene, (2) is toluene.

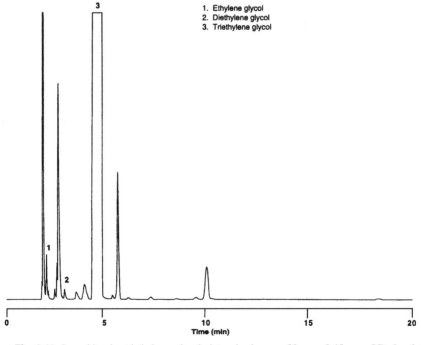

Fig. 9.46. Impurities in triethylene glycol determined on a 30 m × 0.45 mm I.D. bonded methylpolysiloxane column with a 1.27 μm film thickness at 170°C. The neat sample was injected in the split mode. The ethylene glycol and diethylene glycol impurities are noted on the chromatogram.

the increased amount of stationary phase (contributed by both the larger column diameter and the thicker stationary phase film) has greatly increased the "sample-accepting capacity" of the column without affecting the detector sensitivity. Hence the dynamic range of the system has been greatly expanded. The lower phase ratio of this column ($r/2d_f = 265 \, \mu m/10 \mu m = 26.5$) forced an increase in solute retention factors, which was beneficial to the separation of these low-k compounds but would be a severe handicap in the analysis of impurities in higher-boiling solvents. The application of 0.45-mm-ID columns to solvent analysis has also resulted in increased speed of analysis without an observable loss in resolution (see Fig. 9.46).

9.4 Environmental Applications

The environmental analyst may be concerned with the detection and/or quantitation of many different substances in a diversity of matrices. These may range from pesticide chemicals, to polynuclear aromatic hydrocarbons (PAHs), to chlorinated compounds from vinyl chloride to polychlorinated biphenyls, in air, water, and soils. In many cases, the initial problem is one of sample preparation. Water samples, for example, may range from sources of drinking water to industrial wastewater to seepage from chemical dumps and disposal sites.

In many cases, regulatory agencies have specified a "standard" or "accepted" procedure for the analysis of given materials in a given matrix; with the large number of creative scientists engaged in these analytical activities, it is not surprising to find that our capabilities have progressed rapidly. With few exceptions, the "official methods of analysis" are less accurate, less sensitive, and more time-consuming than methods that were developed since their adoption. Regulations of this type change only slowly, but they do change, and the alternative (usually open tubular column) technology is often utilized in the meantime as a confirming technique.

The US Environmental Protection Agency (EPA) first specified procedures for monitoring industrial eluents in 1977 [70]; some of the approved methods were amended in 1980 [71]. In general, US EPA methods numbered in the 500s are for drinking water; the 600s are for wastewater; and the 8000 series is for industrial waste, solids, and sludges. While the initial methods specify the use of packed columns, many of them have been revised for the application of capillary column technology, especially the 500 and 8000 series. Methods 601, 602, 603, and 624 are concerned with volatile pollutants and specify purge-and-trap procedures, discussed briefly in Chapter 3. Methods for establishing the composition of a stationary phase tailored to optimize the separation of a given mixture of compounds, discussed in Section 8.5, permitted designing a new stationary phase, DB-1301, precisely tailored to optimize the relative retention of those solutes [72, 73]. The DB-624 column is coated with that phase, with both α values and the phase ratio β, optimized to yield the best possible separation of these solutes [46]. Further application of phase design technology led to the development of DB-VRX, a proprietary phase optimized for the separation of the 60-compound list from the US EPA. Since that time the compound list for method 524.2 was revised (Rev. 4) [74] and has been increased to 82 plus the necessary internal standards (see Fig. 9.47). To optimize the separation the increased efficiency of a slightly narrower column was necessary while still being able to accommodate the flows from a purge and trap sampler without cryogenic focusing. Thus the inside diameter of the DB-VRX is 0.45 mm instead of 0.53 mm. Both of these columns offer significant advantage over the traditional columns for this analysis in terms of speed and resolution. While method 624 requires the use of an MS detector to confirm peak identities, methods 502, 602, and 8021 make use of an ELCD and PID in series for confirmation. The ELCD is a very sensitive and selective detector for halogenated compounds, and the PID is selective for unsaturated and aromatic compounds. A effective application of method 8021 uses a DB-624 and DB-VRX in a dual-column configuration into the tandem detectors (see Fig. 9.48). The selectivity differences coupled with the data from four detectors on the same sample gives added confidence in the assigned peaks.

While purge-and-trap methods with specified detectors are mandated in the United States, environmental chemists in the rest of the world prefer the use of headspace sampling with an ECD for volatiles determination. Many of these pollutants can be detected at the parts-per-billion and parts-per-trillion levels by direct injections of water or water headspace [76–78]. Solid phase microextraction (SPME), has also been applied to the analysis of drinking water volatiles [79]. SPME was compared to purge and trap and found to have great potential.

Although the concentrating steps associated with volatiles analysis are difficult, the separation challenges are then simplified by virtue of the desorption methods and the way the samples are introduced into the instrument. The columns generally have long lives and the detectors require little maintenance. This is not true for the analysis of the so-called semivolatile environmental pollutants (e.g., PAHs, PCBs, pesticides). These generally require liquid–liquid extraction coupled with a variety of cleanup methods and finally reconcentration in a suitable solvent. The resulting sample is dirty at best and in the worst cases can be damaging to the column. Because of the volatility range and low concentration range of the analytes, splitless injection is common practice. Retention gaps not only help focus the injected analytes but also provide a degree of protection to the column by trapping the nonvolatile residues.

It is for the analysis of semivolatile compounds that the siliarylene phase technology (Chapter 4) has proved to be so valuable. Phases are available with selectivities comparable to 5%, 35%, 50%, and even 65% phenyl polymethylsiloxanes. Additional phases are available with no known siloxane counterpart, i.e., DB-XLB, which possess unique selectivity nonetheless. It is the extreme inertness

1. Dichlorodifluoromethane
2. Chloromethane
3. Vinyl chloride
4. Bromomethane
5. Chloroethane
6. Trichlorofluoromethane
** 7. 1,1-Dichloroethene (m/z 61)
8. Methylene chloride
9. *trans*-1,2-Dichloroethene
10. 1,1-Dichloroethane (m/z 63)
11. *cis*-1,2-Dichloroethene (m/z 96)
12. 2,2-Dichloropropane
13. Bromochloromethane
14. Chloroform
15. 1,1,1-Trichloroethane (m/z 97)
16. Carbon tetrachloride
17. 1,1-Dichloropropene
18. Benzene
19. 1,2-Dichloroethane
20. Trichloroethene
21. 1,2-Dichloropropane
22. Dibromomethane
23. Bromodichloromethane (m/z 83)

24. *cis*-1,3-Dichloropropene (m/z 75)
25. Toluene
26. trans-1,3-Dichloropropene
27. 1,1,2-Trichloroethane
28. Tetrachloroethene
29. 1,3-Dichloropropane
30. Dibromochloromethane
31. 1,2-Dibromoethane
32. Chlorobenzene
33. 1,1,1,2-Tetrachloroethane
34. Ethylbenzene
35. m-Xylene
36. p-Xylene
37. o-Xylene (m/z 106)
38. Styrene
39. Bromoform
40. Isopropylbenzene
41. 1,1,2,2-Tetrachloroethane (m/z 83)
42. Bromobenzene
43. 1,2,3-Trichloropropane
44. n-Propylbenzene
45. 2-Chlorotoluene
46. 1,3,5-Trimethylbenzene

47. 4-Chlorotoluene
48. tert-Butylbenzene
49. 1,2,4-Trimethylbenzene
50. sec-Butylbenzene
51. 1,3-Dichlorobenzene
52. p-Isopropyltoluene
53. 1,4-Dichlorobenzene
54. n-Butylbenzene
55. 1,2-Dichlorobenzene
56. 1,2-Dibromo-3-chloropropane
57. 1,2,4-Trichlorobenzene
58. Hexachlorobutadiene
59. Naphthalene
60. 1,2,3-Trichlorobenzene
61. 4-Bromofluorobenzene
     (surrogate standard)
62. Fluorobenzene
     (internal standard)
63. Acetone
64. Diethyl ether
65. Iodomethane (m/z 142)
66. Acrylonitrile (m/z 53)
67. Allyl chloride

68. Carbon disulfide
69. Methyl-tert-butyl ether
70. Propionitrile (m/z 54)
71. 2-Butanone (MEK)
72. Methacrylonitrile (m/z 41)
73. Methyl acrylate
74. Tetrahydrofuran
75. 1-Chlorobutane (m/z 56)
76. Chloroacetonitrile
77. 2-Nitropropane (m/z 43)
78. Methyl methacrylate
79. 1,1-Dichloro-2-propanone (m/z 43)
80. 4-Methyl-2-pentanone (MIBK)
81. Ethyl methacrylate
82. 2-Hexanone (MBK)
83. trans-1,4-Dichloro-2-butene
84. Pentachloroethane
85. Hexachloroethane
86. Nitrobenzene

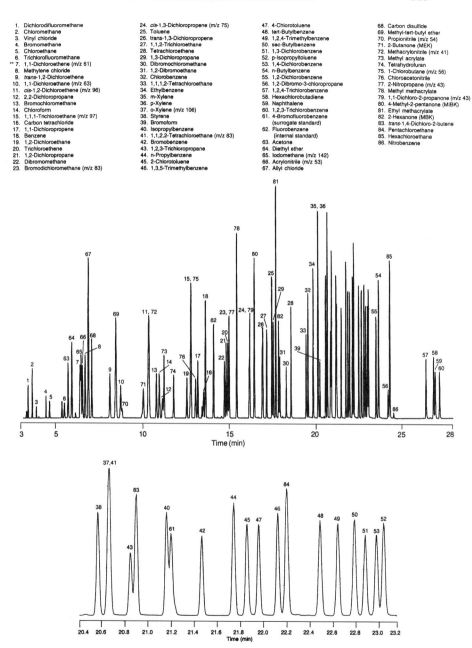

Fig. 9.47. US EPA Method 524.2, Revision 4, volatile organic compounds on a 60 m × 0.25 mm
I.D. DB-VRX with a 1.4 μm film thickness. For this method of drinking water analysis, the purge and
trap unit desorbs into a cryofocusing module (−160°C) at the head of the column. The method calls
for the analysis of 82 compounds plus internal standards. The lower chromatogram is an expanded
portion of the upper chromatogram.

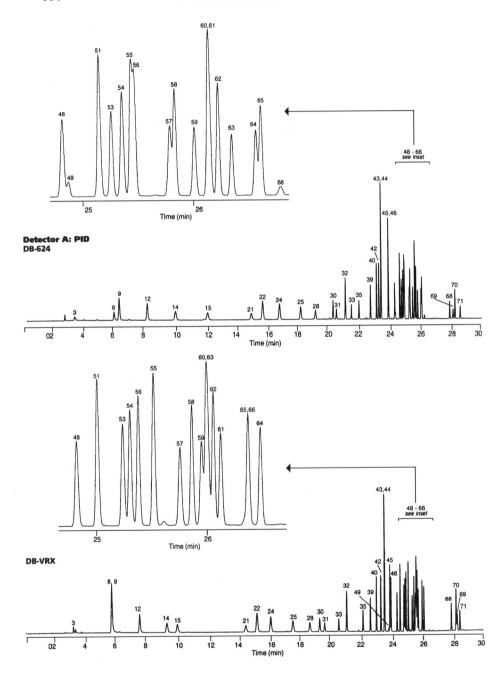

Fig. 9.48. (*continued*)

Compound list for all chromatograms
1. Dichlorodifluoromethane
2. Chloromethane
3. Vinyl chloride
4. Bromomethane
5. Chloroethane
6. Trichlorofluoromethane
7. 2-Chloropropane (IS)
8. 1,1-Dichloroethene
9. Iodomethane
10. Allylchloride
11. Methylene chloride
12. trans-1,2-Dichloroethene
13. 1,1-Dichloroethane
14. Chloroprene
15. cis-1,2-Dichloroethene
16. 2,2-Dichloropropane
17. Bromochloromethane
18. Chloroform
19. 1,1,1-Trichloroethane
20. Carbon tetrachloride
21. 1,1-Dichloropropene
22. Benzene
23. 1,2-Dichloroethane

24. Fluorobenzene (IS)
25. Trichloroethene
26. 1,2-Dichloropropane
27. Dibromomethane
28. Trifluorotoluene (IS)
29. Bromodichloromethane
30. 2-Chloroethyl vinyl ether
31. cis-1,3-Dichloropropene
32. Toluene
33. trans-1,3-Dichloropropene
34. 1,1,2-Trichloroethane
35. Tetrachloroethene
36. 1,3-Dichloropropane
37. Dibromochloromethane
38. 1,2-Dibromoethane
39. 1-Chloro-3-fluorobenzene (IS)
40. Chlorobenzene
41. 1,1,1,2-Tetrachloroethane
42. Ethylbenzene
43. m-Xylene
44. p-Xylene
45. o-Xylene
46. Styrene
47. Bromoform

48. Isopropylbenzene
49. cis-1,4-Dichlorobutene
50. 1,1,2,2-Tetrachloroethane
51. Bromobenzene
52. 1,2,3-Trichloropropane
53. n-Propylbenzene
54. 2-Chlorotoluene
55. 1,3,5-Trimethylbenzene
56. 4-Chlorotoluene
57. tert-Butylbenzene
58. 1,2,4-Trimethylbenzene
59. sec-Butylbenzene
60. 1,3-Dichlorobenzene
61. p-Isopropyltoluene
62. 1,4-Dichlorobenzene
63. Benzyl chloride
64. n-Butylbenzene
65. 1,2-Dichlorobenzene
66. bis (2-Chloroisopropyl) ether
67. 1,2-Dibromo-3-chloropropane
68. 1,2,4-Trichlorobenzene
69. Hexachlorobutadiene
70. Napthalene
71. 1,2,3-Trichlorobenzene

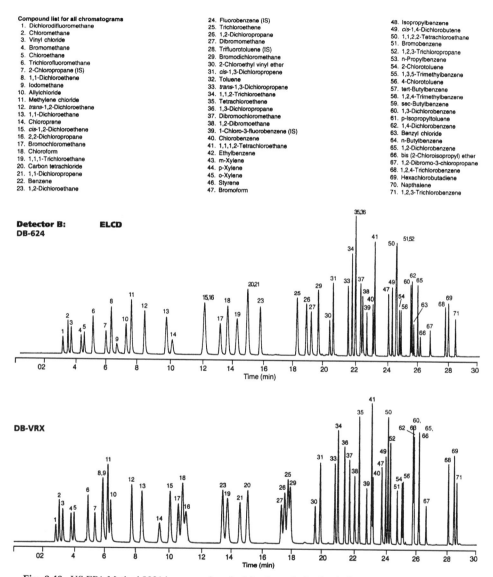

Fig. 9.48. US EPA Method 8021 is a general method for the analysis of volatile priority pollutants in industrial waste. A dual column application with tandem PID-ELCD detection is presented here. The result being four chromatograms from a single sample. The columns are a 75 m × 0.45 mm I.D. DB-VRX and DB-624, with 2.55 μm films. Purge and trap injection without cryofocusing is used to inject the sample. The combination of dissimilar selectivities of *both* stationary phases and detectors provides an ambiguous assignment of peaks.

toward trace levels of active analytes, increased upper temperature limits, and lower bleed for both selective and MS detectors that make these phases effective for semivolatiles analysis.

Figure 9.49 shows the chromatogram of an EPA method 8270 test mix on 5% phenyl equivalent silylarylene capillary column. The complexity of the method demands high resolution and MS detection. An application of EPA method 525.2 for polynuclear aromatic hydrocarbons in drinking water is shown in Fig. 9.50. The combination of column selectivity and a high upper temperature limit lead to a 25-min. run time.

Pesticide concentrations are regulated in food as well as all aspects of the physical environment, including the workplace. Pesticides are analyzed in a manner similar to the semivolatiles and according to their class (i.e., chlorinated, organophosphorus, triazines, and herbicides). The columns for these analyses are generally phenyl polymethylsiloxanes, usually 50% phenyl, as in EPA method 608, or 14% cyanopropylphenyl polymethylsiloxane (i.e., OV-1701 or equivalent). Many of the analytes are active, and some are structurally similar. Columns must be highly inert and possess a high degree of resolving power. Attention to the injection process and sample preparation are also crucial to success. Figure 9.51 shows the analysis of some organochlorine pesticides on a dual-column assembly using the afore mentioned columns and electron capture detection. The best resolution of triazines is accomplished with a 35% phenyl polymethylsiloxane or its silylarylene equivalent (Fig. 9.52).

Polychlorinated biphenyls (PCBs), produced by chlorinating biphenyl to a specified weight percent chlorine, are mixtures containing a number of positional isomers. The most commonly used PCBs were the Aroclors. With the exception of Aroclor 1016 (a PCB mixture containing 41% chlorine), the final two digits specify the percentage (by weight) of chlorine. Because they were relatively stable compounds characterized by a high dielectric potential, PCBs were widely used for purposes as diverse as transformer and hydraulic fluids, and their occurrence in other formulations and oily waste products is not uncommon. Some authorities now argue that these compounds can be carcinogenic; other authorities view at least some of those claims as ill-founded. Nevertheless, the growing concern led to the development of techniques for monitoring uncharacterized fluids and spills (e.g., [80–82]). Aroclor pattern recognition by GC with an ECD is the predominate method of PCB characterization. A dual column approach facilitates positive

---

Fig. 9.49. US EPA Method 8270 is general method for the analysis of semivolatile priority pollutants. This figure shows the analysis of an 8270 standard on a 30 m × 0.25 mm I.D. column with a 0.5 μm film of DB-5ms, a silylarylene equivalent to a 5% phenyl methylpolysiloxane. Because of the necessity of splitless injection and the amount of non-volatile material that these samples can carry the use of a retention gap is strongly recommended. Full scan MS detection over a wide mass range is necessary to characterize these samples.

Compound	m/z	RT
1. N-Nitrosodimethylamine	42	5.95
2. 2-Picoline	93	8.02
3. Methyl methanesulfonate	80	9.10
4. Ethyl methanesulfonate	79	10.70
5. Phenol	94	11.58
6. Aniline	93	11.65
7. bis(2-Chloroethyl) ether	93	11.80
8. 2-Chlorophenol	128	11.90
9. 1,3-Dichlorobenzene	146	12.24
10. 1,4-Dichlorobenzene	146	12.40
11. Benzyl alcohol	108	12.67
12. 1,2-Dichlorobenzene	146	12.73
13. 2-Methyl phenol	108	12.92
14. bis(2-Chloroisopropyl) ether	45	12.97
15. 4-Methyl phenol	108	13.25
16. Acetophenone	105	13.28
17. N-Nitrosodi-n-propylamine	70	13.29
18. Hexachloroethane	117	13.48
19. Nitrobenzene	77	13.65
20. N-Nitrosopiperidine	114	14.00
21. Isophorone	82	14.21
22. 2-Nitrophenol	139	14.35
23. 2,4-Dimethylphenol	122	14.43
24. bis(2-Chloroethoxy)methane	93	14.64
25. Benzoic acid	122	14.83
26. 2,4-Dichlorophenol	162	14.89
27. 1,2,4-Trichlorobenzene	180	15.06
28. Naphthalene	128	15.25
29. (a, a)-Dimethylphenethylamine	58	15.32
30. 4-Chloroaniline	127	15.37
31. 2,6-Dichlorophenol	162	15.39
32. Hexachlorobutadiene	225	15.51
33. N-Nitroso-di-n-butylamine	84	16.11
34. 4-Chloro-3-methylphenol	107	16.40
35. 2-Methylnaphthalene	142	17.01
36. Hexachlorocyclopentadiene	237	17.13
37. 1,2,4,5-Tetrachlorobenzene	216	17.16
38. 2,4,6-Trichlorophenol	196	17.40
39. 2,4,5-Trichlorophenol	196	17.48
40. 2-Chloronaphthalene	162	17.89
41. 1-Chloronaphthalene	162	17.94
42. 2-Nitroaniline	65	18.09
43. Dimethyl phthalate	163	18.49
44. 2,6-Dinitrotoluene	165	18.64
45. Acenaphthylene	152	18.81
46. 3-Nitroaniline	138	19.01
47. Acenaphthene	154	19.20
48. 2,4-Dinitrophenol	184	19.25
49. 4-Nitrophenol	139	19.35
50. Pentachlorobenzene	250	19.49
51. 2,4-Dinitrotoluene	165	19.54
52. Dibenzofuran	168	19.59
53. 1-Naphthylamine	143	19.75
54. 2,3,4,6-Tetrachlorophenol	232	19.84
55. 2-Naphthylamine	149	19.93
56. Diethyl phthalate	149	20.06
57. 4-Chlorophenyl phenyl ether	204	20.32
58. Fluorene	166	20.36
59. 4-Nitroaniline	138	20.39
60. 4,6-Dinitro-2-methylphenol	198	20.45
61. Diphenylamine	169	20.60
62. Decomposed N-Nitrosodiphenylamine	169	20.60
63. 1,2-Diphenylhydrazine	77	20.68
64. Phenacetin	108	21.25
65. 4-Bromophenyl phenyl ether	248	21.42
66. Hexachlorobenzene	284	21.58
67. 4-Aminobiphenyl	169	22.00
68. Pentachlorophenol	266	22.01
69. Pentachloronitrobenzene	295	22.04
70. Pronamide	173	22.09
71. Phenanthrene	178	22.50
72. Anthracene	178	22.52
73. Di-n-butyl phthalate	149	23.65
74. Fluoranthene	202	25.15
75. Benzidine	184	25.38
76. Pyrene	202	25.66
77. p-Dimethylaminoazobenzene	120	26.23
78. Butyl benzyl phthalate	149	27.08
79. 3,3-Dichlorobenzidine	252	28.69
80. bis(2-Ethylhexyl)phthalate	149	28.76
81. Benzo[a]anthracene	228	28.85
82. Chrysene	228	28.98
83. Di-n-octyl phthalate	149	31.41
84. 7,12-Dimethylbenz[a]anthracene	256	33.13
85. Benzo[b]fluoranthene	252	33.16
86. Benzo[k]fluoranthene	252	33.26
87. Benzo[a]pyrene	252	34.20
88. 3-Methylcholanthrene	268	35.34
89. Dibenz[a,j]acridine	279	37.49
90. Indeno[1,2,3-c,d]pyrene	276	38.23
91. Dibenz[a,h]anthracene	278	38.30
92. Benzo[g,h,i]perylene	276	39.34

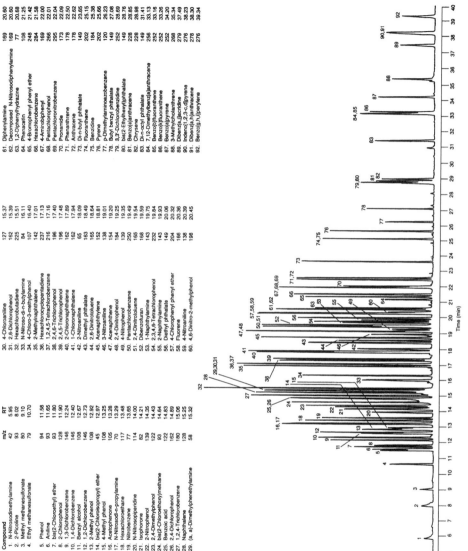

Time (min)

Fig. 9.49

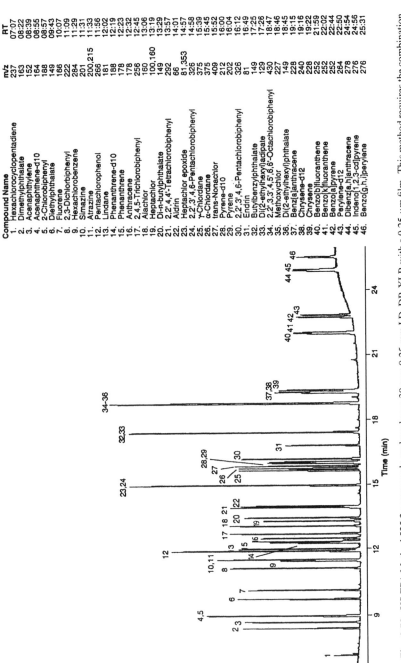

	Compound Name	m/z	RT
1.	Hexachlorocyclopentadiene	237	07:07
2.	Dimethylphthalate	163	08:22
3.	Acenaphthylene	152	08:39
4.	Acenaphthene-d10	164	08:55
5.	2-Chlorobiphenyl	188	08:57
6.	Diethylphthalate	149	09:43
7.	Fluorene	166	10:07
8.	2,3-Dichlorobiphenyl	222	11:09
9.	Hexachlorobenzene	284	11:29
10.	Simazine	201	11:31
11.	Atrazine	200,215	11:33
12.	Pentachlorophenol	266	11:56
13.	Lindane	181	12:02
14.	Phenanthrene-d10	188	12:19
15.	Phenanthrene	178	12:23
16.	Anthracene	178	12:32
17.	2,4,5-Trichlorobiphenyl	256	12:45
18.	Alachlor	160	13:06
19.	Heptachlor	100,160	13:19
20.	Di-n-butylphthalate	149	13:29
21.	2,2',4,4'-Tetrachlorobiphenyl	292	13:57
22.	Aldrin	66	14:01
23.	Heptachlor epoxide	81,353	14:57
24.	2,2',3',4,6-Pentachlorobiphenyl	326	14:58
25.	γ-Chlordane	375	15:39
26.	α-Chlordane	375	15:45
27.	trans-Nonachlor	409	15:52
28.	Pyrene-d10	212	16:00
29.	Pyrene	202	16:04
30.	2,2',3,4,6-Pentachlorobiphenyl	326	16:12
31.	Endrin	81	16:49
32.	Butylbenzylphthalate	149	17:25
33.	Di(2-ethylhexyl)adipate	129	17:26
34.	2,2',3,3',4,5',6,6'-Octachlorobiphenyl	430	18:47
35.	Methoxychlor	227	18:46
36.	Di(2-ethylhexyl)phthalate	149	18:45
37.	Benz[a]anthracene	228	19:15
38.	Chrysene-d12	240	19:16
39.	Chrysene	228	19:22
40.	Benzo[b]fluoranthene	252	21:59
41.	Benzo[k]fluoranthene	252	22:02
42.	Benzo[a]pyrene	252	22:44
43.	Perylene-d12	264	22:50
44.	Dibenz[a,h]anthracene	278	24:54
45.	Indeno[1,2,3-cd]pyrene	278	24:56
46.	Benzo[g,h,i]perylene	276	25:31

Fig. 9.50. US EPA Method 525.2 compounds analyzed on a 30 m × 0.25 mm I.D. DB-XLB with a 0.25 μm film. This method requires the combination of analyte resolution and the higher temperatures necessary to elute the higher boiling PAH's. A ion trap detector was used in this example and the oven was programmed from 60 to 350°C in three ramps.

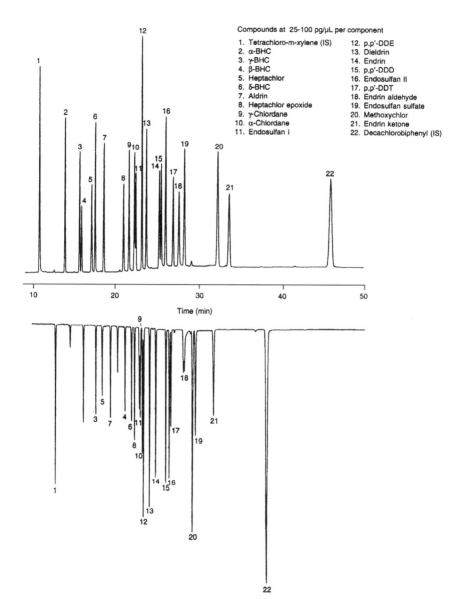

Compounds at 25-100 pg/µL per component

1. Tetrachloro-m-xylene (IS)
2. α-BHC
3. γ-BHC
4. β-BHC
5. Heptachlor
6. δ-BHC
7. Aldrin
8. Heptachlor epoxide
9. γ-Chlordane
10. α-Chlordane
11. Endosulfan I

12. p,p'-DDE
13. Dieldrin
14. Endrin
15. p,p'-DDD
16. Endosulfan II
17. p,p'-DDT
18. Endrin aldehyde
19. Endosulfan sulfate
20. Methoxychlor
21. Endrin ketone
22. Decachlorobiphenyl (IS)

Fig. 9.51. Organochlorine pesticides analyzed on a dual column assembly of a 30 m × 0.53 mm I.D. 50% phenyl methylpolysiloxane, *top*, with 0.83 µm film and a 30 m × 0.53 mm I.D. and a (14% cyanopropylphenyl) methylpolysiloxane, DB-1701 *bottom*, with a 1.0 µm film. The injection was Megabore direct with ECD detection. The dual column gives retention confirmation of the pesticides.

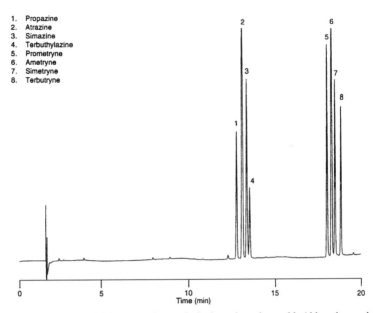

1. Propazine
2. Atrazine
3. Simazine
4. Terbuthylazine
5. Prometryne
6. Ametryne
7. Simetryne
8. Terbutryne

Fig. 9.52. Triazine herbicides are used extensively throughout the world. Although as a class they are easily separated, their resolution from each other requires a stationary phase with a high phenyl content. The resolution of eight triazines by a 30 m × 0.25 mm I.D. column with a 0.25 μm film of 35% phenyl methylpolysiloxane is shown here.

identifications. Figure 9.53 shows the chromatograms of Aroclor 1254 on DB-5, DB-1701, and DB-17 columns. The DB-5/DB-17 pair is the most common in the dual-column application. Studies of PCB's have shown that some species of the 209 possible are more toxic than the others. Separating the individual components, congeners, of this complex mixture is, at this writing, best accomplished with a DB-XLB. Figure 9.54 shows a chromatogram of a mixture of Aroclors 1232, 1248, and 1262 with ECD detection. The unique silylarylene chemistry of this phase provides the best congener resolutions to date for resolving PCB congeners on a single column [83]. MS detection in the SIM mode is very helpful to resolve these species with a single column.

Fig. 9.53. Aroclor determination in environmental samples is improved when the sample is chromatographed on two phases of dissimilar selectivity. The resulting patterns are more readily identified and assigned. Shown here is a sample of Aroclor 1254 chromatographed on three columns of different selectivity. The top chromatogram is from a 5% phenyl methylpolysiloxane, the middle chromatogram is from a (14% cyanopropylphenyl) methylpolysiloxane and the bottom chromatogram is from a 50% phenyl methylpolysiloxane. All columns were 30 m × 0.32 mm I.D.

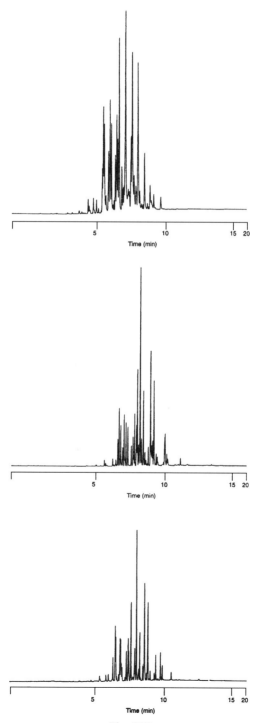

Fig. 9.53

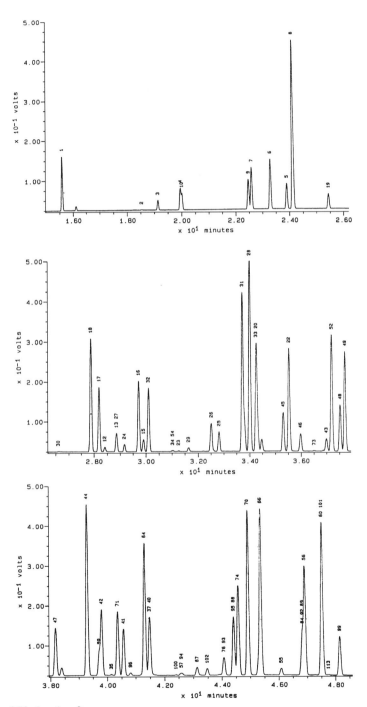

Fig. 9.54. (*continued*)

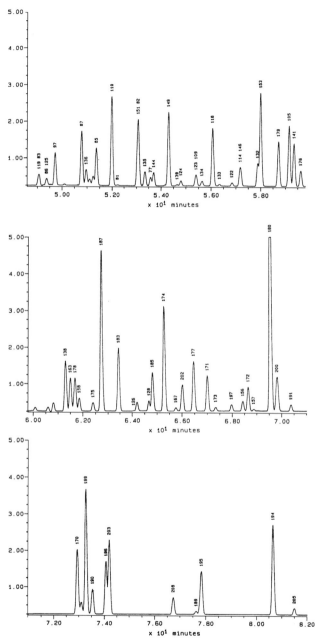

Fig. 9.54. A chromatogram of poly chlorinated biphenyls, listed by IUPAC #, on a 60 m × 0.18 mm I.D. DB-XLB with a 0.18 μm film. The oven was programmed from 75°C (2 min.) at 15°C/min to 150°C, then at 1.5°C/min to 285°C. The carrier was H₂ with an average linear velocity of 40 cm/sec. The sample was a dilute mixture of Aroclor 1232, 1248 and 1262. *Courtesy of Jack Cochran, Hazardous Waste Research and Information Center, Champaign, IL, USA.*

An abundance of isomers also complicates the analysis of dioxins [(tetra-chlorodibenzodioxins (TCDDs)]. There are reports that the 2,3,7,8-isomers are both mutagenic and carcinogenic, but such claims are understandably difficult to authenticate with human subjects. Again, analytical techniques have developed. Suspect samples are first screened by EPA method 625 (which also isolates other contaminants), and positive samples are then subjected to EPA method 613. Safe handling procedures, important for all toxic substances and of critical importance for 2,3,7,8-TCDD, are also discussed in the latter document. The gas chromatographic separation of the 2,3,7,8-isomer is not trivial, and method 613 uses high resolution mass spectrometric detection for its differentiation. The gas chromatographic separation requires a polar stationary phase, a large number of theoretical plates, and relatively high temperatures. Although specialty phases exist for this method, the application of a 5% phenyl polymethylsiloxane in conjunction with a 50% cyanopropylphenyl 50% methyl polysiloxane are most commonly used.

The seepage of fuel from underground storage tanks raises the possibility of groundwater contamination and has become an active area of environmental concern. As tanks age, fuel leaks into the adjacent soil and may access water supplies. Many of these tanks have been replaced with tanks of more inert materials but the sites must be remediated. The adjacent soils, as well as water supplies are analyzed for contamination by "gasoline-range organics" (GROs), or "diesel-range organics" (DROs). While the diesel components are analyzed *en toto*, some degree of speciation is desired for the GRO (usually BTEX). Figure 9.55 shows a BTEX analysis of water with an FID and PID in tandem.

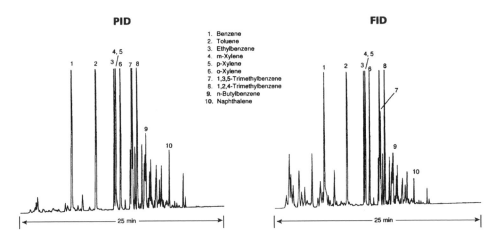

Fig. 9.55. The analysis of a contaminated water (5 ppm gasoline) sample for BTEX using a PID and FID in tandem. The PID is selective for unsaturated and aromatic compounds and the FID detects all of the hydrocarbons in the sample. The sample was chromatographed on a 30 m × 0.53 mm I.D. DB-624 column with a 3 μm film. The sample was collected and injected with a purge and trap concentrator.

Amendments to the US 1990 Clean Air Act are concerned with the analysis of ambient air. Method TO-14 is for the analysis of polar and nonpolar compounds present in air in the low to sub-ppb range [84]. In this method, air is drawn into passivated metal canisters (SUMMA canisters are used predominately). The contents of the canister are then concentrated on a glass bead packed cryogenic trap, desorbed, and backflushed into a cryofocusing module (see Chapter 3) prior to GC analysis.

9.5 Biological and Medical Applications

The low-level detection of compounds significant in this field poses stringent requirements for well-deactivated analytical systems; this is, of course, also true for some other active solutes, such as pesticides. These high levels of solute activity may necessitate not only the selection of well-deactivated columns but also attention to other parts of the analytical system; in particular, user deactivation of the sample injection area may be desirable (see Chapter 3). Where quantitation is important, techniques explored by Schomburg [85] and by Dandeneau and Zerrener [86] should be employed to establish that the reliability of the system extends to the levels measured for the solute of concern. For thermally labile solutes, cold on-column injection may also be required. Developments in this application area as they relate to gas chromatography have been reviewed [87].

It is often desirable to monitor the level of an anesthetic, both in the air (or oxygen) stream provided to a patient and sometimes (as a safeguard against leakage) in the room atmosphere. Rapid analysis is essential, and thermal conductivity is the most practical means of detection. Figure 9.56 illustrates the use of a very thick-film large-diameter open tubular column to achieve the separation of several anesthetics in less than 7 min. Vigh has recently reported the chiral separation of the anesthetics desflurane, isoflurane and enflurane (see Fig. 9.57) [88]. Studies in progress are aimed at bulk chiral separations of anesthetics to support studies on the anesthetic efficacy of specific enantiomers [89].

Poisoning by volatile organic solvents is sometimes evidenced by the odor of the patient's clothing or breath, but in other cases the diagnosis is facilitated by blood headspace analysis [90]. The more common agents of abuse, whose sources can include solvents (used per se or from sniffing glue, lacquer, etc.), aerosol propellants, or anesthetics, may include bromochlorodifluoromethane, n-butane, carbon tetrachloride, chlorobutanol, cryofluorane (Halon 114), ethyl acetate, halothane, isobutane, isopropanol, isopropyl nitrate, methyl ethyl ketone (MEK), propane, tetrachloroethylene, toluene, 1,1,1-trichloroethane, 2,2,2-trichloroethanol, trichloroethylene, and trichlorofluoromethane (Halon 11). Ramsey and Flanagan used headspace injections on two dissimilar packed columns and simultaneous flame ionization–electron capture detection (FID-ECD) to facilitate rapid detection and identification of these solutes in blood samples of acutely poi-

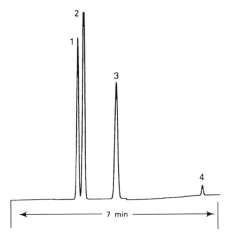

Fig. 9.56. Separation of a mixture of anesthetics on a 30 m × 0.53 mm column, 5.0 μm film of dimethylpolysiloxane; 50°C (3.75 min) to 125°C@ 30°/min, TCD. Solutes: 1) isoflurane; 2) enflurane; 3) halothane; 4) methoxyflurane.

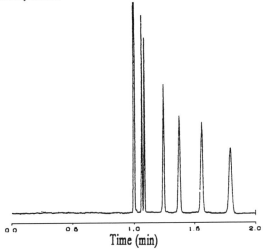

Fig. 9.57. The chiral separation of three anesthetics on a 30 m × 0.25 mm I.D. GTA column with a 0.25 μm film [88]. The enantiomers of desfluran, isoflurane and enflurane were resolved in under 2 minutes at 50°C.

soned patients [90]. It is probable that the DB-624 column could accomplish such analyses even more rapidly. The separation of those solutes that typically occur in an analysis for blood alcohols is shown in Fig. 9.58; analyses of this type can also be performed by headspace sampling procedures, with great benefits to column lifetime.

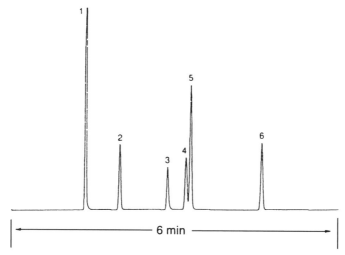

Fig. 9.58. Blood alcohol on a 30 m × 0.53 mm column, 1 μm film of DB-WAX. Injection, 0.2 μL water containing 0.1% of each solute; 40°C (2 min) to 80°C @ 10°/min. Solutes: 1) acetaldehyde; 2) acetone; 3) methanol; 4) isopropanol; 5) isopropanol; 6) ethanol; 7) *n*-propanol.

Drugs are often among the more active solutes, and derivatization is normally required for their analysis on packed columns. The increased inertness of the fused silica open tubular column may allow direct analysis of the underivatized drug. In some cases (e.g., some of the antiepileptics), derivatization is employed to impart thermal stability and is required for both packed and open tubular columns. These analyses would probably benefit from columns that incorporated silarylene stationary phases. A separation of some tricyclic antidepressant drugs is shown in Fig. 9.59; the other drug analyses shown (Figs. 9.60–9.65) all utilized the same large-diameter open tubular column.

In studies of the derivatization of difunctional amines (e.g., β-adrenergic blocking drugs), Jacob *et al.* established that with the exception of ephedrine, all drugs examined produced two stereoisomeric derivatives that were resolved to well-formed peaks under the conditions employed [91]. Alm *et al.* used simultaneous injection of drug mixtures onto two dissimilar columns, one employing FID and the other nitrogen/phosphorus detection (NPD), to generate chromatographic data of the type shown in Fig. 9.66 [92]. A computerized system of identification, based on solute retentions relative to those of two internal standards, was also described. A number of underivatized barbiturate and alkaloid drugs, neat and in spiked urine, were separated by Plotczyk and Larson [93], who also assigned retention indices and studied the effects on column lifetime of injections of polar solutions of such solutes.

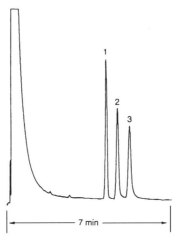

Fig. 9.59. Underivatized antidepressant drugs on a 15 m × 0.53 mm column, 1 μm 50%-dimethyl-50%-diphenylpolysiloxane; 150 to 220°C @ 20°/min. Solutes: 1) amitriptyline; 2) nortriptyline; 3) protriptyline.

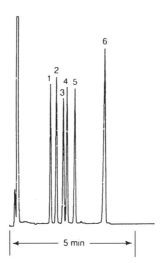

Fig. 9.60. Underivatized barbiturates on a 30 m × 0.53 mm column, 1.5 μm film of dimethyl-polysiloxane; 175°C to 210°C @ 8°/min. Direct injection of one μL methanol containing one μg each analyte. Solutes: 1) aprobarbital; 2) butabarbital; 3) amobarbital; 4) pentobarbital; 5) secobarbital; 6) phenobarbital.

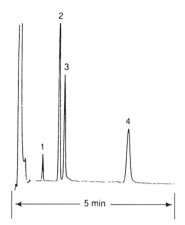

Fig. 9.61. Alkaloid drugs, conditions as in Fig. 9.60 except 250 ng each component, and isothermal at 250°C. Solutes: 1) cocaine; 2) codeine; 3) morphine; 4) quinine.

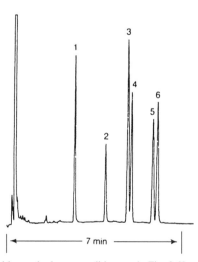

Fig. 9.62. Sedative and hypnotic drugs; conditions as in Fig. 9.60 except 70 ng per component, 150°C to 250°C @ 10°/min. Solutes: 1) aprobarbital; 2) meprobamate; 3) dipenhydramine; 4) mephobarbital; 5) methapyrilene; 6) chlorpheniramine.

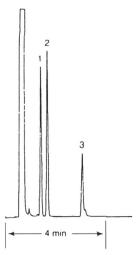

Fig. 9.63. Amphetamines; conditions same as Fig. 9.60, except 500 ng per component, 150°C isothermal. Solutes: 1) amphetamine; 2) methamphetamine; 3) ephedrine.

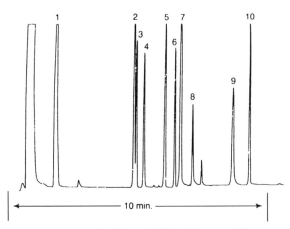

Fig. 9.64. Anticonvulsant drugs. Conditions as in Fig. 9.60, except 200 ng per component, 160°C (2 min) to 275°C @ 15°/min. Solutes: 1) ethosuximide; 2) methsuximide; 3) phensuximide; 4) *n*-desmethylmethsuximide; 5) phenytoin; 6) 5-methyl-5-phenylhydantoin; 7) phenylethylmalonamide; 8) phenobarbital; 9) primidone; 10) carbamazephine.

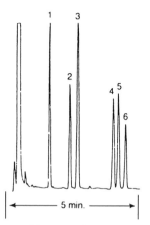

Fig. 9.65. Antihistamine drugs, conditions as in Fig. 9.60 except 85 ng per component, 215 to 275°C @ 5°/min. Solutes: 1) pheniramine; 2) tripelennamine; 3) cyclizine; 4) pyrilamine; 5) triprolidine; 6) promethazine.

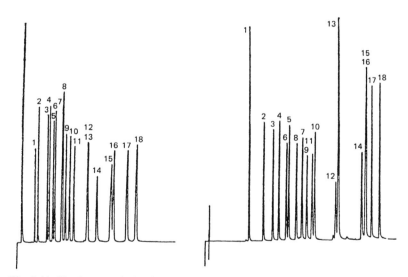

Fig. 9.66. Simultaneous dual column, dual detector analysis of an illicit drug mixture [92]. Left, 11 m × 0.2 mm column, 0.26 μm film 5%-phenyl-95%-methylpolysiloxane, FID. Right, 10 m × 0.2 mm column, 0.2 μm film of trifluropropyl-methy-polysiloxane, NPD. Both columns 75° (1 min) to 280°C @ 10°/min. Solutes: 1) metarbital; 2) barbital; 3) allobarbital; 4) aprobarbital; 5) propylamphetamine (internal standard #1); 6) butylbarbital; 7) amobarbital; 8) nealbarbital; 9) pentobarbital; 10) vinbarbital; 11) secobarbital; 12) brallobarbital; 13) hexobarbital; 14) trioctylamine (internal standard #2); 15) phenobarbital; 16) cyclobarbital; 17) alphenal; 18) heptabarbital.

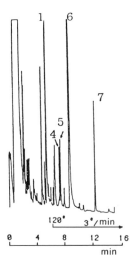

Fig. 9.67. Extract of human plasma following oral administration of isosorbide dinitrate [94]. Column, 30 m × 0.32 mm, 0.25 μm film of 5%-phenyl-95%-methylpolysiloxane; 60° during injection, ballistic rise to 120°C (3 min) to 195°C @ 3°/min. Solutes: 1) isosorbide-2-mononitrate; 4) isomannide mononitrate; 5) isoidide mononitrate; 6) isosorbide-5-mononitrate; 7) isosorbide dinitrate. Reprinted with permission of Huethig Publishing Co.

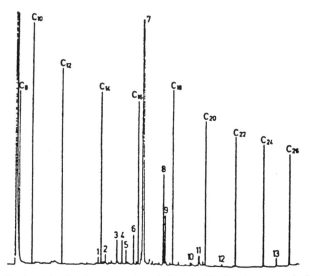

Fig. 9.68. Total acid fraction from urine. Sample from healthy female, extracted with chloroform and reacted with diazomethane. Column, 30 m × 0.33 mm coated with dimethylpolysiloxane; 80° (5 min) to 280°C @ 2°/min, FID. Numbered solutes are mono-, di-, or tri-methyl esters of the following acids: 1) 3-methoxyphenylacetic; 2) citric; 3) 2,4, methyloctanedioic; 4) nonanedioic; 5) N-caetyl-2-aminooctanoic; 6) 3,4-dimethoxyphenylacetic; 7) hippuric; 8) indoleacetic; 9) N-methylhippuric; 10) palmitic; 11) phthalic; 12) stearic; 13) di(n-butyl)phthalate. From [95].

Figure 9.67 shows results presented by Lutz *et al.* [94], who developed an automated procedure for the determination of isosorbide dinitrate and its metabolites in blood plasma, using capillary gas chromatography and ECD. A chromatogram of the total acid fraction of a urine sample from a healthy female is shown in Figs. 9.68, and 9.69 examines the less abundant components of the same sample following removal of the major components by liquid chromatography [95]. Chromatographic parameters yielding the separation of the trimethylsilyl (TMS) derivatives of several steroids were established by Knorr *et al.* [96] and employed to demonstrate differences in the patterns of urinary steroids from normal and diseased patients (Fig. 9.70). Lochner *et al.* [97] suggested that a noninvasive technique, based on GC/MS patterns generated by saliva extracts, could be used in the detection of certain pathological states; patterns from saliva extracts of a normal and a diseased patient are shown in Fig. 9.71.

The plasma concentration of free and total fatty acids in patients suffering from medium chain acyl–CoA dehydrogenase (MCAD) defficiency was studied by gas chromatography (see Fig. 9.72) [98]. Patients with MCAD were found to have increased levels of several fatty acids relative to control and fasting patients.

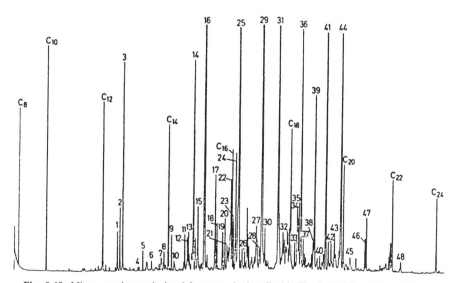

Fig. 9.69. Minor constituents isolated from sample described in Fig. 9.68 [95], same conditions. Solutes represent mono-, di-, or tri-methyl esters of the following acids: 1) 2-methylhexanedioic; 2) 3-methylhexanedioic; 3) 2-hydroxyphenylacetic; 15) 3,5-dimethyloctanedioic; 16) nonanedioic; 17) 3,4-dimethyloctanedioic; 24) hexanedioic; 25) 5-decynedioic; 29) 3-methylfuran-2,5-dipropanoic; 31) 2-(3-carboxy-4-methyl-5-*n*-propylfuranyl)-*n*-propanoic; 41) 2-(3-carboxy-4-methyl-5-*n*-pentylfuranyl)-*n*-propanoic.

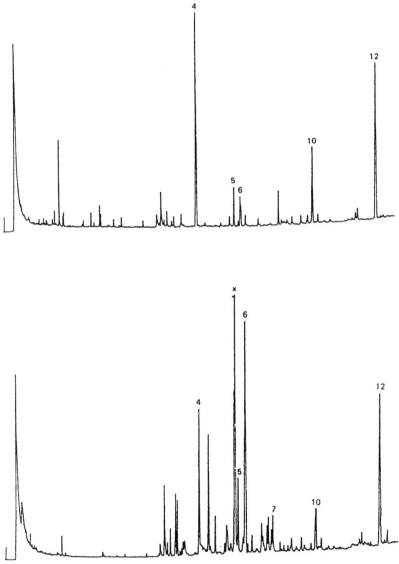

Fig. 9.70. Chromatograms of urinary steroids as trimethylsilyl ethers [96]. Column, 35 m × 0.22 mm, dimethylpolysiloxane; 190-290°C @ 2°/min, FID. Top, normal adult woman; selected solutes 5) pregnanediol; 6) pregnanetriol; 10) tetrahydrocortisone. Bottom, insufficiently treated 12-year old girl with 21-hydrolase deficiency: 4) internal standard; 5) 17-hydroxypregnanolone; 6) pregnanetriol.

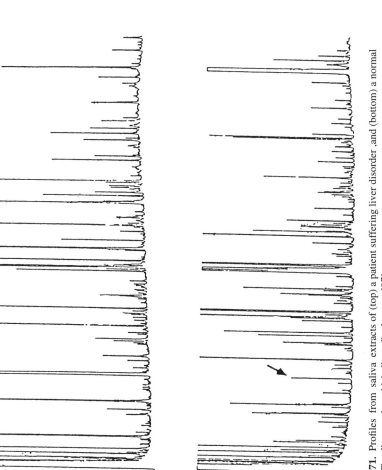

Fig. 9.71. Profiles from saliva extracts of (top) a patient suffering liver disorder, and (bottom) a normal patient. The diagnostic peak is indicated by the arrow [97].

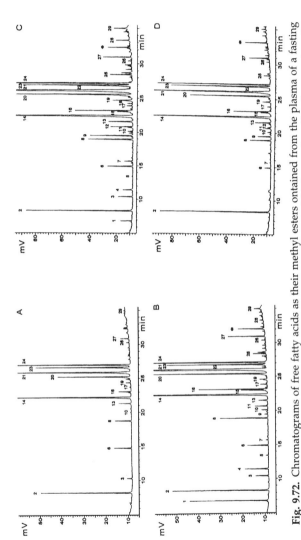

Fig. 9.72. Chromatograms of free fatty acids as their methyl esters obtained from the plasma of a fasting control patient A, a hypoglycemic MCAD patient B, A symptomatic MAD-m patient C and a symptomatic VLCAD patient D. The separation was performed on a 25m×0mm I.D.CP Sil 88 column with a 0.2 μm film. Quantitative results were based on C9:0 and C19:0 internal standards. *Cis* and *trans* isomers were resolved. From [98].

References

1. J. Yancey, *J. Chromatogr. Sci.* **32**:403 (1994).

2. T. Chamblee and B. Clark, *Bioactive Volatile Compounds from Plants*, ACS Symposium Series 525 (R. Teranishi, R. Buttery, and H. Sugisawa, eds.) p. 97, American Chemical Society, Washington, D.C., 1993.

3. G. Takeoka, S. Ebeler, and W. Jennings, *Characterization and Measurements of Flavor Compounds* (D. D. Bills and C. J. Mussinan, eds.) p. 95, American Chemical Society Symposium Series 289, (1984).

4. R. Kaiser, *Bioactive Volatile Compounds from Plants*, ACS Symposium Series 525 (R. Teranishi, R. Buttery, and H. Sugisawa, eds.) (1993).

5. W. Coleman, *J. Chromatogr. Sci.* **30**:159 (1992).

6. A. Padrayuttawat, H. Tamura, and M. Yamao, *J. High Res. Chromatogr.* **19**:365 (1996).

7. J. Field, G. Nickerson, D. James, and C. Heider, *J. Agric. Food Chem.*, **44**:1768 (1996).

8. K. Miller, C. Pool, and T. Pawlowski, *Chromatographia* **42**:639 (1996).

9. C. Author, L. Killam, K. Buchholtz, J. Pawliszvn, and J. Berg, *Anal. Chem.* **64**:1960 (1992).

10. E. Guenther, *The Essential Oils*, Vol. 3. Van Nostrand-Reinhold, Princeton, New Jersey, 1949.

11. J. W. Gramshaw and K. Sharpe, *J. Sci. Food Agric.* **31**:93 (1980).

12. M. M. Bradley and J. W. Gramshaw, *J. Sci. Food Agric.* **31**:99 (1980).

13. W. Jennings, *J. Chromatogr. Sci.* **22**:129 (1984).

14. W. Jennings and T. Shibamoto, *Qualitative Analysis of Flavor and Fragrance Volatiles by Glass Capillary Gas Chromatography*, Academic Press, New York, 1980.

15. T. Shibamoto, in *Capillary Gas Chromatography in Essential Oil Analysis* (P. Sandra and C. Bicchi, eds.). Huethig Verlag, Heidelberg, 1987.

16. R. Adams, *Identification of Essential Oil Components by Gas Chromatography/Mass Specrometry*, Allured Publishing, Carol Stream, Illinois, 1995.

17. L. Nykanen, P. Savolahti, and I. Nykanen, *Topics in Flavor Research* (R. Berger, S. Nitz, and P. Schreier, eds.) p. 109.199 Eichhorn, Marzling-Hangenham, 1985.

18. W. Mick and P. Schreier, *Agric. Food Chem.* **32**:924 (1984).

19. A. Kobayashi, K. Kubota, and M. Yano, *Bioactive Volatile Compounds from Plants*, ACS Symposium Series 525 (R. Teranishi, R. Buttery, and H. Sugisawa, eds.). American Chemical Society, Washington, D.C., 1993.

20. I. Kubo, *Bioactive Volatile Compounds from Plants*, ACS Symposium Series 525 (R. Teranishi, R. Buttery, and H. Sugisawa, eds.). American Chemical Society, Washington, D.C., 1993.

21. M. Guntert, J. Bruning, R. Emberger, R. Hopp, M. Kopsel, H. Surberg, and P. Werkhoff, in *Flavor Precursors-Thermal and Enzymatic Conversions*, ACS Symposium Series 490 (R. Teranishi, G. Takeoka, and M. Guntert, eds.) American Chemical Society, Washington D.C., 1992.

22. C. Macku, *J. Agr. Food Chem.* **35**:845 (1987).

23. M. Guentert, A. Rapp, G.R. Takeoka, and W. Jennings. *Z. Lebensm. Unters. Forsch.* **182**:200 (1986).

24. D. Goldberg, J. Yan, E. Ng, E. Diamandis, A. Karumanchiri, G. Soleas, and A. Waterhouse, *Anal. Chem.* **66**:3959 (1994).

25. G. Soleas, personal communication, 1995.

26. W. Konig, A. Kruger, D. Icheln, and T. Runge, *J. High Res. Chromatogr.* **15**:184 (1992).

27. W. Konig, A. Rieck, I. Hardt, B. Gehrcke, K. Kubeczka, and H. Muhle, *J. High Res. Chromatogr.* **17**:315 (1994).

28. G. Takeoka, R. Flath, T. Mon, R. Buttery, R. Teranishi, M. Guntert, R. Lautamo, and J. Szejtli, *J. High Res. Chromatogr.* **13**:202 (1990).

29. P. Werkhoff, S. Brennecke, W. Bretschneider, M. Guntert, R. Hopp, and H. Surburg, *Z. Lebensm. Unters. Forsch.* **196**:307 (1993).

30. V. Schurig, J. Jauch, D. Schmalzing, M. Jung, W. Bretschneider, R. Hopp, and P. Werkhoff, *Z. Lebensm. Unters. Forsch.* **191**:28 (1990).

31. P. Werkhoff, S. Brennecke, and W. Bretschneider, *Chem. Mikrobiol. Technol. Lebensm.* **13**:129 (1991).

32. E. Bayer, *Z. Naturforsch.* **38**:1281 (1983).

33. W. Christie, *Gas Chromatography and Lipids—A Practical Guide*, The Oily Press, Ayr, Scotland, 1989.

34. G. Pilhofer, H. Lee, M. McCarthy, P. Tong, and J. German, *J. Dairy Sci.* **77**:55 (1994).

35. J. B. German, personal communication (1996).

36. B. A. Ingelse, F. M. Evergerts, J. S. Sevcik, Z. Stransky, and S. Fanali *J. High Res. Chromatogr.* **18**:348 (1995).

37. W. Ratnayake, and J. Beare-Rogers, *J. Chromatogr. Sci.* **28**:633 (1990).

38. R. Wolff, *J. Chromatogr. Sci.* **30**:17 (1992).

39. W. Ratnayaka and G. Pelletier, *J. Am. Oil Chem. Soc.* **69**:95 (1992).

40. C. E. Kientz and A. Verweij, *J. Chromatogr.* **355**:229 (1986).

41. B. Kolb, P. Pospisil, and M. Auer, *Chromatographia* **19**:113 (1984).

42. B. Page and G. Lacroix, *J. Chromatogr.* **648**:199 (1993).

43. D. M. Wyatt, *J. Chromatogr. Sci.* **21**:508 (1983).

44. G. Reineccius, personal communication (1985).

45. C. Furst and P. Furst, in *Capillary Gas Chromatography in Food Control and Research* (R. Wittkowski and R. Matissek, eds.) B. Behr's Verlag, Hamburg, 1992.

46. R. R. Freeman and R. M. A. Lautamo, *Am. Lab.* **18**:60 (1986).

47. S. A. Mooney, F. P. Di Sanzo, and C. J. Lowther, *J. High Res. Chromatogr.* **5**:684 (1982).

48. C. J. Cowper and P. A. Wallis, *Chromatographia* **19**:85 (1985).

49. R. Pauls, in *Advances in Chromatography*, Vol. 35 (P. Brown and E. Churika, eds.), p. 259. Marcel Dekker, New York, 1995.

50. N. Johansen, *Determination of Individual Components in Gasoline by High Efficiency Open Tubular Column Gas Chromatography*, N.G. Johansen, Inc., Boulder, Colorado, 1993.

51. T. Berger, *Chromatographia* **42**:63 (1996).

52. J. Diehl, J. Finkbiner, and F. Di Sanzo, *Anal. Chem.* **64**:3202 (1992).

53. F. Di Sanzo, *J. Chromatogr. Sci.* **28**:73 (1990).

54. R. Shearer, D. O'Neal, R. Rios, and M. Baker, *J. Chromatogr. Sci.* **28**:24 (1990).

55. B. Chawla and F. Di Sanzo, *J. Chromatogr.* **589**:271 (1992).

56. R. Hutte and J. Ray, *Detectors for Capillary Chromatography* (H. Hill and D. McMinn, eds.) John Wiley & Sons, New York, 1992.

57. V. Reeve, J. Jeffery, D. Weihs, and W. Jennings, *J. Forensic Sci.* **31**:479 (1986).

58. K. Furton, J. Bruna, and J. Almirall, *J. High Res. Chromatogr.* **18**:625 (1995).

59. S. R. Lipsky and M. L. Duffy, HRC CC, *J. High Res. Chromatogr. Chromatogr. Commun.* **9**:376 (1986).

60. G. A. Warburton and J. E. Zumberge, *Anal. Chem.* **55**:123 (1983).

61. R. Hwang, *J. Chromatogr. Sci.* **28**:109 (1990).

62. H. Borwitzky and G. Schomburg, *J. Chromatogr.* **240**:307 (1982).

63. B. P. Tissot and D. H. Welte, *Petroleum Formation and Occurrence*, Springer-Verlag, Berlin, New York, 1978.

64. W. Bergmann, W. Heller, A. R. Hernanto, M. Schallies, and E. Bayer, *Chromatographia* **19**:165 (1985).

65. T. A. Michalske and S. W. Freiman, *Nature* (London) **295**:511 (1982).

66. D. Villalonti (Triton Analytical), personal communication (1995).

67. ASTM method D2887E.

68. J. Danielson, R. Snell, and G. Oxborrow, *J. Chromatogr. Sci.* **28**:97 (1990).

69. H. Rasmussen, A. Pinto, M. DeMouth, P. Touretzky, and B.McPherson, *J. High Res. Chromatogr.* **17**:593 (1994).

70. *Fed. Regist.* **44**:(233) (1979).

71. *Fed. Regist.* **45**:(98) (1980).

72. M. F. Mehran, W. J. Cooper, and W. Jennings, *J. High Res. Chromatogr.* **7**:215 (1984).

73. M. F. Mehran, W. J. Cooper, R. Lautamo, R. R. Freeman, and W. Jennings, *J. High Res. Chromatogr.* **8**:715 (1985).

74. J. Eichelberger and W. Budde, *Method 524.2 Measurement of Purgable Compounds in Water by Capillary Column Gas Chromatography/Mass Spectrometry.* US EPA, Cincinnati, Ohio, 1989.

75. US EPA Method 524.2 Revision 4.

76. K. Grob and A. Habich, *J. High Res. Chromatogr.* **6**:11 (1983).

77. K. Grob, *L. Chromatogr.* **299**:1 (1984).

78. M. F. Mehran, W. J. Cooper, M. Mehran, and W. Jennings, *J. Chromatogr. Sci.* **24**:142 (1986).

79. T. Nilsson, F. Pelusio, L. Montanarella, B. Larsen, S. Facchetti, and J. Madsen, *J. High Res. Chromatogr.* **18**:617 (1995).

80. *The Determination of PCBs in Transformer Fluid and Waste Oil*, EPA 600/4-81-045. Environmental Monitoring and Support Laboratory, Office of Research and Development, USEPA, Cincinnati, Ohio, 1982.

81. *Methods for Organic Chemical Analysis of Municipal and Industrial Wastewater*, EPA 600/4/82-057. Environmental Monitoring and Support Laboratory, Office of Research and Development, USEPA, Cincinnati, Ohio, 1982.

82. *Interim Methods for the Sampling and Analysis of Priority Pollutants in Sediments and Fish Tissue.* Environmental Monitoring and Support Laboratory, Office of Research and Development, USEPA, Cincinnati, Ohio, 1980.

83. G. Frame, personel communication (1996).

84. J. Winberry, B. Carhart, A. Randall, and D. Decker, "Method TO-14," *Compendium of Methods for the Determination of Toxic Organic Compounds in Ambient Air*, EPA-600/4-89-017, U. S. Environmental Protection Agency, Research Triangle Park, North Carolina, 1988.

85. G. Schomburg, *J. High Res. Chromatogr.* **2**:461 (1979).

86. R. D. Dandeneau and E. H. Zerrener, *J. High Res. Chromatogr.* **2**:451 (1979).

87. R. Clement (ed.), *Gas Chromatography-Biochemical, Biomedical and Clinical Applications*, John Wiley & Sons, New York, 1990.

88. A. Shitangkoon, D. Staeer, and G. Vigh, , *J. Chromatogr.* **657**:387 (1993).

89. G. Vigh, personel communication, (1996).

90. J. D. Ramsey and R. J. Flanagan, *J. Chromatogr.* **240**:423 (1982).

91. K. Jacob, G. Schnabl, and W. Vogt, *Chromatographia* **19**:2161 (1985).

92. S. Alm, S. Jonson, H. Karlsson, and E. G. Sundholm, *J. Chromatogr.* **254**:179 (1983).

93. L. L. Plotczyk and P. Larson, *J. Chromatogr.* **257**:211 (1983).

94. D. Lutz, J. Rasper, W. Gielsdorf, J. A. Settlage, and H. Jaeger, *J. High Res. Chromatogr.* **7**:58 (1984).

95. G. Spiteller, *Angew. Chem., Int. Ed. Engl.* **24**:451 (1985).

96. D. Knorr, F. Bidlingmaier, and U. Kuhnle, *Horm. Res.* **16**:201 (1982).

97. A. Lochner, S. Weisner, A. Zlatkis, and B. S. Middleditch, *J. Chromatogr.* **378**:267 (1986).

98. W. Onkenhout, V. Venizelos, P. van der Poel, M. van den Heuvel, and B. Poorthuis, *Clin. Chem.* **41**:1467 (1995).

CHAPTER 10

TROUBLESHOOTING

10.1   General Considerations

Elementary troubleshooting, as based on baseline trace and the shape and behavior of solute peaks, has been well covered by Hinshaw [1]. The present chapter builds on a basic troubleshooting knowledge to considerations of other chromatographic problems and methods for their rectification.

New-Column Installation

Most practitioners tend to condemn the column for defects such as an excessively rising or unsteady baseline, for ghost, malformed, or tailing peaks, and for many other chromatographic anomalies. These problems can also be due to extracolumn factors (e.g., deposits or residues in the inlet). Active sites within the inlet can produce tailing peaks on chromatograms generated on even the best-deactivated column.

A new column that produces excellent results when first installed sometimes undergoes rapid deterioration, resulting in decreased resolution, tailing peaks, and/or a high rate of bleed. High-quality columns are both material- and labor-intensive. Column manufacturing costs can be reduced by using lower-quality supplies (stationary phases, tubing), by devoting less time and effort to deactivating and coating the tubing and individually testing each finished column, and by lowering the minimum quality control performance standards so that a higher percentage of the finished columns can be shipped and sold. A given lot of columns produced in this manner will usually still include a few columns that are excellent, a number that are mediocre, and some that are of very poor quality. The manufacturing costs and initial purchase price of higher-quality columns are generally higher, but the columns can be expected to produce superior results for longer times when given proper care. The lifetime of a lower-quality column is usually preordained by manufacturing parameters and is less affected by operational conditions. The lifetime of a high-grade column, on the other hand, is less limited by conditions of its manufacture and is very much influenced by the operational conditions. One of the more critical circumstances that can jeopardize column lifetime is the use of contaminated carrier gas.

Gas scrubbers and purifiers should be used on even high-purity gases. The presence of contaminants in the carrier gas reaching the column testifies to either exhausted (or improperly installed) gas scrubbers and filters or subsequent contamination of the clean gas by postfilter passage through dirty lines or soiled inlets. Gas scrubbers should be positioned *vertically, and never horizontally*. Packings inevitably settle, and in the horizontal position, settling opens a path of least resistance through the open space over the packing; the scrubber is totally ineffective.

Prior to the installation of a new column, it is good general practice to recheck in-line gas scrubbers and filters, to clean the inlet liner and the injection block cavity in which it is housed, and to replace the septum. The column should require only brief conditioning (see Chapter 6), and the manufacturer's test chromatogram should be duplicated under the conditions originally used to generate that test chromatogram. If the regenerated test chromatogram is inferior to that enclosed with the column (in terms of component separation, peak symmetry and conformation, and/or bleed), it indicates inadequacies in the equipment, in column installation, or in operational techniques. One or more of these factors is degrading chromatographic performance and may jeopardize column lifetime. Particularly in the case of a high bleed, the cause should be ascertained and corrected before the column is subjected to routine use.

10.2   Use of Test Mixtures

Polarity test mixtures include solutes that embrace a range of functional groups. Hydrocarbon solutes are almost always included. By evaluating the numbers of theoretical plates developed on hydrocarbon solutes and the shapes of those peaks, the analyst can establish the behavior of a nonactive solute under a particular set of conditions on that apparatus. Malformed hydrocarbon peaks testify to analytical inadequacies that may include poor injection technique and gas flow problems. The latter may be related to gross departures from the optimum carrier gas velocities or to poorly swept volumes in the gas flow path, such as excessive or dead volumes in the inlet or detector, an inadequate split ratio (split injection), a cracked injector liner, flashback, insufficient makeup gas, or improper positioning of the column ends in the inlet and detector.

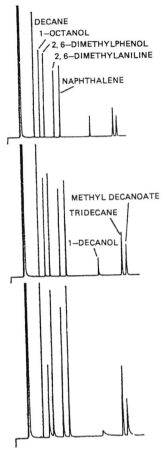

Fig. 10.1. Test chromatograms generated on three different columns in the same instrument and under the same test conditions. All columns 30 m × 0.32 mm, coated with dimethylpolysiloxane. All runs at 105°C isothermal. Top, a well deactivated column. Center, the non-hydrocarbon solutes all exhibit some tailing, and while the octanol peak seems normal, the decrease in the decanol peak indicates irreversible adsorption. *Note: sorption is strongly influenced by test temperature conditions. It is probable that if the test temperature were lowered 10 to 15°, both alcohols would fail; if it were raised 10 to 15°, both alcohols would pass.* Bottom, excessive activity; note that the hydrocarbon peaks are still acceptable.

Well-formed hydrocarbon peaks serve as "standards" to which the peak shapes of active solutes can be compared. Figure 10.1 shows a typical column test mixture on three different columns. The top chromatogram, in which the peak shapes for hydrocarbon solutes are essentially the same as those for active solutes, is indicative of a well-deactivated column properly operated in good equipment. In the center chromatogram, both reversible and irreversible adsorption are evident

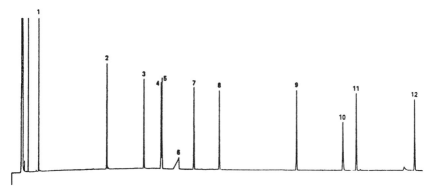

Fig. 10.2. Chromatogram of the Grob test mixture [2] on a 30 m × 0.25 mm 5%-diphenyl 95%-dimethyl polysiloxane column. Hydrogen carrier, t_M 60 sec @ 40°C. 40 to 140°C @ 1.67°C/min. Solutes: (1) 2,3-butanediol; (2) n-decane; (3) 1-octanol; (4) nonanal; (5) 2,6-dimethylphenol; (6) 2-ethylhexanoic acid; (7) 2,6-dimethylaniline; (8) n-dodecane; (9) methyl decanoate; (10) dicyclohexylamine; (11) methyl undecanoate; (12) methyl dodecanoate. See text for discussion.

on the two alcohol peaks. Reversible adsorption is evidenced as tailing, and the decrease in the areas of those peaks, relative to those of their hydrocarbon neighbors (compare to top chromatogram), testifies to irreversible adsorption and signals an alert against using this system for quantitation of these active solutes. In addition, peaks generated by the phenol, the aniline, and the ester are unsatisfactory. Because the same unit was used to generate the test chromatograms on all three columns, the activity can probably be blamed on the column rather than on the inlet or other parts of the chromatographic system.[1] The bottom chromatogram shows a column with extreme activity where the peak shapes of active solutes, relative to those generated by hydrocarbons, are extremely poor. If this is a new column containing methyl siloxane or methylphenyl siloxane, it is possible that its performance may improve with resilylation (see below). Chromatograms of this type can also be generated by injection residues in the inlet or in the column. Careful cleaning or removal of a few meters of column at the inlet end may resolve the problem (see below).

A chromatogram of a test mixture proposed by Grob *et al.* [2] is shown in Fig. 10.2. Under their standardized conditions (discussed in Ref. 3), each peak should extend to reach a curved line drawn through the maxima of the non-sorbed solutes. In the example shown, abstraction of a portion of each of the solutes represented by peaks 3 (octanol), 4 (nonanal), 7 (2,6-dimethylaniline), and 10 (dicyclohexylamine) is obvious. The fronting evidenced by peak 6 is due not to

---

[1]This assumption must be viewed with some caution. In the case of a wandering baseline, ghost peaks and similar anomalies that may originate in the inlet and eventually percolate through the column, installation of a new column will seemingly solve the problem, and the analyst feels justified in blaming the problem on the old column. However, if the real source of the problem was a dirty injector, then the installation of a new column is only a short-term solution. In time, the new column will become similarly contaminated, and the problem will resurface.

activity but to overload (see below) and testifies to the unsuitability of the methyl siloxane phase for carboxylic acid analysis (see also Section 6.2). The diminution of peaks 4 and 7 is probably due to aldehyde–aniline interaction producing the Schiff base product in the test solution, a phenomenon that shows the danger of storing premixed test solutions, and that led to the recommendation that, where both solutes are required, they should be presented in separate test mixtures [4].

10.3   Column Bleed

It is important to distinguish between "perceived bleed" and "bleed." True column bleed is evidenced as a temperature-dependent rise in baseline. Because the signal-eliciting materials are being generated throughout the entire column simultaneously and there is no "point source," the baseline signal rises to a level that remains constant (steady state signal) until the column temperature is changed. *Peaks must originate with a point source, and column bleed cannot produce peaks.* What is perceived as bleed may in some cases be detector-specific. Aspiration of a small amount of phosphate-containing leak detection fluid into a column can cause that column to produce baselines that are acceptable by flame ionization detection (FID) but wildly erratic with nitrogen/phosphorus detection (NPD). Contamination of the detector itself at the column attachment fitting can also lead to high noise levels and/or temperature-related signal that can be confused with column bleed.

In cases where the oven is turned off and carrier flow is maintained over night, analysts sometimes observe a dramatic baseline rise on the first run of the next day. This is often attributable to contaminants in the carrier gas or contained in the heated injector that become concentrated at the head of the cooler column [5]. As the column is heated on the first run of the day, these contaminants chromatograph through the column and can appear as discrete peaks and/or a rising baseline that slowly returns to normal levels (Fig. 10.3). Typically, the phenomenon becomes less acute with each successive injection, and eventually disappears—until the column is again allowed to stand at ambient temperature for some period of time [6]. In one case involving GC/MS, an analyst reported a high-level "silicone signal" (m/z 73, 207, 281, 355), typical of column bleed. While these fragments are typical of siloxane degradation products, *they do not necessarily result from column bleed.*

Whereas true column bleed does not produce peaks, bleed from the septum can produce peaks. Typically, septum bleed is characterized by the elution of one or more peaks, sometimes accompanied by a broad hump, during the GC run. The predominant ions associated both with column bleed and with the degradation of polysiloxane stationary phases include m/z 73, 207, 281, and 355, *but these same ions are produced by GC septa.* Both septa and siloxane-based stationary phases are made from polysiloxane polymers. Septa also contain plasticizers that impart a degree of elasticity and the ability to reseal after needle penetration, and these substances are slowly released at higher temperatures. These contaminants migrate through the heated injection port, and are prone to recondense on the

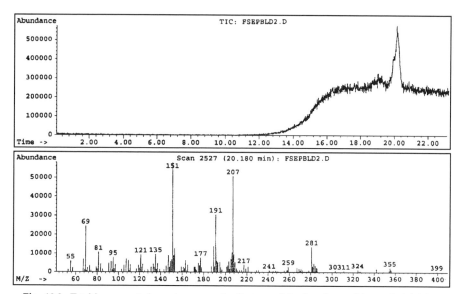

Fig. 10.3. Total ion current chromatogram on the first run of the day. In stand by mode (night), the injector and detector are left on, the oven is turned off, and a slight flow of carrier is maintained. Both the abnormal baseline rise and the discrete peaks result from the elution of materials that have condensed on the unheated column. See text for discussion. *From [6]; reprinted with permission of Huethig Publishing.*

cooler column. The longer the column is maintained at a lower temperature, the greater the accumulation of such contaminants.

Low-bleed septa are usually produced by a high-temperature treatment or solvent extraction to remove some of the plasticizing materials. However, these treatments can imperil the ability of the septum to reseal, and septum leakage (which, because of backdiffusion, also allows oxygen contamination of the carrier gas) may become more of a problem.

To establish what portion of the signal (if any) is attributable to the septum, first complete a high-temperature run to leave the column clean. Remove the septum, wrap it in aluminum foil prior to reinstallation, and then generate a chromatogram to determine what peaks are now absent. As a partial solution to the problem, septa can be baked at an oven temperature 25°C higher than the injector temperature for 4–8 hr; vacuum oven treatments can also be useful. This will help correct problems of septum bleed, but septum lifetimes will be shorter.

Figure 10.4 shows a chromatogram with a serious bleed problem. This relatively common phenomenon was formerly attributed to the distillation of stationary phase from the column. Today we recognize that although bleed attests to material entering the detector, there are several possible origins of that material and several plausible explanations for its presence in or generation from some of those sources.

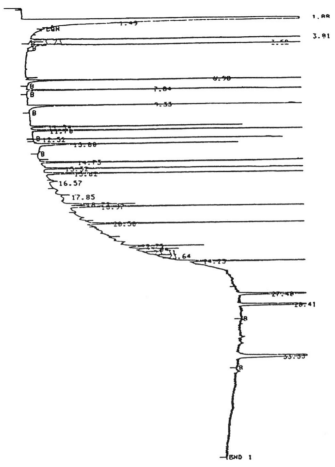

Fig. 10.4. Excessive bleed, due to stationary phase degradation by silanol groups, generated by oxygen exposure. While this type of behavior can also be generated by residues in the injector, deposits in the detector, and the associated gas lines and filters, it is rarely this extreme.

Isolating Bleed Problems

When a bleed problem becomes evident, the column should be removed from the oven and the column attachment fittings at both the inlet and the detector should be closed with approved solid plug fittings. The detector should then be activated, the detector signal trace energized, and the oven temperature increased to the level where problems were previously experienced. An increase in the baseline level indicates that the signal is extracolumn. In this case, the column is not responsible for the "bleed." What the user perceives as "bleed" may actually result from a dirty detector, or from residues that have accumulated, sometimes over extended periods of time, in the lines supplying makeup gas and/or combustion hydrogen to the detector. The detector should be disassembled and cleaned according to the manufacturer's instructions, and all gas lines, in-line valves, and flow controllers supplying the detector should be carefully cleaned, and in-line gas scrubbers should be replaced. Finally, the efficacy of the cleaning procedures should be evaluated by repeating the trace generated only by the detector, that is, with the column removed and the column fitting at the detector capped off.

Even if the signal has now been reduced to a satisfactory level, it is important to realize that the original problem may have been a dirty inlet, and if so, it discharged those contaminants to the detector via the column. If that was the case and the column is simply reinstalled after cleaning the detector, the detector will be recontaminated. Contaminated columns should be cleaned as described below before they are reinstalled.

The condition of the inlet can be checked by installation of a dummy column. (A few meters of clean stainless steel or fused silica tubing will usually suffice. *Metal tubing should first be rinsed with a succession of solvents of varying polarity, and dried with a flow of clean gas.*) With the detector energized, the high-temperature signal trace should be reevaluated; excessive baseline rise indicates contamination originating from the inlet side. The source may be a dirty injector or deposits in the carrier gas line, flow controllers, regulators, or expended gas scrubbers.

Sometimes a heat gun, judiciously used, can identify the culprit with some precision. Shielding other portions of the instrument with small pieces of heavy cardboard or the like, direct the heat at, such as an inline flow controller, a valve, or a length of gas line, while the signal trace is running. Warming a contaminated device or area *just a few degrees* is usually sufficient to cause a significant increase in the signal level; *too vigorous an application of the heat gun can damage some in-line devices.* Once the cause has been identified, that device should be replaced, or removed and thoroughly cleaned, reinstalled, and the system retested.

Flashback, which occurs when the abrupt increase in sample volume cannot be accommodated by the volume of the vaporization chamber (and the rate at which the carrier gas empties the vaporization chamber by sweeping the vaporized sample to the column), is usually evidenced by a massive tail on the solvent peak. Quantitation, especially of higher-boiling solutes, is also affected adversely. Flashback can also contaminate with residues of everything in that injection any area into which the rapid expansion forces the solvent–solute mixture. These include the inner face of the septum, the septum seat, the liner housing, and even the carrier gas line and filters therein. These deposits can cause malformed and tailing peaks, ghost peaks, "perceived bleed," and cause qualitative and quantitative errors in analyses that are performed weeks later. "Memory effects" (discussed below) are not uncommon.

Flashback is aggravated by (1) the injection of large sample volumes, (2) lower carrier gas flows (higher flows more rapidly carry the volatilizing sample to the column), and (3) injector temperature that are much higher than the boiling point of the solvent. Again, thorough cleaning is required. Inlet liners should be removed and all surfaces and cavities within the injector block cleaned, including those that would appear to be out of the main flow stream.

It can be informative to check a given chromatograph for "memory effects," Using a vaporizing mode of injection (split, splitless, PTV or hot on-column) and an injection temperature of 250–300°C, inject 6–8 μL of a high-purity low-boiling solvent such as methylene chloride. This will normally generate a massive solvent peak that exhibits severe tailing. Extraneous peaks on that tail indicate that the solvent vapors, expanding out of the vaporization area, have picked up residues from previous injections, which are now introduced to the column. Obviously, this can have adverse effects on both qualitative and quantitative analyses. Again, the solution is cleaning. The injector, injector gas supply lines, and anything that might come in contact with the flashback vapors, should be dissembled, cleaned, and retested as described above until the extraneous peaks no longer appear on the solvent tail.

Once a satisfactory trace is generated on the dummy column, attention can be directed to the original column. Some columns with crosslinked surface-bonded stationary phases can be cleaned by rinsing with several column volumes of solvent (see below). In other cases, the column can be connected to the inlet with the detector end disconnected. Detector gases should be turned off and the detector connection plugged. The column should be reconditioned at a relatively high gas flow (unless provisions are made to discharge the carrier outside the oven, hydrogen carrier is not recommended), at a temperature at or approaching that where bleed was previously experienced. Less column damage will usually occur if conditioning is done for longer times at lower temperatures rather than vice versa. Following conditioning, the column should be reconnected to the detector, the detector energized, and the entire system (inlet–column–detector) evaluated by signal trace.

If neither the trace from the detector nor that from the inlet–dummy column–detector exhibited a high-level signal, then the observed bleed must have originated with the column and could have come from several causes. Sources of column bleed include (1) materials dispersed through the stationary phase polymer that become volatile at higher temperatures (usually oligimers that failed to polymerize) and degradation products of the polymer, produced by; (2) depolymerization triggered by residual catalyst; (3) reaction with active sites (e.g., —OH) on the solid support; (4) reaction with —OH groups within the stationary phase; (5) reaction of the stationary phase with injected materials or injection residues; (6) reaction of the stationary phase with impurities and contaminants (e.g., residual catalyst, as in (2), above) in the carrier gas; and (7) the eventual emergence of low-volatility materials deposited on the column from other sources, including higher-boiling solutes that were present in previously injected samples or inadequately cleaned carrier gas. These often produce a wavy baseline.

Any oligimeric materials that are present, either as residues from incomplete polymerization or as products of a depolymerization triggered by residual catalyst, will, of course, contribute to column bleed, but bleed contributed by factors 1 and 2 is seldom a problem with the higher-quality stationary phases, and the contribution of factor 3 is minimized by careful deactivation procedures. Control of these three factors is a responsibility of the column manufacturer. The manufacturer should also have taken steps to minimize bleed due to factor 4 by end-capping the terminal —OH groups of the polysiloxane (Chapter 4). Factor 6 can lead to reappearance of this problem. Exposing the column to even traces of oxygen at temperatures above 250°C cleaves Si—CH_3 bonds and regenerates Si—OH entities [7]. These engender degradative reactions that are discussed in greater detail below [8–11]. An example of factor 5 occurs with the injection of ethereal extracts of alkaline solutions, which can lead to degradation of phase in the front portion of siloxane-coated columns and to severe bleed problems [12].

Residues from the injection of "dirty" or "real world" samples accumulate at points of sample vaporization. With properly executed split and splitless injection, these residues accumulate in the inlet liner, which should periodically be inspected, cleaned, and replaced. With on-column injections, injection residues accumulate on the column per se and are commonly evidenced as peak broadening (loss of separation efficiency). Other peak-related anomalies such as tailing or splitting have also been attributed to injection residues [13]. Where the residues are of a carbohydrate or proteinaceous character, polar solutes are usually more affected. Residues from waxes, plastics, and lubricating oils tend to exercise their greatest effect on apolar solutes [13]. Residues from previous injections can also interact with solutes in subsequent injections vaporized from that soiled surface. Degradation can lead to fragmentation, and silyl derivatives can decompose and contribute additional nonvolatile residues [14].

10.4 Temperature and Oxygen Effects

Polysiloxane Stationary Phases

Bulk polysiloxanes exposed to high temperatures and/or oxygen undergo degradation. Dimethyl polysiloxane subjected to programmed heating in vacuum produced volatile degradation products, primarily the cyclic trimeric oligimer, which was detectable at 343°C and reached a maximum at 443°C [15]. The reaction, which was accompanied by crosslinking, reportedly occurred in a stepwise manner from the hydroxyl-terminated ends of the polymer chain, was accelerated by oxygen, and was strongly accelerated by KOH. The polymer was much more stable when the terminal hydroxyls were replaced by trimethylsilyl groups (end-blocking). Later work demonstrated that the substitution of phenyl for some of the methyl resulted in a more thermally stable polysiloxane [16]. Larson *et al.* [17] used columns coated with OV-1 (polydimethylsiloxane) and SE-54 (5% phenyl, 1% vinyl, 94% dimethyl polysiloxane) with hydrogen, helium plus 110 ppm O_2 and helium plus 525 ppm O_2 as carrier gases at 325°C and 400°C in their studies of column stress parameters. Their results also indicated that phenyl substitution lent both thermal and oxidative stability to the polysiloxane. Evans [18] measured the retention indices of solutes on packed columns containing several different stationary phases that had been subjected to air flow at 225°C for up to 60 hr. Among the polysiloxanes, the largest changes occurred with the dimethyl siloxane OV-101 and the trifluoropropylmethyl siloxane OV-210. The most resistant to change was the 50% phenyl OV-17, while the cyanopropylphenylmethyl siloxane OV-225 was intermediate.

Polyethylene Glycol Stationary Phases

The PEG-based phases are in general less stable than the polysiloxanes. It has been suggested that three factors are involved in their decomposition: oxygen in the carrier gas, the catalytic action of the solid support, and catalysis by acidic products of the degradation [19]. Among the decomposition products of Carbowax 20M are acetaldehyde and acetic acid [20]. The thermal decomposition of the Carbowax stationary phases is encouraged by even trace levels of oxygen. Conder *et al.* [19] reported that inclusion of an oxygen scrubber that reduced the O_2 content of their commercial carrier gas by three orders of magnitude decreased the rate of Carbowax 20M decomposition by a factor of 5 and raised the temperature at which decomposition was first evidenced from 160 to 200°C.

10.5   Column Rejuvenation

Sources of oxygen contamination were considered earlier, and the material explored above indicates that polysiloxane-coated columns can be severely damaged by exposure to oxygen at high temperatures (Fig. 10.4). The polysilarylenesiloxane stationary phases (Chapter 4) are more resistant to heat and oxidative degradation. Oxidative cleavage of Si—C bonds regenerates hydroxyl groups on the polysiloxane chain, which then engage in "backbiting" reactions to produce cyclic trimers (primarily) of Si—O from the end of the chain. It is this material that generates the "bleed" signal. With each such reaction, a new —OH is left on the chain to perpetuate the reactions. Capping of the exposed hydroxyl groups, which would block those degradative reactions, has been the subject of several studies (e.g., [8–11]). More recently, "chain stiffening groups" have been inserted into stationary phase polymers to inhibit the chain flexibility that can encourage the back-biting reaction.

Bleed levels that are quite acceptable with FID or even MSD become intolerable with the more bleedsensitive ion trap detector (ITD). It is important to recognize that for a given quality of stationary phase under a given set of conditions (gas quality, gas flow rate, temperature), bleed is always a function of the mass of stationary phase that lies in the flow path [12]. Any change the analyst can make

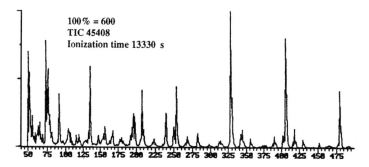

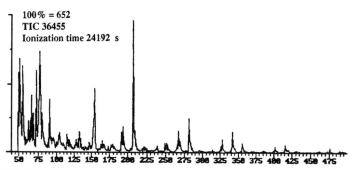

Fig. 10.5. Electron impact spectra generated on an ITD. Top, the end of a 5%-diphenyl-95%-dimethylpolysiloxane was threaded through the heated transfer line into the ion source. Bottom, a length of uncoated deactivated fused silica tubing was attached to the column and threaded through the transfer line. In the ITD, ionization times vary inversely with the molecular population in the analyzing section. Note the increased ionization time (and the improved spectral quality) in the bottom figure.

to reduce the amount of stationary phase—shorter column, smaller-diameter column, thinner film of stationary phase—will reduce the bleed level. The newer polyarylenesiloxane phases have been very beneficial to the ITD (Fig. 10.5).

Some columns can be successfully rejuvenated by cleaning, followed by resilylation. The procedures give better results with crosslinked, surface-bonded, high-methyl polysiloxane columns (e.g., DB-1, DB-5). Grob and Grob [8] specify that the column must first be thoroughly cleaned. Otherwise the silylation reactions will confer volatility on column contaminants that originated from injection residues. This can create an entirely new bleed problem that will be difficult to correct. The contaminated column should be rinsed with several column volumes of an appropriate solvent, and it may be desirable to use both polar and nonpolar solvents. Depending on the nature of the sample residues expected, the first rinse might be water, followed by methanol and acetone. Methylene chloride is a good final rinse and in some cases may be the only solvent required. *It is important to consider the intended use of that column. If it is destined for use with an ECD, it would be wise to avoid the use of chlorinated solvents.* The column should then be filled with methylene chloride (or hexane) and allowed to stand flooded overnight to allow materials within the stationary phase time to migrate into the solvent. The column is flushed with fresh solvent, drained, and dried at room temperature with a stream of clean, dry, nitrogen. The heavily solvated stationary phase may now appear as a thick puffy film. At temperatures above the boiling point of the solvent, violent solvent evaporation can tear the film to pieces. Instead, the column should be subjected to a nitrogen flow stream and held at a temperature just below the boiling point of the solvent for several hours. The temperature should then be increased at a rate of 1–2°/min to a point ~20°C above the boiling point of the solvents, and held for 2 or 3 hr. After allowing the column to cool, vacuum is used to draw a 10% pentane solution of diphenyltetramethyldisilazane (DPTMDS) into the column until about one-tenth of the column is filled. A very slow flow of clean dry nitrogen is passed through the column to encourage evaporation of the pentane. The flow should not be so fast that plugs of liquid are blown from the end of the column. When the pentane appears to have evaporated, the nitrogen flow is increased slightly (2–3 mL/min for 0.25–0.32-mm columns; 10–20 mL/min for 0.53-mm columns) and the column is held at about 75°C to ensure total evaporation of the solvent (approximately 1 hr), leaving the DPTMDS in the column. Both ends of the column are then flame-sealed in such a manner as to ensure that it is nitrogen-filled, and the column is heated to 300°C for 30–60 min, cooled, flushed with solvent, and dried.

The 100% polydimethylsiloxane and 5% phenyl–95% dimethylpolysiloxane phases give the most satisfactory results on rejuvenation. With most other stationary phases, the results are spotty and rarely justify the effort required. Rejuvenated columns sometimes demonstrate aggressiveness toward active solutes [21]. Berezkin and Korolev [22] reported that the contribution of adsorption to the retention behavior of polar solutes increases as the degree of crosslinking increases, and resilylation could also encourage more extensive crosslinking of the stationary phase.

10.6   Peak Distortion

Injection technique can affect peak conformation and in extreme cases can result in misshaped, malformed, or split peaks (Chapter 3). The relative polarities of solvent, solutes, and stationary phase have also been shown to play roles [23, 24]. Other general causes of peak distortion and asymmetry have been addressed by Conder [25], who described the two extremes in peak skewing as (1) a simple skewing of the whole peak without a long tail and (2) tailing (or "fronting") where the (very) upper part of the peak is nearly symmetrical, but is followed (or preceded) by a distinct, drawn-out tail of exponential form.

Fronting, where the leading edge exhibits a slow rise and the drop from the peak maximum is much more abrupt, can sometimes be attributed to injector or detector problems, but is more generally due to column "overload." In the case of solutes that are compatible with the stationary phase, the phenomenon has been attributed to localized decrease in the column phase ratio [26] (Fig. 10.6). A chromatographing solute engages in a highly dynamic equilibrium partitioning

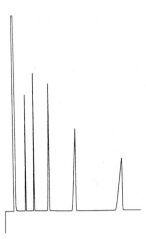

Fig. 10.6. An isothermal chromatogram. Note the "fronting" on the last two peaks, typical of "overload" [26]. See text for discussion.

between the two phases, in accordance with its distribution constant K_c, where K_c is the ratio of the solute concentration in stationary phase relative to the concentration in mobile phase. The solute retention factor k is governed by the relationship $K_c = \beta k$, where β is the ratio of the column volume occupied by stationary phase (V_S) relative to the column volume occupied by mobile phase (V_M). When the amount of solute in (or on) the stationary phase is comparatively large (e.g., a large injection, or an incompatible or low volatility solute), that portion of the solute that is condensed also acts like stationary phase. In this localized section of the column, the "apparent" V_S is now higher, the "apparent" V_M is now lower, and the "apparent" (and localized) column phase ratio is lower. Solute molecules in the mobile phase are now affected differently, depending on their position in that solute band. Those at the front of the chromatographing band are retained in a normal fashion and exhibit a "normal" k, as the carrier sweeps them forward over virgin stationary phase where the column phase ratio has not yet been affected. Solute molecules in the mobile phase that are not at the advancing front of the band continue to chromatograph in localized regions where the "apparent" phase ratio is lower, their k becomes larger, and they progress more slowly. The detector sees a progressively increasing concentration of the less retained molecules, gradually tapering into a higher concentration of more retained molecules, followed by an abrupt return to baseline as the last of the latter elute from the column. The defect can be corrected by decreasing the sample size (which would decrease the amount of that solute in both phases), increasing the column temperature (which would increase the solute vapor pressure, and so decreasing the amount of solute in stationary phase), or increasing the column phase ratio [$K_c = \beta k = (r/2d_f)k$].

Skewing that takes the form of tailing may result from flashback (see above), from gas flow problems, or from active sites that may be present in the inlet, the column, or the detector itself. Tailing of hydrocarbon peaks generally testifies to flashback or to inadequacies in the gas flow stream. The split ratio may be too low, or the flow of makeup gas may be low or misdirected. Activity in the flow stream is indicated if hydrocarbons yield well-formed peaks while active solutes exhibit tailing. Column substitution can help determine whether the problem is extracolumn. The inlet should be inspected for residues and cleaned if necessary, and positioning of the outlet end of the column should be verified. Deactivation of the inlet liner may be necessary. If the problem seems to be associated with the column, it may help to backflush the column with solvent to remove residues. The removal of one to several meters of column from the inlet end may also correct the problem. A different troubleshooting case that involved poor pentachlorophenol quantitation was solved this way (Fig. 10.7) [27]. Because resolution varies with

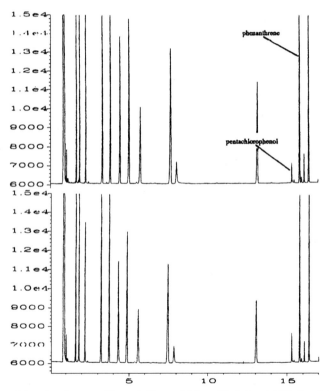

Fig. 10.7. Selected analytes on a 30 m × 0.32 mm, d_f0.5μm 5%-diphenyl-95%-dimethyl polysiloxane column. 140°C (12 min) to 280°C @ 20°C/min. Helium carrier @ 35 cm/sec. Top, pentachlorophenol peak (active solute) is 6.8 of the phenanthrene peak (comparatively inactive solute). Bottom, one meter of column removed from the inlet end. Note that the response of the pentachlorophenol peak has more than doubled, to 15% of the phenanthrene peak. *From [27], and reproduced with permission of Huethig Publishing.*

Fig. 10.8. The "Christmas Tree" effect. Peak splitting artificially generated in pentadecane by short term fluctuations in the limb of a low thermal mass column [30].

the square root of column length, shortening the column rarely has a pronounced effect on the separation efficiency, but it will affect retention times. Column resilylation may correct activity problems in some cases, but in other cases it may accentuate tailing (see above).

Peak splitting and "Christmas tree" effects are discussed in Chapter 7. These interrelated problems are usually attributed to exposure of a portion of the low thermal mass columns to nonconstant thermal gradients or radiant heat sources [28–31] and are related to acceleration of restricted portions of the chromatographing band. Reed and Hunt [32] studied the phenomena by subjecting the detector limb of the column to a pulsating heat source (Fig. 10.8).

10.7   Other Sorptive Residues

Any of the phenomena discussed above (split and malformed peaks and memory effects that result in ghost peaks) can also be caused by chromatographic conditions that permit the sample to come in contact with materials that are capable of interacting with sample components. These interactions can range from adsorption to solution. Chapter 7 discussed risks associated with graphite particles that are scraped from the ferrule by the column end to lodge in the flow stream. The series of ECD chromatograms shown in Fig. 10.9 illustrates a related problem [33]. Analysis I shows the programmed temperature separation of a series of pesticides. Analysis II was run 1 hr later and exhibits peak distortions. Analysis III is the same mixture, injected immediately after analysis II. In IV, V, and VI, injections of the solvent only (methanol) establish that these anomalies are attributable to memory effects (ghosting). In analysis VII the column was programmed without injection, and VIII was run 15 min after VII. Both indicate a source of sample in

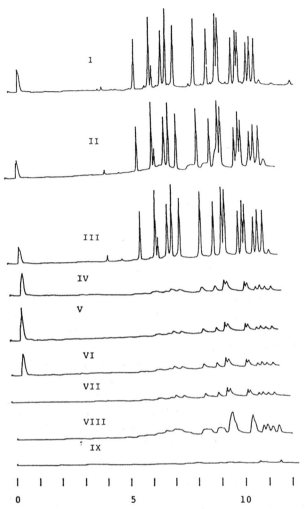

Fig. 10.9. Effects of sorptive material in the inlet, programmed runs with ECD. (I) normal separation. (II) second analysis, immediately after completion of I. (III) analysis after a one hour pause. (IV–VI), traces engendered by methanol injections. (VII, VIII) column programmed without injection. (IX) programmed baseline after removing a septum particle from the inlet.

the flow stream. A postcolumn source would yield a high background signal, but the components would not be separated. Because the "peaks" exhibit retentions comparable to those exhibited by normal injections, the source must precede the column or be situated in the very first section of the column. Careful examination of the injector block with the liner removed revealed that a minute crumb of silicone rubber, presumably chewed from the septum by the syringe needle, had lodged within the carrier gas line at the point of entry to the injector. The bottom trace shows the baseline generated under programmed conditions after removal of that small particle. Other sorptive materials in the flow stream can give rise to entirely different problems (see below).

10.8   Column Coupling and Junction Problems

Because of the advantages they offer (delivering focused bands to the column, serving as a guard column), many laboratories routinely use retention gaps. Usually a meter of uncoated but deactivated fused silica tubing is sufficient, but those required to analyze particularly dirty samples sometimes use retention gaps of 5 m and longer. Residues from these dirty samples usually accumulate in the first few centimeters of the tubing, and as they accumulate, the front end of the retention gap can be removed and the system reactivated with minimum down time.

Junctions are required to attach retention gaps to columns, to couple columns, or in the use of "Y"s and "T"s, where (e.g.), the injection is split to two columns, or the column effluent is split to two detectors. Especially with higher program rates, the thermal mass of the fitting may result in its temperature lagging behind that of the column and the oven, and higher-boiling solutes can be cold-trapped in the fitting. An homologous series of hydrocarbons can exhibit good conformation of earlier peaks, but a sudden break in their elution sequence signals trouble. As the temperature of a high-thermal-mass fitting slowly increases, trapped solutes are eluted as split and malformed peaks. "Christmas tree" effects may also be apparent. One solution is to wire the metal fitting tightly against the base of the heated detector. Another is to accelerate transfer of heat into the high thermal mass fitting by enclosure in a separately heated "oven."

"Press-fit" connectors, usually of borosilicate glass, sometimes of fused silica, offer convenient low thermal mass fittings that, properly installed, can be highly dependable. These are available as "sized," or as "universal" fittings. The latter—capable of accepting 0.25-, 0.32-, and 0.53-mm-OD tubing, are usually less satisfactory than those designed for one specific size. The universal fitting presents a steeper sealing surface, and seals over a shorter length than the more gradually tapered single-size fitting. Tubing ends should be freshly cut square, wiped with an acetone or methanol wetted tissue to remove any fragments, and examined under suitable magnification. If the tubing is clean and square, immerse the end in acetone or methanol for 5–10 min, and quickly force the wet tube into a connector that has been wetted with the same solvent. For more critical connections, scribe the tubing with a suitable cutter, and cover the scribe mark with a thick coat of polyimide resin (available from several supply houses). Allow the resin to tack up for 10–15 min., and snap the tubing. This leaves a thick area of polyimide at the tubing end, while the flow path remains uncontaminated. Place another droplet of

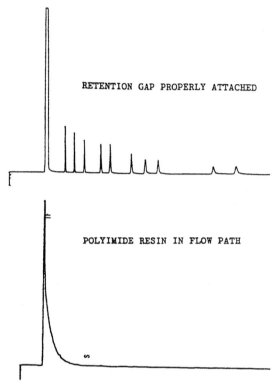

Fig. 10.10. On-column injections of a test mixture through a 1-meter retention gap attached to the column with a press-fit connector. Top, normal results with proper press-fit installation. Bottom, polyimide in the flow-stream.

polyimide resin on the tubing about 5–8 mm from the end, and carefully insert into a dry press-fit connector, ensuring that resin does not contaminate the flow path. Install the assembled column, and maintain a gentle gas flow (0.2–1 mL/min) for an hour or longer. Too high a gas flow can cause "blow out" of the still uncured polyimide seal. Too short a time, and the column will still contain air, which will cause problems during the thermal cure. The column is then heated to 150°C for 30 min to set the resin [34]. Figure 10.10 shows the results of a poor seal, where the absorptive polyimide has entered the flow stream.

10.9   Flame Jet Problems

The small diameter and flexibility of the fused silica column can be used to good advantage at the detector end, because the column end can be sited so that solutes are eluted within 2–4 mm of the point of detection, eliminating the possibilities

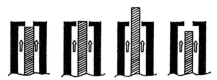

Fig. 10.11. Positioning the outlet end of the column inside the flame jet of an FID. (I) and (II) can restrict the gas flow and make difficult ignition and flame maintenance. (III) generates an FID signal shown in Fig. 10.12. (IV), proper positioning.

of adsorptive sites in transfer tubing, flame jets, and so on. Some of the problems that can occur are illustrated in Fig. 10.11, which shows a fused silica column terminating within a flame jet. If the diameter of the column exceeds (I) or is similar to (II) that of the jet orifice, the flow of combustion hydrogen (and makeup gas) may be restricted to such a degree that it becomes difficult to ignite or maintain the flame. It is also possible to thrust a smaller column completely through the jet (III) and into the flame. This condition is usually characterized by wild, fluctuating signals (Fig. 10.12). Ideally, the column should terminate 2–4 mm below the jet orifice (IV), in such a position that solute contact with the flame jet is precluded by the flow path of the surrounding gases and the end of the column does not restrict free passage of gas from the jet orifice.

10.10   Miscellaneous Chromatographic Problems

Separation of a solute occurring at trace levels is usually a problem if the trace component is immediately followed by a major component.

The trailing edge of the minor component behaves as if the major component were stationary phase. The phase ratio into which the trace component is moving is large, and the phase ratio in the area it is leaving is small ($\beta = V_M/V_S$). Since $K_c = \beta k$, the front of this band has a lower k than the rear of the band, the band lengthens and the peak is destroyed (Fig. 10.13). It is also evident from Fig. 10.13 that this same phenomenon benefits a minor component following a major component by sharpening the peak [34, 35].

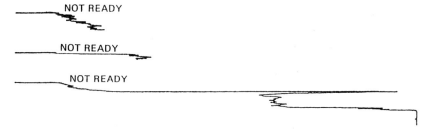

Fig. 10.12. Signal generated by an FID with the column positioned as in III, Fig. 10.11.

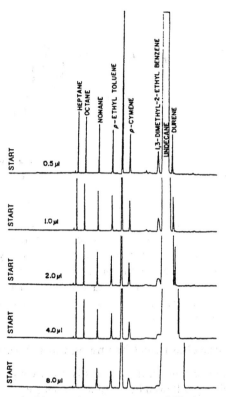

Fig. 10.13. Effects exercised by a major component on neighboring trace solutes. Note that as the injection size increases, peak shapes of preceding solutes are affected adversely, and the envelope of influence is influenced by the amount injected. Trace components following the major solute are sharpened [34, 35].

A more detailed approach to troubleshooting GC systems appears in Ref. 36.

References

1. J. Hinshaw, *LC/GC* **11**:274 (1993).
2. K. Grob, G. Grob, and K. Grob, Jr., *J. Chromatogr.* **156**:1 (1978).
3. W. Jennings, *Comparisons of Fused Silica and Other Glass Columns in Gas Chromatography.* Huethig, Heidelberg, 1981.
4. I. Temmerman and P. Sandra, *J. High Res. Chromatogr.* **7**:332 (1984).
5. W. Jennings, *J. High Res. Chromatogr.* **17**:682 (1994).
6. W. Jennings, *J. High Res. Chromatogr.* **19**:127 (1996).
7. A. E. Coleman, *J. Chromatogr. Sci.* **13**:198 (1973).
8. K. Grob and G. Grob, *J. High Res. Chromatogr.* **5**:349 (1982).
9. G. Schomburg, R. Dielmann, H. Borwitzky, and H. Husmann, *J. Chromatogr.* **167**:337 (1978).

10. V. Paramasigamani and W. A. Aue, *J. Chromatogr.* **168**:202 (1979).

11. S. Hoffmann, L. G. Blomberg, J. Buijten, K. Markides, and T. Wannman, *J. Chromatogr.* **302**:95 (1984).

12. W. Jennings, *J. High Res. Chromatogr.* **19**:176 (1996).

13. K. Grob, Jr., *J. Chromatogr.* **287**:1 (1984).

14. M. Donike, *Chromatographia* **6**:190 (1973).

15. N. Grassie and 1. G. MacFarlane, *Eur. Polym. J.* **14**:875 (1978).

16. N. Grassie, 1. G. MacFarlane, and K. F. Francie, *Eur. Polym. J.* **15**:415 (1979).

17. P. Larson, T. Stark, and R. Dandeneau, *Proc. Int. Symp. Capillary Chromatogr., 4th, 1981*, p. 727 (1981).

18. M. B. Evans, *Chromatographia* **15**:355 (1982).

19. J. R. Conder, N. A. Fruitwala, and M. K. Shingari, *J. Chromatogr.* **269**:171 (1983).

20. J. Debraurwere and M. Verzele, *J. Chromatogr. Sci.* **14**:296 (1976).

21. W. Jennings, *J. Chromatogr. Sci.* **21**:337 (1983).

22. V. G. Berezkin and A. A. Korolev, *Chromatographia* **20**:482 (1985).

23. R. G. Jenkins, *Proc. 4th Int. Symp. Capillary Chromatography*, p. 803 (1981).

24. L. Ghaoui, F. S. Wang, H. Shanfield, and A. Zlatkis, *J. High Res. Chromatogr.* **6**:497 (1983).

25. J. R. Conder, *J. High Res. Chromatogr.* **51**:341, 397 (1982).

26. M. Hastings, *J. High Res. Chromatogr.* **17**:53 (1994).

27. W. Jennings, *J. High Res. Chromatogr.* **19**:243 (1996).

28. S. A. Mooney, *J. High Res. Chromatogr.* **5**:507 (1982).

29. G. Schomburg, *J. Chromatogr. Sci.* **21**:97 (1983).

30. F. Munari and S. Trestianu, *J. Chromatogr.* **279**:457 (1983).

31. F. J. Schwende and J. D. Gleason, *J. High Res. Chromatogr.* **8**:29 (1985).

32. G. D. Reed and R. J. Hunt, *J. High Res. Chromatogr.* **9**:341 (1986).

33. M. F. Mehran, W. J. Cooper, M. Mehran, and F. Diaz, *J. High Res. Chromatogr.* **7**:639 (1984).

34. R.J. Miller and W. Jennings, *J. High Res. Chromatogr.* **2**:72 (1979).

35. W. Jennings, *J. High Res. Chromatogr.* **17**:567 (1994).

36. D. Rood, *Care, Maintenance, and Troubleshooting of Capillary GC Systems.* Huethig, Heidelberg, 1995.

ABBREVIATIONS, TERMS, AND NOMENCLATURE

A.1 Generic Terms and Nomenclature

The wide variety of terms and symbols used by gas chromatographers can lead to confusion. Systems of gas chromatographic nomenclature have been proposed by both the International Union of Pure and Applied Chemistry (IUPAC) and the American Society for Testing and Materials (ASTM). This book utilizes the most recent recommendations of the latter, which appeared in *Pure Appl. Chem.* **65**:819–872 (1993). Other terms listed herein reflect commonly accepted usage, modified only where extensive experience in teaching gas chromatography indicates that an alternative would be more logical to the student.

A	Packing factor term of the van Deemter (packed column) equation.
α	Separation factor; Ratio of the adjusted retention times of two solutes, measured under identical conditions. By convention, α is never less than 1; hence it is the function of the more retained solute relative to the same function of the less retained solute.
B	Longitudinal diffusion term of the van Deemter (packed column) and Golay (open tubular column) equations.
β	Column phase ratio. The column volume occupied by mobile (gas) phase relative to the volume occupied by stationary liquid phase. In open tubular columns, $\beta = ([r_c/2d_f])/2d_f \cong r_c/2d_f$.
c_M, c_S	Solute concentrations in mobile and stationary phases, respectively.

C — Resistance to mass transport (or mass transfer) in the van Deemter (or Golay) equations; C_M and C_S denote mass transport from mobile to stationary and from stationary to mobile phases, respectively.

d_c — Inner diameter of the column. Both millimeters and micrometers are commonly used. The latter, while consistent with the units used for d_f, implies three significant figures, which is rarely true.

d_f — Thickness of the stationary phase film, usually in micrometers.

D — Diffusivity; D_M and D_S denote solute diffusivities in the mobile and stationary phases, respectively; usually given in square centimeters per second.

ECD — Electron capture detector.

EPC — Electronic pneumatic control. A system in which the pressure (or flow) of the gas supply is electronically (computer-) controlled.

F — Volumetric flow of the mobile phase, usually in cubic centimeters per minute. Many practical chromatographers assume equivalency with (and hence employ) milliliters per minute.

FID — Flame ionization detector.

FPD — Flame photometric detector.

GC/MS — The combination of gas chromatography and mass spectrometry, usually a single integrated unit in which fractions separated by GC are sequentially introduced to the MS.

H — Height equivalent to a theoretical plate: $H = L/N$. More precisely, the length of column occupied by one theoretical plate. When measured at u_{opt}, the result is termed H_{min}.

$j_1, j_2, \ldots, j_n,$ — Gas compressibility (or pressure drop) correction factors.

k — Retention factor (formerly termed the *partition ratio* or the *capacity factor*). Ratio of the amounts of a solute (or time spent) in stationary and mobile phases, respectively. Because all solutes spend t_M time in the mobile phase, $k = [(t_R - t_M)]/t_M$; $k = t'_R/t_M$.

K_c — Distribution constant; formerly K_D. Ratio of solute concentrations in stationary and mobile phases, respectively: $K_c = c_S/c_M$.

L — Length of the column, usually expressed in meters for column length per se, in centimeters for the determination of $\bar{u}$, and in millimeters for the determination of H.

N — Theoretical plate number; $N = (t_R/\sigma)^2$, where σ is the standard deviation of the peak.

N_{req} — Number of theoretical plates required to separate two solutes of a given α and given retention factors to a given degree of resolution: $N_{req} = 16Rs^2[(k+1)/k]^2[\alpha/(\alpha-1)]^2$.

NPD	Nitrogen/phosphorus detector. There are two types: in the thermionic detector, an alkali metal bead (Rubidium is favored) is heated in the flame of an otherwise conventional FID; the "plasma type detector" heats the bead electrically in an unlighted detector.
OD	Outer diameter of the column.
p_i	Column inlet pressure.
p_o	Column outlet pressure.
Δp	Pressure drop through the column; $\Delta p = p_i - p_o$.
P	Relative pressure; $P = p_i/p_o$.
r_c	Inside radius of column. Both millimeters and micrometers are commonly used. The latter implies three significant figures, which is rarely true.
R_s	Peak resolution. A measure of separation as evidenced by both the distance between the peak maxima and by the peak widths. ASTM and IUPAC definitions are based on w_b (peak width at base) measurements, which require extrapolation. If peaks are assumed to be Gaussian, $R_S = 1.18 \times [t_{R(B)} - t_{R(A)}]/[w_{h(A)} + w_{h(B)}]$.
σ	Standard deviation of a Gaussian peak.
t_M	Gas holdup time. The time (or distance) required for a nonretained substance (e.g., mobile phase) to transit the coulmn.
t_R	Retention time. The time (or distance) from the point of injection to the peak maximum.
t'_R	Adjusted retention time. Equivalent to the residence time in stationary phase; therefore the difference of the solute retention time and the gas holdup time: $t'_R = t_R - t_M$.
$\bar{u}$	Average linear velocity of the mobile (gas) phase: $\bar{u}_{cm/sec} = L_{cm}/t_{M,sec}$
v	Linear velocity of the solute band.
V	Volume; V_M and V_S represent volumes of the mobile and stationary phases, respectively.
w_b	Peak width at base. Determined by measuring the length of baseline defined by intercepts extrapolated from the points of inflection of the peak; this is equivalent to four standard deviations in a Gaussian peak.
w_h	Peak width at halfheight. Measured across the peak halfway between baseline and peak maximum; this can be measured directly without extrapolation, and is equal to 2.35 standard deviations in a Gaussian peak.

A.2 Proprietary Terms

Trademark	Company
Aroclor	Monsanto
Carbopack	Supelco
Carbosieve	Supelco
Carbowax	Union Carbide
Chirasil Val	Alltech Associates
Chromosorb	Johns Manville Corp.
Connex	J&W Scientific
DB	J&W Scientific
Hall	Tracor Instruments
Megabore	J&W Scientific
Microbore	J&W Scientific
Minibore	J&W Scientific
Quickfit	Quadrex
Silar	Silar
Snoop	Nupro Co.
Swagelok	Crawford Fitting Co.
Teflon	E. I. Du Pont Nemours & Co.
Vespel	E. I. Du Pont Nemours & Co.

INDEX